# Molecular Phylogeny of Microorganisms

Aharon Oren
Department of Plant and Environmental Sciences
The Hebrew University of Jerusalem
Israel

and

R. Thane Papke
Department of Molecular and Cell Biology
University of Connecticut
USA

Caister Academic Press

Caister Academic Press
Norfolk, UK

www.caister.com

British Library Cataloguing-in-Publication Data
A catalogue record for this book is available from the British Library

ISBN:978-1-904455-67-7

Description or mention of instrumentation, software, or other products in this book does not imply endorsement by the author or publisher. The author and publisher do not assume responsibility for the validity of any products or procedures mentioned or described in this book or for the consequences of their use.



Cover image: courtesy of Olga Zhaxybayeva

Printed and bound in Great Britain

# Contents

# Contributors

**Cheryl P. Andam**
Department of Molecular and Cell Biology
University of Connecticut
Storrs, CT
USA

cheryl@andam@uconn.edu

**Jongsik Chun**
School of Biological Sciences and Institute of Microbiology
Seoul National University
Gwanak-gu
Seoul, Republic of Korea

jchun@snu.ac.kr

**Dion G. Durnford**
Department of Biology
University of New Brunswick
Fredericton, NB
Canada

durnford@unb.ca

**Gregory P. Fournier**
Department of Biological Engineering
NASA Astrobiology Institute
Massachusetts Institute of Technology
Cambridge, MA

g4nier@mit.edu

**J. Peter Gogarten**
Department of Molecular and Cell Biology
University of Connecticut
Storrs, CT
USA

gogarten@uconn.edu

**Radhey S. Gupta**
Department of Biochemistry
McMaster University
Hamilton, Canada

gupta@mcmaster.ca

**Soon Gyu Hong**
Polar Biocenter
Korea Polar Research Institute
KORDI
Songdo Techno Park
Yeonsu-gu
Incheon, Republic of Korea

polypore@kopri.re.kr

**Christopher E. Lane**
Department of Biological Sciences
University of Rhode Island
Kingston, RI
USA

clane@mail.uri.edu

**Wolfgang Ludwig**
Lehrstuhl für Mikrobiologie
Technische Universität München
Freising, Germany

ludwig@mikro.biologie.tu-muenchen.de

**Maureen A. O'Malley**
Egenis
University of Exeter
Exeter, UK

M.A.O'Malley@exeter.ac.uk

**Aharon Oren**
Department of Plant and Environmental Sciences
The Institute of Life Sciences and The Moshe Shilo Minerva Center for Marine Biogeochemistry
The Hebrew University of Jerusalem
Jerusalem
Israel

orena@cc.huji.ac.il

**R. Thane Papke**
Department of Molecular and Cell Biology,
University of Connecticut
Storrs, CT
USA

robertson.papke@uconn.edu

**Pablo Vinuesa**
Centro de Ciencias Genómicas
Universidad Nacional Autónoma de México
Cuernavaca, Morelos
Mexico

vinuesa@ccg.unam.mx

**David Williams**
Department of Molecular and Cell Biology
University of Connecticut
Storrs, CT
USA

david.williams@uconn.d-dub.org.uk

# Preface

## From a 'Protoplasmal Primordial Atomic Globule' to a Wealth of 'Scientific Names of Beings Animalculous'

*Don't mention it. I am, in point of fact, a particularly haughty and exclusive person, of pre-Adamite ancestral descent. You will understand this when I tell you that I can trace my ancestry back to a protoplasmal primordial atomic globule. Consequently, my family pride is something inconceivable.*

This statement was made by Poo-Bah, the 'Lord High Everything Else' of the Town of Titipu in *The Mikado,* the comic opera with lyrics by William Schwenck Gilbert and music by Arthur Sullivan, first produced in 1885. It is a surprising statement, as in that period hardly anybody thought about the phylogeny of those ancient 'protoplasmal primordial atomic globules' that may have had some resemblance to the microorganisms found in nature today. Charles Darwin, who introduced the concepts of phylogeny and evolution to the scientific world, was indeed optimistic that some time in the future, the secrets of phylogeny will be revealed. 'The time will come, I believe, though I shall not live to see it, when we shall have very fairly true genealogical trees of each great kingdom of Nature', thus Darwin wrote in a letter to Thomas Henry Huxley (26 September 1859). It is also noteworthy that the only illustration to appear in *The Origin of Species* was a tree-like representation showing how life may have diverged from one or more common ancestors to result in the current diversity of living organisms on Earth. However, it is interesting to note that Darwin did not often mention the microbial world in his writings. Darwin surely knew about the diversity of bacteria, protozoa, etc., known since the days when Antonie van Leeuwenhoek saw the first bacteria through the lens of this primitive microscope. The diversity of protozoa was documented in detail in the first decades of the nineteenth century by Christian Ehrenberg and others. What Darwin did not know was how genetic information was inherited, though Mendel sent him a manuscript of his work with peas. Despite the fact that Fredrick Griffith's 1928 experiments demonstrating DNA as the molecule of genetic information was clearly a case of lateral gene transfer, acceptance of this mode of inheritance as an import mechanism for the evolution of prokaryotes had to wait until the creation of inexpensive DNA sequencing and computational power. Today, lateral gene transfer is acknowledged as a relevant source of genetic variation in each domain of life, but the debate still rages as to whether phylogenetic reconstruction of the 'tree of life', especially for prokaryotes, represents organism or gene evolution, which has an enormous impact on all facets of evolution and taxonomy. Here, we explore phylogeny and its usefulness with respect to microorganisms.

The first chapter by Aharon Oren provides an historical overview of how scientists looked at the phylogeny of microbes, and especially of prokaryotic microorganisms, in the past – from the nineteenth century morphology-based schemes to schemes based on physiological properties and information on the cell wall, to the 16S rRNA gene-based phylogenetic trees used nowadays by nearly all workers in the field to establish order in the complex world of the prokaryotes. This molecular approach, pioneered by Carl Woese in the 1970s, leading to the three-domain model (Archaea–Bacteria–Eucarya), has revolutionized our thinking about evolution in the microbial world.

The wealth of sequence information that is accumulating at an ever-increasing speed can only be used for phylogenetic reconstruction when appropriate computational tools are available to use the information to reconstruct phylogenetic trees. Chapter 2 by Jongsik Chun and Soon Gyu Hong provides an overview of the methods and programs for the calculation of phylogenetic relationships and trees. Even though we may never know the true phylogeny, phylogenetic analysis provides best assumptions, thereby providing a framework for various disciplines in microbiology. Due to the technologically innovative burst of modern molecular biology and the rapid advancement in computational science, inferring the true phylogeny of a gene or organism seems possible in the near future. The chapter reviews major steps in phylogenetic analysis and introduces relevant computer software tools with respect to their accuracy, efficiency and availability.

Chapter 3 by Pablo Vinuesa presents a review of the critical factors that have to be considered and evaluated in multilocus sequence analysis, a technique increasingly used in order to make robust estimates of bacterial species phylogenies when the 16S rRNA alone does not provide the full information. The chapter presents a number of case studies, reviews criteria for marker selection and provides practical advice on the computational aspects, potential pitfalls, and software choices available for each step in multilocus sequence analysis, emphasizing the importance of using multiple isolates per species/lineage, proper model selection and thorough tree searches to get a good estimate of a multispecies phylogeny.

In Chapter 4, Wolfgang Ludwig asks whether the small subunit RNA sequence is still a useful marker to reconstruct the molecular phylogeny of microorganisms. The current taxonomy of prokaryotes as well as modern probe and chip based identification methods are mainly based upon rRNA derived phylogenetic conclusions. The significance of single gene based phylogenetic inference was evaluated by including alternative global markers such as elongation and initiation factors, RNA polymerase subunits, DNA gyrases, heat shock and recA proteins. Although the comparative analyses are hampered by the generally low phylogenetic information content, a domain and prokaryotic phyla concept is globally supported.

The result of the 16S rRNA sequence based phylogenetic tree reconstruction is summarized in Chapter 5, in which Aharon Oren presents the major phyla of prokaryotes, based on cultured and named species as well as on 'environmental' sequences recovered from DNA isolated from different environments. To obtain the organisms harbouring these sequences and to study their properties is a major challenge of microbiology today. This type of classification, in which the now over 8200 species of prokaryotes with validly published names are classified in respectively 27 phyla of Bacteria and two phyla of Archaea (as of November 2009), was adopted in the latest edition of *Bergey's Manual* of *Systematic Bacteriology*. It should, however, be stressed that there is no official classification of prokaryotes. For the higher taxa there even is no official nomenclature: the rules of the International Code of Nomenclature of Prokaryotes do not cover taxa above the rank of class. Alternative classifications have been proposed, based, for example, on the structure of the cell wall.

The approaches towards the rooting of the universal tree of life are discussed in Chapter 6 by Greg Fournier. Identifying the location of the root of the tree corresponding to the most recent common ancestor is a challenging and distinct problem that has yet to be completely solved. To date, many investigations have proposed various roots, using a wide diversity of biological data and techniques. A survey of the most promising of these models illustrates the difficulty faced in reaching a scientific consensus on the issue, as well as the additional philosophical complications posed by our emerging understanding of the role of horizontal gene transfer in genome evolution.

Comparative analysis of genome sequences is leading to discovery of large numbers of novel molecular markers that are proving very helpful in understanding many important aspects of microbial phylogeny. Of these molecular markers, conserved inserts or deletions (indels) in protein sequences provide a means for identifying different groups of microbes in molecular terms and for understanding how they have may have evolved from a putative common ancestor. Chapter 7 by Radhey Gupta discusses the applications of conserved indels for analysing microbial phylogeny, and shows how genetic and biochemical studies of these markers should also lead to identification of novel properties that are unique to different groups of microbes.

Elucidation of an ever increasing number of genome sequences of Bacteria and Archaea shows how important lateral gene transfer has been in the evolution of the prokaryotic world. Realization of the significance of lateral gene transfer and other non-vertical processes has reconceptualized and reoriented attempts to construct the universal phylogeny. Chapter 8 by Maureen O'Malley discusses the occurrence of paradigm shifts as construction, deconstruction and reconstruction of the tree of life changed, based on our increasing understanding of lateral gene transfer phenomena.

The impact of gene transfer on the formation of groups of organisms is described in further depth in Chapter 9 by David Williams, Cheryl Andam and Peter Gogarten. Gene transfer can make it more difficult to define and determine relationships. In those cases where many genes have been transferred between partners, the majority of genes in a genome may reflect biased gene acquisition, and as a consequence, if a coherent signal is detected, one nevertheless might not be sure that the signal is due to shared organismal ancestry. Lateral gene transfer may have positive aspects as well when trying to reconstruct phylogenetic trees. The presence of a particular transferred gene was shown, in several cases, to constitute a shared derived character useful in classification. Gene transfer can put together new metabolic pathways that open up new ecological niches, and consequently, the transfer of an adaptive gene might create a new group of organisms

In Chapter 10, finally, Christopher Lane and Dion Durnford discuss the role of endosymbiosis and the evolution of plastids. Photosynthesis is one of the most successful energy production strategies on the planet and has been co-opted numerous times throughout evolutionary history via the uptake and retention of photosynthetic cells by non-photosynthetic eukaryotic heterotrophs. The result of this process is clear, but the mode and tempo of plastid movement among eukaryotes, particularly plastids of red algal derivation, are largely unknown. Recent changes in our understanding of the relationships between eukaryotic supergroups have only served to complicate the picture further.

The different approaches applied today to elucidate the molecular phylogeny of the prokaryotes (and of eukaryotic protists as well) have taken us a long way toward illuminating the complex world of the microbes that Linnaeus described under a single taxon, 'Chaos infusorium'. The methods available to the microbiologist today will rapidly place a newly isolated strain or an environmental small subunit rRNA gene sequence within the existing framework of the universal tree, even if some details of the exact topology of the tree may still be unknown.

The number of isolated, characterized, and named species of prokaryotic and eukaryotic microorganisms – presumably all descendants of the above-mentioned 'protoplasmal primordial atomic globule' that supposedly was the beginning of all life of Earth – is ever increasing. A proper understanding of the diversity, systematics and nomenclature of the microbes is important in many branches of biological science. Having opened this preface with a quote from a Gilbert & Sullivan comic opera, we close with another quotation from their work, this time from *The Pirates of Penzance* (1879), that shows that knowledge of microbial systematics, nomenclature and phylogeny can be of importance to many professionals, even to a Major-General:

> *I'm very good at integral and differential calculus;*
> *I know the scientific names of beings animalculous:*
> *In short, in matters vegetable, animal, and mineral,*
> *I am the very model of a modern Major-General.*

Aharon Oren and R. Thane Papke

# Current Books of Interest

|  |  |
|---|---|
| *Streptomyces*: Molecular Biology and Biotechnology | 2011 |
| Alphaherpesviruses: Molecular Virology | 2011 |
| Recent Advances in Plant Virology | 2011 |
| Vaccine Design: Innovative Approaches and Novel Strategies | 2011 |
| *Salmonella*: From Genome to Function | 2011 |
| PCR Troubleshooting and Optimization: The Essential Guide | 2011 |
| Insect Virology | 2010 |
| Environmental Microbiology: Current Technology and Water Applications | 2010 |
| Sensory Mechanisms in Bacteria: Molecular Aspects of Signal Recognition | 2010 |
| Bifidobacteria: Genomics and Molecular Aspects | 2010 |
| Molecular Phylogeny of Microorganisms | 2010 |
| Nanotechnology in Water Treatment Applications | 2010 |
| Iron Uptake and Homeostasis in Microorganisms | 2010 |
| Caliciviruses: Molecular and Cellular Virology | 2010 |
| Epstein-Barr Virus: Latency and Transformation | 2010 |
| Anaerobic Parasitic Protozoa: Genomics and Molecular Biology | 2010 |
| Lentiviruses and Macrophages: Molecular and Cellular Interactions | 2010 |
| Microbial Population Genetics | 2010 |
| *Borrelia*: Molecular and Cellular Biology | 2010 |
| Influenza: Molecular Virology | 2010 |
| RNA Interference and Viruses: Current Innovations and Future Trends | 2010 |
| Retroviruses: Molecular Biology, Genomics and Pathogenesis | 2010 |
| Metagenomics: Theory, Methods and Applications | 2010 |
| *Aspergillus*: Molecular Biology and Genomics | 2010 |
| Environmental Molecular Microbiology | 2010 |
| *Neisseria*: Molecular Mechanisms of Pathogenesis | 2010 |
| Frontiers in Dengue Virus Research | 2010 |
| ABC Transporters in Microorganisms | 2009 |
| Pili and Flagella: Current Research and Future Trends | 2009 |
| Lab-on-a-Chip Technology: Biomolecular Separation and Analysis | 2009 |
| Lab-on-a-Chip Technology: Fabrication and Microfluidics | 2009 |
| Bacterial Polysaccharides: Current Innovations and Future Trends | 2009 |
| Microbial Toxins: Current Research and Future Trends | 2009 |
| *Acanthamoeba* | 2009 |
| Bacterial Secreted Proteins: Secretory Mechanisms and Role in Pathogenesis | 2009 |
| *Lactobacillus* Molecular Biology: From Genomics to Probiotics | 2009 |
| *Mycobacterium*: Genomics and Molecular Biology | 2009 |
| Real-Time PCR: Current Technology and Applications | 2009 |
| Clostridia: Molecular Biology in the Post-genomic Era | 2009 |
| Plant Pathogenic Bacteria: Genomics and Molecular Biology | 2009 |
| Microbial Production of Biopolymers and Polymer Precursors | 2009 |
| Plasmids: Current Research and Future Trends | 2008 |
| *Pasteurellaceae*: Biology, Genomics and Molecular Aspects | 2008 |

# Concepts About Phylogeny of Microorganisms – an Historical Overview

1

Aharon Oren

Abstract

When at the end of the nineteenth century information began to accumulate about the diversity within the bacterial world, scientists started to include the bacteria in phylogenetic schemes to explain how life on Earth may have developed. Some of the early phylogenetic trees of the prokaryote world were morphology-based; others were based on the then-current ideas on the presumed conditions on our planet at the time that life first developed. Around 1950 many leading microbiologists had become pessimistic with respect to the possibility of ever reconstructing bacterial phylogeny. The concept of the prokaryote–eukaryote dichotomy did little to clarify phylogenetic relationships. The developing technology of nucleic acid sequencing, together with the recognition that sequences of building blocks in informational macromolecules (nucleic acids, proteins) can be used as 'molecular clocks' that contain historical information, led to the development of the three-domain model (Archaea – Bacteria – Eucarya) in the late 1970s, primarily based on small subunit ribosomal RNA sequence comparisons. The information currently accumulating from complete genome sequences of an ever increasing number of prokaryotes now once more will lead of modifications of our views on microbial phylogeny.

*'The time will come I believe, though I shall not live to see it, when we shall have very fairly true genealogical trees of each great kingdom of nature.'*

(letter from Charles Darwin to Thomas Huxley, 26 September 1857)
(Burkhardt and Smith, 1990)

*'There is no other one thing so important in systematic biology as the fact that grouping of organisms reflects and expresses their true relationships.'*

(Edwin Bingham Copeland, 1927)

## Introduction

When Darwin wrote the above-quoted prophetic statement in 1857, he did not especially refer to microorganisms. In fact, Darwin hardly mentioned the microorganisms in his writings. It must also be stressed that in Darwin's days the tremendous diversity within the bacterial world was not yet recognized, and at the time no more than five or six genera of prokaryotes had been named (not including the cyanobacteria, whose diversity was already well known and had been extensively described by nineteenth-century botanists).

In the second half of the nineteenth century the extent of the diversity within the bacterial world started to become clear. Robert Koch and his fellow medical microbiologists succeeded in isolating many pathogenic bacteria, and Sergei

Winogradsky and Martinus Beijerinck established the basis for our current understanding of the role of microorganisms in the biogeochemical cycles in soils and in aquatic environments. Already in those early years did the microbiologists ask questions about the nature of the evolutionary processes that led to prokaryote diversity, and the first attempts were made to establish the place of the bacteria within the general scheme of the evolution of life on Earth.

When in 1905 Beijerinck received the Leeuwenhoek Medal of the Dutch Royal Academy of Sciences, his address to the Academy included the following statement in which he clearly explained his views:

> *Die richting is kort te omschrijven als het onderzoek van de Oekologie der mikroben, dat is van het verband tusschen bepaalde levensvoorwaarden en bepaalde levensvormen die daaraan beantwoorden. Daar het mijn overtuiging is, dat deze bij den tegenwoordigen stand der wetenschap de meest noodzakelijke en meest vruchtbare richting is om orde te brengen in onze kennis aangaande dat deel van het natuurlijke stelsel, dat the laagste grens omvat van de organische wereld, en dat ons aanhoudend het grote vraagstuk naar den oorsprong van het leven zelve in scherpe trekken voor ogen stelt, is het mij tot groote voldoening, dat de Akademie blijkbaar de beoefening daarvan in den beoefenaar wil bekronen.*
>
> *(This direction can briefly be described as the study of microbial ecology, i.e. the relation between certain environmental conditions and certain corresponding forms of life. I am convinced that, at the current state of science, the most necessary and most fruitful direction is to establish order in our knowledge of that part of the natural system that deals with the lowest limits of the organic world, and which constantly reminds us very clearly of the great question of the origin of life itself. Therefore it is a great pleasure to me that the Academy apparently wishes to promote this line of science by presenting an award to its practitioner.)*
>
> (Translation: A.O.)

Beijerinck thus understood that the study of the microbial world and its diversity would eventually lead to a deeper insight into the origin of life and its evolution. Already then Beijerinck outlined the microbiology of today – an integrated microbiology that synthesizes organisms, ecology, and evolution (Woese and Goldenfeld, 2009).

This chapter provides an attempt to trace the development of the concepts on microbial and especially prokaryotic phylogeny from the early times when the extent of microbial diversity started to become clear until the present day. To illustrate the different ideas proposed by many of the leading scientists of the past century I included many quotations from their original publications.

The evolution of the theories on prokaryote phylogeny has been discussed earlier by several authors, notably by Carl Woese (1994, 2004) and by Jan Sapp (2005, 2006). Much of the information provided below is therefore based on their analyses and their views.

## Bacterial classification and nomenclature in the eighteenth and the nineteenth centuries

The different theories about bacterial phylogeny as these developed in the course of the twentieth century all depended on the then current ideas about the morphological and physiological diversity of the group and on the concepts of how to define bacterial species, genera, families, etc. Therefore the discussions will start with a short overview of the earliest attempts toward a bacterial classification and nomenclature.

When Antonie van Leeuwenhoek first viewed bacteria through his primitive microscopes in the 1670s, he did not assign names to the novel types of small organisms he observed, and neither did several subsequent generations of naturalists. The first attempts to classify and name the small bacteria were made by the Danish biologist Otto Frederik Müller (1730–1784), who described two genera: *Monas* [*Vermis inconspicuus, simplicissimus, pellucidus, punctiformis* (an inconspicuous, extremely simple, translucent, dot-shaped worm)] and *Vibrio* [*Vermis inconspicuus, simplicissimus, teres, elongatus* (an inconspicuous, extremely simple, curved elongated worm)]. Within the genus *Monas* he discriminated three species – *M. termo*, *M. lens* and M. *mica* – and he provided descriptions of no fewer than 15 species of *Vibrio*: *V. lineola*, *V. bacillus*, *V. anguillula*, *V. serpentulus*, *V. vermiculus*, *V. undula*, *V. intestinum*, *V. proteus*, *V. falx*, *V. anser*, *V. cygnus*, *V. malleus*, *V. utriculus*, *V. fascicola*, and *V. colymbus* (Müller, 1786).

The first true classification of the bacteria (at the time considered as belonging to the animal kingdom) can probably be attributed to the German Christian Gottfried Ehrenberg (1795–1876). His scheme published in 1838 contained five genera:

1 *Bacterium*, inflexible rod forms;
2 *Vibrio*, flexible rod forms;
3 *Spirillum*, spiral, inflexible;
4 *Spirochaeta*, spiral, flexible,
5 *Spirodiscus*, flattened spirals.

These were the only genera of heterotrophic prokaryotes that had been named until the middle of the nineteenth century. In addition, many types of cyanobacteria had already been described and named, as well as some long, conspicuous filamentous non-phototrophic bacteria such as *Beggiatoa* and a few large phototrophic sulfur bacteria (e.g. *Monas okenii*, now *Chromatium okenii*, and *Ophidomonas jenensis*, now *Thiospirillum jenense*, both already described by Ehrenberg). In the famous evolutionary scheme of Ernst Haeckel (1866), the bacteria were placed in the order Moneres (later Monera) at 'the lowest stage of the protist kingdom'. Bacteria were unique, he argued, because unlike other protists, they possessed no nucleus. They were as different from nucleated cells as 'a hydra was from a vertebrate' or 'a simple alga from a palm' (Sapp, 2005).

The starting point of modern systematic bacteriology is often considered to be Ferdinand Cohn's *Untersuchungen über Bakterien* (Cohn, 1872a). Morphological differences were the primary properties used to delineate genera and higher groups. Physiological differences would indicate only varieties and races; they originated from the same germ, but through constant natural or artificial culture under the same conditions these were thought to become hereditary (Sapp, 2006). Cohn named four tribes and six genera:

| | |
|---|---|
| Tribe I. | Sphaerobacteria (sphere bacteria) |
| | Genus 1. *Micrococcus* (spherical cells) |
| Tribe II. | Microbacteria (rod bacteria) |
| | Genus 2. *Bacterium* (isolated short rods) |
| Tribe III. | Desmobacteria |
| | Genus 3. *Bacillus* (long rigid rods or filaments) |
| | Genus 4. *Vibrio* (long flexible rods or filaments) |
| Tribe IV. | Spirobacteria |
| | Genus 5. *Spirillum* (spiral, rigid cells) |
| | Genus 6. *Spirochaeta* (spiral, flexible cells) |

Cohn based his type concept of a bacterium primarily on the properties of a relatively homogeneous group: unicellular (Eu)bacteria that multiply by binary fission. He considered these bacteria to belong to the first and simplest division of living organisms, and they seemed to him naturally differentiated from the different groups of higher plants, even though their characteristic attributes are more of a negative than a positive nature (Cohn, 1875).

Cohn was the first to realize the close structural similarities between the known bacteria and the cyanobacteria ('blue–green algae'). He recognized that the bacteria had no relation to yeasts or moulds, as was then often held, and referred them to a new group, the Schizosporeae, a name he subsequently changed to the more suitable name Schizophyta. The latter comprised the Schizophyceae (Myxophyceae; cyanobacteria) and the Schizomycetes (Bacteria). In a communication on the 'Verwandtschaftsbeziehungen der Bakterien' (1875), Cohn provided a detailed account of his views. The bacteria were considered to be near, though primitive allies of the Myxophyceae, adapted to a saprotrophic or parasitic mode of life. While refuting any relation to animals or fungi, Cohn stated that the bacteria are at the lower end of the line of myxophycean evolution, although a closer relation between certain forms of the two classes was assumed' (Cohn, 1853, 1872a,b, 1875; Pringsheim, 1949).

Walter Migula (1863–1938) extended Cohn's morphology-based classification in this two-volume *System der Bakterien*, published in 1897–1900, in which he integrated most of the new acquired information (Migula, 1897–1900). In Migula's classification, emphasis was placed upon motility, the type of flagellation, and endospore formation.

## The first half of the twentieth century: morphology-based phylogenetic trees

The two most influential papers on bacterial evolution and phylogeny written in the 1930s and 1940s were authored by three of the most famed microbiologists of the time: Albert Jan Kluyver, Cornelis van Niel, and Roger Stanier. Their attempts to reconstruct bacterial phylogeny resulted in schemes based primarily on cell morphology. Phylogenetic classification was thus first and foremost based on increased morphological complexity. In view of the great impact that these articles had at the time, a more extensive description of these publications is justified here. Kluyver and his student van Niel introduced their 'natural system of classification of bacteria' with the following words (Kluyver and van Niel, 1936):

> *... the only truly scientific foundation of classification is to be found in appreciation of the available facts from a phylogenetic point of view. Only in this way can the natural interrelationships of the various bacteria be properly understood. ... it cannot be denied that the studies in comparative morphology made by botanists and zoologists have made phylogeny a reality. Under these circumstances it seems appropriate to accept the phylogenetic principle also in bacteriological classification ...*

Thus, the principles used were derived from zoological classification: morphological characters have the greatest impact, and the morphologically simplest forms are the ancestral ones. Only morphological properties were used for the demarcation of larger units, tribes, families, etc., whereas both morphological and biochemical characters would be used to define genera. The system presented in the 1936 paper was extended in a later paper by Stanier and van Niel (1941). The authors gave the following rationale for their proposed phylogenetic classification scheme:

> *... a clear recognition of the larger natural groups of bacteria, their characteristics and relationships, would seem an indispensable basis for more detailed work. ...*
>
> *In most biological fields it is considered a truism to state that the only satisfactory basis for the construction of a rational system of classification is the phylogenetic one. ... However, the mere fact that a particular phylogenetic scheme has been shown to be unsound by later work is not a valid reason for total rejection of the phylogenetic approach.*
>
> *... It is our belief that such pessimism is not entirely justified, and that at present some relationships can be recognized which can well be incorporated in a system of classification.*

Even granting that the true course of evolution can never be known and that any phylogenetic system has to be based on some extent of hypothesis, there is good reason to prefer an admittedly imperfect natural system to a purely empirical one.

> *... In the classification of higher plants and animals, systematicists have relied almost exclusively on morphology. ...*

The chief stumbling block in attempting to draw up a phylogenetic system on a primarily physiological basis is the necessity of making a large number of highly speculative assumptions as to what constitute primitive and advanced metabolic types.

In spite of the comparative simplicity of bacteria it is rather naïve to believe that in the distribution of their metabolic characters one can discern the trend of physiological evolution. For these reasons, a phylogenetic system based solely on physiological grounds seems unsound. It is our belief that the greatest weight in making the major subdivisions on the *Schizomycetes* should be laid on morphological characters, although correlative physiological characters may also be used.'

Stanier and van Niel thus opposed the physiology-based schemes of Orla-Jensen and others (see below), while denouncing problems inherent to any phylogenetic system based primarily on physiology. The proposed scheme was again primarily based on morphology. The result of their attempts to use only morphological characters to distinguish larger units of classification looked as follows:

> *Starting from the hypothetical primitive coccus type, the first line leads through the micrococci to the sarcinae, culminating in the spore-forming sarcinae. ...*
>
> *The second line consists of the polarly flagellated rods, starting with the pseudomonas type and leading through the vibrios to the spirilla. ....*
>
> *The third line is morphologically highly diverse. It takes its origin in the streptococci, and passes through the lactic acid bacteria, the propionic acid and other corynebacteria, and the mycobacteria to the actinomycetes. ...*
>
> *The fourth line ... comprises all peritrichously flagellated rods ...*
>
> *It is at least certain that morphologically the Myxophyta resemble the true bacteria far more closely than they do any of the other algal groups.*
>
> *The foregoing pages have, we feel, justified the claim that a phylogenetic approach to the taxonomic problems of the bacteria is capable of yielding fruitful results. ... All other bacteria can be readily subdivided into three large groups whose sharp and easy separation on the basis of one or two fundamental morphological characters is possible.*

Morphological similarities also guided Pringsheim (1949) in his attempts to classify filamentous colourless sulfur bacteria such as *Beggiatoa* and *Thiothrix*. The superficial resemblance of the shape of their filaments with some filamentous cyanobacteria (such as *Oscillatoria*) was sufficient at the time to classify these colourless microorganisms as 'apochlorotic Myxophyceae', i.e. cyanobacteria that had lost their pigmentation and therefore were leading a non-photosynthetic mode of life.

Today, in retrospect, we know that morphological characters such as cellular shape, motility, as well as general physiological properties, tell us nothing whatsoever on the true phylogenetic relationships among the prokaryotes. In their later lives both Stanier and van Niel acknowledged that their earlier efforts toward constructing a natural bacterial system had not led to the goal.

## Phylogeny reconstructions based on physiological properties

Quite early in the history of microbiology some taxonomists, unhappy with a morphology-based phylogeny of the bacteria, proposed using physiological properties to reconstruct phylogenetic schemes. The idea behind this was quite straightforward: if it is possible to make educated guesses regarding the environmental conditions on early Earth at the time that life originated and estimate how those conditions have subsequently changed, then one can, based on the properties known from the current-day forms of life, make assumptions about the types of organisms that thrived in the earliest times, and deduce how other types then could have evolved from the first primitive bacteria.

The Oparin–Haldane hypothesis that 'chemical evolution' had preceded the origin of life and biological evolution dominated the thoughts behind many of these physiology-based phylogenetic schemes. If indeed there was a wealth of simple organic molecules in the 'primordial soup' and molecular oxygen was not available, then the first type of metabolism may have resembled the simple fermentation pathways known today e.g. from the homolactic fermentation, based on the Embden–Meyerhof–Parnas glycolytic pathway. Such pathways yield the ATP necessary for life in a few relatively simple enzymatic steps without the use of an external electron acceptor such as oxygen, while degrading energy-rich organic substances to simpler compounds.

However, long before Oparin and Haldane independently launched their ideas in the 1930s, an altogether different phylogenetic scheme for the bacteria based on physiological properties rather than on cell shape was proposed by the Danish bacteriologist Sigurd Orla-Jensen

(1870–1949). Orla-Jensen's ideas, first published in 1909, were based on the assumption that the first organisms on Earth must have developed in the dark in an environment devoid of organic matter, and of course they should have been independent of the presence of other life forms. The only organisms known today to be capable of that are chemosynthetic bacteria. Chemolithotrophy as a mode of life was already well known thanks to the pioneering work of Sergei Winogradsky on chemoautotrophic sulfur bacteria (*Beggiatoa*), bacterial nitrification (*Nitrosomonas, Nitrobacter*), and iron oxidation (*Gallionella*) in the last two decades of the nineteenth century. Orla-Jensen therefore assumed that such organisms may have been the first living forms on earth; the rest of his system was based on an evaluation of increasing physiological and biochemical complexity. Thus, his scheme includes propositions for the establishment of bacterial families such as the Alkalibacteriaceae (aerobic forms producing ammonia from proteins), Putribacteriaceae (anaerobic forms decomposing proteins), Thiobacteriaceae (colourless sulfur bacteria), Luminibacteriaceae (phosphorescent bacteria), Reducibacteriaceae (actively reducing bacteria), and others (Orla-Jensen, 1909; Sapp, 2006).

Orla-Jensen's system was at the time highly regarded, by some even until the mid-1940s. A good illustration can be found in the comments made by Charles-Edward Winslow, president of the Society of American Bacteriologists (now the American Society for Microbiology) during an address in Montreal, Canada (Winslow, 1914):

> *Whatever minor criticisms may be made of Professor Jensen's scheme, I believe that no one who has thought seriously about bacterial relationships can study it carefully without feeling that it is by far the most successful attempt yet made at a real biological classification of the group and that future progress will probably consist in its modification and extension rather than in any profound reversal of its basic principles.*

In later years, the Oparin-Haldane idea of the organic-rich 'prebiotic soup' for the origin of life gained popularity. As van Niel (1956) wrote:

> *Haldane's hypothesis, that the first living entity was an organism with little or no biosynthetic activity, is much easier for present-day biologists to accept than the earlier 'autotroph hypothesis'. ... The conclusion that chemo- and photoautotrophy represent the final stages of biochemical evolution leads to the problem of the order of appearance of these processes. There are, I believe, sound reasons in favor of the thesis that anaerobic chemoautotrophy preceded photoautotrophy, the latter also starting from an anaerobic pattern.*

Different phylogenetic models were based on this and on similar modes of reasoning. Such models always begin with the simplest fermentative anaerobes and progress to the photosynthetic bacteria and from these to the cyanobacteria. Oxygen production by the latter then opens the way for the evolution of aerobic respiration (Hall, 1971; Horvath, 1974; Uzzell and Spolsky, 1974). In his 1975 book *The Evolution of the Bioenergetic Processes*, Engelbert Broda presented a detailed rationale for such a scheme, but he also realized the limitations of his speculations:

> *Clearly, the scheme will need revision on the basis of further biochemical, physiological and morphological data. It will be particularly interesting to check the proposed evolutionary sequences, or improved versions, against the data for compositions of DNA and of selected proteins.*

Today's advanced insights into the microbial phylogeny, as based on sequence analyses of small subunit rRNA and other information-rich molecules, have not yet led to a full understanding of what type of metabolism may have been the first to emerge on our planet. It rather appears that quite soon after life started, many different types of metabolism were already active. However, examination of the rooted universal rRNA sequence-based tree shows how many hyperthermophilic sulfur- and/or hydrogen-dependent chemoautotrophs are found close to the presumed root. Whether indeed chemoautotrophy may have been the first mode of microbial life cannot be ascertained, but in any case the 'chemoautotrophs-first' hypothesis launched over a hundred year ago by Orla-Jensen now looks surprisingly modern (see also Martin and Russell, 2003, for recent views on this topic).

## 'The concept of a bacterium'

'The distinctive property of bacteria and blue–green algae is the procaryotic nature of their cells.' This sentence is one of the keys to the understanding of Stanier and van Niel's famous 1962 paper on 'The concept of a bacterium'.

Before discussing the importance of the paper and its impact at the time, some information on the origin of the terms 'prokaryote' and 'eukaryote' may be useful. In their essay, Stanier and van Niel referred to the earlier writings of the Swiss protozoologist Edouard Chatton (1883–1947), who in the 1930s had coined the terms. The names 'procaryote' and 'eucaryote' were recommended to Stanier in 1961 by his teacher André Lwoff, who had been a student of Chatton and had used Chatton's terms in some of his own publications. It was often suggested that Chatton already had a profound insight into the true nature of the bacterial world ('singular prescience' in the words of Stanier and van Niel). However, the in-depth historical research by Jan Sapp (2005) requires a significant change of this view.

Chatton first used the terms in his 1925 paper on '*Pansporella perplex*: reflections on the biology of the Protozoa'. This article dealt with the life history of an amoeba that lives as a parasite in the intestines of *Daphnia*. The words 'procaryotes' and 'eucaryotes' only appear in two figures at the end of that paper, discussing the place of *Pansporella* among the amoebae. In his first figure ('Essai de Classification des Protistes'), Chatton grouped Cyanophyceae, Bacteriaceae, and Spirochaetaceae as 'procaryotes'. In his second figure ('Essai sur la Phylogénie des Protistes'), he placed the Cyanophyceae (blue–green algae, cyanobacteria) at the base of the tree leading to the 'eucaryotes', the earliest of which were promastigotes – primitive flagellated forms. Spirochaetes and Bacteriaceae were drawn as side branches from the Cyanophyceae. The term 'procaryotes' was thus used in this figure as a label for the bacteria. It was not included in the text. Chatton appears not to have used 'procaryote' to connote a common structural organization or a common ancestry of bacteria, but rather to suggest that the schizomycetes (procaryotes) preceded nucleated cells (eucaryotes) in the evolutionary sequence.

Also in Chatton's 1938 compendium *Titres et Travaux Scientifiques*, sometimes considered a landmark publication in biology, very little is actually said about prokaryotes. The only reference to the term states:

> *Protozoologists agree today in considering the flagellated autotrophs the most primitive of the Protozoa possessing a true nucleus, Eucaryotes (a group which also includes the plants and the Metazoans), because they alone have the power to completely synthesize their protoplasm from a mineral milieu. Heterotrophic organisms are therefore dependent on them for their existence as well as on chemotrophic Procaryotes and autotrophs (nitrifying and sulphurous bacteria, Cyanophyceae).*
>
> (Translation: J. Sapp)

Thus, Chatton's writings from the 1930s did not provide any clear statements and no 'singular prescience' about a fundamental prokaryote–eukaryote dichotomy, as so often suggested in later works.

About twenty years after Chatton's work, and probably independent from it, the prokaryotic–eukaryotic dichotomy reappeared in a 1957 paper by Ellsworth Dougherty. After he had earlier endorsed the kingdom Monera (Dougherty, 1955), now the terms 'eukaryon' and 'prokaryon' appeared to designate the complex, well-organized nucleus of higher organisms (animals, plants, protists) and the more primitive nucleus of the Monera.

When Stanier and van Niel wrote their highly influential 1962 paper, they no longer believed that it could be possible to reconstruct a true phylogeny of the bacteria; at the time they considered the search for bacterial evolutionary relationships to be a futile worthless effort:

> *Our first joint attempt to deal with this problem was made 20 years ago (Stanier and van Niel, 1941). At that time, our answer was framed in an elaborate taxonomic proposal, which neither of us cares any longer to defend. But even though we have become skeptical about the value*

> *of developing formal taxonomic systems for bacteria (see van Niel, 1946, for an explanation of the reasons), the problem of defining these organisms as a group in terms of their biological organization is clearly still of great importance, and remains to be solved.*

Therefore they made an effort to provide a definition of what bacteria are:

> *Any good biologist finds it intellectually distressing to devote his life to the study of a group that cannot be readily and satisfactorily defined in biological terms; and the abiding intellectual scandal of bacteriology has been the absence of a clear concept of a bacterium.*

The long list of definitions of what a prokaryote is, as given in the Stanier and van Niel paper, is all based on negative properties: absence of cytoplasmic organelles, absence of a membrane-surrounded nucleus, absence of the eukaryotic type of cilia and flagella, etc. All bacteria, it was asserted, are prokaryotes, sharing a 'prokaryotic' organization as ascertained by electron microscopic studies. As originally formulated, the prokaryote–eukaryote distinction only described cellular organization, and this distinction has only seldom been questioned. An interesting exception is found in a short paper published in Nature in 1973 by K.A. Bissett of the University of Birmingham. The author argued that the inner layer of the bacterial cell envelope was actually comparable to the nuclear membrane of protists, and he therefore proposed that 'the structure which now represents the cell membrane of bacteria may have originated in the nuclear membrane of an ancestral form'. The same author had earlier made some unusual statements about prokaryote phylogeny, for example: 'Although there is little direct evidence, I am inclined to the opinion, that the spiral water bacteria, with their polar flagella, are the ancestral bacterial form' (Bissett, 1950).

Although not intended as such when first proposed, the meaning of the prokaryote–eukaryote dichotomy quickly changed so as to signify a phylogenetic distinction. It was soon assumed that all organisms on Earth were derived from a common prokaryotic ancestor (Stanier *et al.*, 1963). This idea quickly became generally accepted and remained influential for a long time (e.g. Murray, 1974; Harold, 2001). When later the prokaryote–eukaryote dichotomy model was replaced by the three-domain Archaea–Bacteria–Eucarya model, and it had become clear that the prokaryotes do not form a phylogenetically coherent taxon, Woese (2004) made the following comments on Stanier van van Niel's 1962 paper:

> *There was one occasion (perhaps the only one) on which the 'lack of a concept of a bacterium' was recognized and denounced as the 'abiding scandal' of bacteriology [Stanier and van Niel, 1962]. But, rather than use this insight to begin a much-needed dialog within the field, the authors concocted a guesswork solution to settle the matter then and there, thereby removing the question/problem from the arena of discourse.*

## The phylogeny of chloroplasts and mitochondria

Since the work of Lynn Margulis, the idea is well known that mitochondria and chloroplasts were derived from the prokaryote world, and had originally been incorporated into the primitive eukaryotic cell as endosymbionts (Margulis, 1970). However, it is seldom realized that the concept that the eukaryotic cytoplasmic organelles are related to bacteria already originated in the nineteenth century. Sapp (2005) has provided an excellent overview of the old literature on this topic.

The possible origin of the chloroplasts from a symbiotic association was probably first suggested by Andreas Schimper in 1883. It was further discussed by Haeckel in 1905, and was developed most prominently by the Russian botanist Constantin Merezhkowsky (1855–1921). Merezhkowsky coined the word 'symbiogenesis' for the synthesis of new organisms by symbiosis (Merezhkowsky, 1905, 1910), and he provided arguments to prove that chloroplasts had originated from symbiotic cyanobacteria in the remote past (Merezhkowsky, 1920).

At about the same time the idea of the origin of mitochondria from bacterial symbionts was developed. The French biologist Paul Portier (1866–1962), working at the Institut Océanographique de Monaco, had developed a theory of symbiosis as a fundamental aspect of life. He argued that mitochondria originated as symbionts and had been transformed from free-living bacteria to their current function as intracellular respiratory organelles in the course of the evolution of the eukaryotic cell (Portier, 1918). Similar ideas were published in the 1920s by Ivan Wallin of the University of Colorado (1883–1969). In his book *Symbionticism and the Origin of Species* (1927), Wallin proposed that bacterial symbionts were the source of new genes, and that such symbiotic associations were one of the primary mechanisms for the origin of species. He even claimed that he had cultured mitochondria to prove their actual bacterial nature (Wallin, 1927).

Analysis of the 16S rRNA of these organelles has fully confirmed their bacterial nature, the mitochondria cluster within the Proteobacteria branch of the Bacteria and the chloroplasts cluster within the cyanobacterial lineage.

## 'The Dark Age'

Until the 1940s there still was a general feeling of optimism that the major questions of microbial phylogeny could be solved and that the construction of a universal tree of life may be possible. The next two decades, called 'The Dark Age' by Woese (1994), were a period of deep pessimism. At this time many of the leading microbiologists had realized that the classical morphology- and physiology-based approaches to establish a natural bacterial taxonomy could not crack the problem. This even led to a general feeling that bacterial phylogenies are inherently unknowable (Stanier *et al.*, 1957; van Niel, 1955). An impasse was reached, which lasted until the late 1970s when Woese revealed the first results of this small subunit rRNA sequencing work and analysed its implications. To illustrate the atmosphere of the pre-1977 period, here are some ideas expressed by four leading microbiologists of those days: Pringsheim, Winogradsky, van Niel and Stanier.

The Czech microbiologist Ernst Georg Pringsheim (1881–1970) had already very early realized that the concepts of bacterial systematics and the attempts to reconstruct bacterial phylogeny were entirely based on negative characteristics. Early in his career he contended that similarities between certain groups may well have arisen by convergent evolution rather than by a true evolutionary relationship (Pringsheim, 1923). This also included the question of the possible ancestor of the cyanobacteria (Myxophyceae). He argued that it was entirely possible, perhaps likely, that the similarities between the blue–green algae and the bacteria resulted from convergent evolution. Arguments in favour of a true relationship as presented by Stanier and van Niel (1941), such as absence of true nuclei, absence of sexual reproduction, and absence of plastids were all negative properties, and therefore the possibility of convergent evolution of the two classes should be seriously considered (Pringsheim, 1949). Fifteen years later he made the following highly pessimistic statement (Pringsheim, 1964):

> *... müssen wir uns ... durchaus von der Vorstellung freimachen, als könne man die uns heute umgebenden Lebensformen stammbaumartig miteinander verknüpfen. Die heute lebenden Organismen sind nur die verschwindend wenigen übrig gebliebenen Abkömmlinge einer unvorstellbar großen Zahl verschiedenartiger Lebensformen, deren weitaus größter Teil spurlos verschwunden ist... es ist unwahrscheinlich, daß eine Art sich über lange Zeiten unverändert fortgepflanzt hat, nachdem eine andere durch Mutation aus ihr hervorgegangen ist.*
>
> *(... we should altogether abandon the idea that it may be possible to connect the currently found life forms in the form of a family tree. The organisms living today are no more than the negligibly few remaining descendants of an unthinkably large number of life forms of different nature, of which by far the greatest part has disappeared without a trace. ... It is improbable that a species has propagated itself unchanged over long periods after another species had originated from it by mutation.)*
>
> (Translation: A.O.)

Towards the end of his long and productive life, Sergei Winogradsky (1856–1953) commented in his very last publication that principles of phylogenetic classification that work well for plants and animals cannot be applied to bacteria. Comparable efforts in the realm of the bacteria (and blue–green algae/cyanobacteria) are doomed to failure because it does not appear likely that criteria of truly phylogenetic significance can be devised for these organisms (Winogradsky, 1952). He also reminded bacteriologists that the naming of species, genera, tribes, families, and orders in *Bergey's Manual* was only a façade, and had nothing to do with any possible true relationships.

Cornelis van Niel, who earlier had co-authored two highly influential papers with proposed morphology-based natural systems of classification of bacteria (Kluyver and van Niel, 1936; Stanier and van Niel, 1941), had by 1946 ceased to believe in the value of his earlier models. He first expressed his pessimistic views on bacterial classification at the Cold Spring Harbor Symposium of 1946 (van Niel, 1946). He concluded that criteria of truly phylogenetic significance cannot be devised for these organisms (van Niel, 1955):

> *Even for a general outline along phylogenetic lines, the available information is entirely inadequate. Much of this is, of course, the result of the paucity of characteristics, especially those of a developmental nature.*

Roger Stanier, one of the very few microbiologists who at the time still maintained any interest at all in bacterial evolution, wrote in his textbook *The Microbial World* that 'the ultimate scientific goal of biological classification cannot be achieved in the case of bacteria' (Stanier *et al.*, 1963), and a few years later he came with the following statement (Stanier, 1970):

> *Evolutionary speculation constitutes a kind of metascience, which has the same intellectual fascination for some biologists that metaphysical speculation possessed for some mediaeval scholastics. It can be considered a relatively harmless habit, like eating peanuts, unless it assumes the form of an obsession, then it becomes a vice.*

Thus, in the 1950s Stanier and van Niel had become disappointed in their search for bacterial relationships, and neither one wished to put further serious thought or effort into the problem. Woese (1994) noted there were 'two C.B. van Niels and two Roger Staniers, for their careers spanned and reflect both sides of the watershed separating the Classical Period from the ensuing Dark Age'. And Sapp (2005) commented: 'Woese saw Stanier as a tragic figure who had given up on his hope for phylogenetics and who had discouraged its pursuit.' However, others had a different view. Thus, John Ingraham and Horishi Nikaido (1994) wrote:

> *Those of us who lived through the 1960s and 1970s as professional microbiologists know that this description does not reflect what actually happened. First Roger Stanier did not destroy the enthusiasm of microbiologists for phylogeny; he stimulated it. Perhaps his principal goal throughout his professional life was to make sense of the microbial world by organizing microorganisms into phylogenetically related groups and thereby to integrate microbiology into the rest of biology.*

Ingraham and Nikaido thus were convinced that Stanier had not really abandoned phylogeny but rather 'sometimes let out cries of despair', frustrated by the inadequacy of the methods then available.

## The origins of the three-domain universal phylogenetic tree

In the late 1970s, Woese's three-domain model with the Archaebacteria, Eubacteria and Eukaryote branches started replacing the earlier generally accepted phylogenetic prokaryote–eukaryote dichotomy. The story of the events that led to the new concepts has been told many times (e.g. Woese, 1987, 1994, 2004), and therefore a few comments will suffice here.

The emergence first of the technology to determine amino acid sequences in proteins and later nucleotide sequences in RNA and DNA has revolutionized all biological disciplines. The prospect that it may become possible to track the course of evolution from analysis of such sequences was already predicted more than 50 years ago by Francis Crick (1958):

> *Biologists should realize that before long we shall have a subject which might be called 'protein taxonomy' – the study of amino acid sequences of proteins of an organism and the comparison*

*of them between species. It can be argued that these sequences are the most delicate expression possible of the phenotype of an organism and that vast amounts of evolutionary information may be hidden away with them.*

The theoretical framework for such evolutionary reconstructions was established in the now classic paper by Zuckerkandl and Pauling (1965), who defined biological macromolecules as documents of evolutionary history, which enables the reconstruction of phylogeny from comparisons of molecular structures. The authors introduced the concept of the 'molecular clock': random changes in the sequence of amino acids or nucleotides may occur at a more or less constant rate, so that the number of changes in a certain molecule with a conserved function should be approximately proportional to evolutionary time. The technology to read the information came first with the development of methods to assess the amino acid sequences in proteins, and later of sequencing nucleic acids. In the mid-1960s, Frederick Sanger and his co-workers had devised a method for partially characterizing RNA sequences, based on a two-dimensional paper electrophoretic method called oligonucleotide cataloguing (Sanger *et al.*, 1965). Woese and his colleagues then applied this method of T1 RNAse cataloguing to the small-subunit ribosomal RNA (Fox *et al.*, 1977a) and they started comparing 16S rRNA molecules of different prokaryotic microorganisms and 18S rRNA from eukaryotes. Unexpected results were obtained when in June 1976 for the first time a methanogen, supplied by the laboratory of Ralph Wolfe, was subjected to this kind of analysis (Fox *et al.*, 1977b). In his scientific autobiography, Wolfe (1991) described the events as follows:

*When I asked Woese about the results of the first attempt to label the 16S rRNA of a methanogen, he replied that something had gone wrong with the extraction – perhaps they had isolated the wrong RNA. The experiment was repeated with special care, and this time Carl's voice was full of disbelief when he said, 'Wolfe, these things aren't even bacteria'.*

In May 1977 similar sequence information was obtained for the first extreme halophile (Magrum *et al.*, 1978). Sequences deduced from 16S rRNA oligonucleotide catalogues of the hyperthermophiles *Thermoplasma* and *Sulfolobus* completed the dataset on which the three-domain model was originally based (Fox *et al.*, 1977b; Olsen *et al.*, 1985). Based on the sequence information it was obvious that 'These 'bacteria' appear to be no more related to typical bacteria than they are to eukaryotic cytoplasms' (Woese and Fox, 1977). Woese immediately realized that the newly acquired capacity to read the cell's historical information (by the sequencing of the genes) would drastically alter the way we look at all of biology (Woese, 1987).

A more in-depth investigation into the properties of the 'unusual' groups of prokaryotes – methanogens, extreme halophiles and thermophiles/thermoacidophiles – quickly showed that these organisms share many more special features in which they differ from the 'normal' prokaryotes. In fact, the presence of isoprenoid ether lipids in *Halobacterium* had already been known since the early 1960s (Sehgal *et al.*, 1962), but only now did the implications of the finding become clear when it was ascertained that similar lipids occur in the methanogens and the thermoacidophiles as well. To further stress that the newly established domain is basically unrelated to the other prokaryotes, Woese and his collaborators proposed the terms Archaea and Bacteria to replace the earlier names Archaebacteria and Eubactera, to avoid all possible misunderstanding (Woese *et al.*, 1990).

When the three-domain model was first published, Woese and his colleagues showed unrooted trees. In 1989, the groups led by Peter Gogarten and by Naoyuki Iwabe independently used ancient gene duplications to root the tree by means of an outgroup (Gogarten *et al.*, 1989; Iwabe *et al.*, 1989).

Although the division of the life forms on Earth into Archaea, Bacteria and Eucarya is now commonly accepted, not all biologists have been equally happy with the three-domain tree, and alternative views have been expressed as well (Margulis and Guerrero, 1991; Cavalier-Smith, 1992). In 1990, Ernst Mayr noted that the establishment of the Archaebacteria/Archaea as a new domain was grossly misleading, and he claimed that in Woese's arrangement the previous imbalance is replaced by a different one, as the ranking in the classification suggested by Woese is based entirely on the amount of difference in ribosomal RNA, ignoring the enormous evolutionary step from the prokaryotes to the eukaryotes. Instead, Mayr (1990) proposed the domain Prokaryota with two subdomains, Eubacteria and Archaebacteria, and the domain Eukaryota with two subdomains: Protista and Metabionta. In a later paper Mayr considered Woese's three-domain model absurd as evolution was 'an affair of phenotypes' and 'all archaebacteria are nearly indistinguishable'. Even if one took prokaryotes as a whole, he argued, it 'does not reach anywhere the size and diversity of eukaryotes' (Mayr, 1998).

Whatever model one may prefer, phylogenies derived from sequence analysis have to be accepted for what they minimally are: hypotheses, to be tested and either strengthened or rejected on the basis of other kinds of data (Jensen, 1985).

## Phylogeny and Bergey's Manual

*Bergey's Manual of Determinative Bacteriology* (nine editions since 1923) and its successor *Bergey's Manual of Systematic Bacteriology* (first edition 1984–1989, second edition from 2001 onwards) has been the major source of information on the properties of prokaryotes for over 80 years now. The editors of each new edition of the Manual had to cope with the question how to arrange the ever increasing number of prokaryotic microorganisms in a more or less logical scheme. Although the handbook never claimed to be a guide on bacterial phylogeny, it was understandable that the editors chose to use the at the time current insights when setting up their classification and identification schemes. With the notable exception of the most recent edition of the Manual, this resulted in different arbitrary classification systems that used all kinds of properties: morphological, physiological, nomenclatural, medical, and others. An overview of the history of *Bergey's Manual* can be found in volume 1 of the last edition (Murray and Holt, 2001).

The first edition was prepared in the early 1920s by a committee of the Society of American Bacteriologists, headed by David Bergey (1845–1932). From the beginning the editorial board of *Bergey's Manual* opted for a reasonably stable, unnatural classification instead of a phylogenetic classification. In their preface to the first edition, the editors wrote (Bergey *et al.*, 1923):

'The Committee [on Characterization and Classification of the Society of American Bacteriologists] does not regard the classification of species offered here as in any sense final, but merely a progress report leading to more satisfactory classification in the future.'

Many of the leading microbiologists found the classification of the bacteria in the different editions of *Bergey's Manual* highly unsatisfactory. Stanier and van Niel (1941) wrote: 'A more inadequate definition than that given by Bergey would be hard to conceive'. Kluyver and van Niel (1936) had made a similar statement. The editors, however, argued that natural relationships were obscure and that a natural system would be highly speculative and constantly changing. Thus, Robert Breed, one of the editorial board members of *Bergey's Manual*, commented: 'Realistic workers have on their side been impatient with idealists who have introduced many unjustified speculations regarding relationships between the various groups of bacteria.' (Breed, 1939). The classification used in *Bergey's Manual* also influenced the arrangement of taxonomic material in textbooks. Thus Henrici's textbook *The Biology of the Bacteria* (1939) followed the classification of the 5th edition of *Bergey's* in its chapters discussing the diversity in the bacterial world.

The introductory chapter to the 7th edition of *Bergey's Manual*, published in 1957, makes some interesting statements on prokaryote phylogeny and on the philosophy of the editors of the manual at the time (Breed, 1957):

> *'With this thought in mind, some students of the systematic relationships of living things have thought of the chemoautotrophic bacteria that still exist as being more like primordial living things than are other types of bacteria.*[1] *It is true that the chemoautotrophic organisms are able to live on simple inorganic foods that were, in*

1 This was possibly based on the above-discussed scheme by Orla-Jensen, proposed in 1909.

*all probability, available to living things under early conditions in the development of the earth. However, it does not necessarily follow that chemoautotrophic forms are the only ones that could have existed in the beginning. It seems even more reasonable to assume that early living forms developed a pigment like chlorophyll that enabled primordial bacterial to utilize the sun's energy in synthesizing organic matter. Such photosynthetic pigments are found in purple or green bacteria. These photoautotrophic forms could have existed on the simple foods available when life began as readily as could chemoautotrophic forms. ... In the present edition of Bergey's Manual, the classification used has been rearranged on the assumptions that the photoautotrophic bacteria extant today presumably are the living organisms that are most nearly like the primordial types of bacteria.*

Thus, a new artificial scheme arose that, without any obvious reason, placed the anoxygenic phototrophic bacteria first on the basis of some new assumptions that were nowhere documented in further detail.

The following edition of the *Manual* no longer tried to link its classification to certain models of bacterial phylogeny. The editors wrote (Buchanan and Gibbons, 1974):

*The Manual is meant to assist in the identification of bacteria. No attempt has been made to provide a complete hierarchy, as in previous editions, because a complete and meaningful hierarchy is impossible.*

And in an introductory chapter to the book, Murray (1974) wrote:

*The future will have to bring a regrouping of higher taxa to express a more coordinate view. This example is constructed without hierarchical prejudice (e.g. no attempt is made to indicate that non-photosynthetic procaryotes must have preceded the photosynthetic groups on an evolutionary scale) and using common names or tentative names in parentheses, as a suggestion of a possible arrangement.*

But still the prominent placement of the anoxygenic phototrophs, as given in the 7th edition, was retained, so that the 8th edition of *Bergey's Manual* was structured as follows (Murray, 1974):

Kingdom *Procaryotae*
Division I: Phototrophic procaryotes
('Photobacteria')
Class I: Blue–green photobacteria
Class II: Red photobacteria
Class III: Green photobacteria
Division II: Procaryotes indifferent to light
('Scotobacteria')
Class I: The bacteria
Class II: Obligate intracellular Scotobacteria in eucaryotic cells – Rickettsias
Class III Scotobacteria without cell walls – *Mollicutes*

The author then added the following comments:

*No doubt time will settle the problems involved in making a reasonable arrangement within the Procaryotae. Haste is unwise; all previous classifications seem to have suffered infinite rearrangement due to insufficient information. The groupings suggested above, even if the rank designations may need adjustment, might appear to be stable until a new level of understanding overtakes us. The new insights are likely to come from a clear understanding, on a comparative level, of the components of the genome of procaryotic cells. This may allow us to understand why and how vastly different phylogenetic groupings of phototrophic, lithotrophic and organotrophic organisms have remarkably similar and complex structural components (as exemplified by the wide range of spirilla). Surprises may well be in store for us; hence, our adherence to simplicity.*

The predicted surprises indeed came very soon: just three years after the publication of the 8th edition, Woese first revealed his three-domain tree of life!

When in 1984 the first volume of a new edition of the manual was published (Krieg and Holt, 1984), the title was changed from *Bergey's Manual of Determinative Bacteriology* to *Bergey's Manual of Systematic Bacteriology* to reflect the change in scope (Krieg, 1984; Staley and Krieg, 1984):

> *One important goal of the Manual is to assist in the identification of bacteria, but another goal, equally important, is to indicate the relationships that exist between the various kinds of bacteria. The methods of molecular biology have now made it possible to envision the eventual development of a comprehensive classification of bacteria based on their relatedness to one another ... Such a general scheme, however, cannot yet be perceived fully. The relatedness within and between some bacterial groups has been intensively studied, but for other groups very little work has been done. Moreover, the relatedness studies that have been done often have involved the use of one or another method without confirmation by other methods.*
>
> *As recently as the seventh edition of Bergey's Manual [i.e. 1957] ... the view was expressed that bacteria were a primitive group of organisms, and the classification scheme presented in that edition of the Manual claimed to be a natural scheme in which the photosynthetic bacteria were treated first, because they were regarded as the most primitive bacterial group. However, because of the lack of objective evidence for this (or any other) phylogenetic scheme, the eighth edition of the Manual abandoned all attempts at a phylogenetic approach to bacterial classification. ... Phylogenetic information has increased since the eighth edition, however, largely through the increasing use of methods for measuring genetic relatedness (i.e. DNA/DNA hybridization, DNA/rRNA hybridization, rRNA oligonucleotide cataloging, and protein sequencing). ... Unfortunately, the phylogenetic information is still in a fragmentary form, and it seems probable that the interpretation of the data is still not entirely clear ...*

Indeed, although the editors dreamt of a phylogenetic arrangement of the organisms, this was not yet completely feasible at the time. Their edition of the manual was published in four subvolumes: (a) the Gram-negatives of general, medical or industrial importance; (b) the Gram-positives other than actinomycetes; (c) the archaeobacteria, cyanobacteria and remaining Gram-negatives; and (d) the actinomycetes. In some cases the arrangement of chapters was quite curious indeed. Thus, the (archaeal) family *Halobacteriaceae* appeared as Family V in the section on Gram-negative aerobic rods and cocci in the first published subvolume, after the families *Pseudomonadaceae*, *Azotobacteriaceae*, *Rhizobiaceae*, and *Methylococcaceae*, and was followed by the *Acetobacteraceae* and the *Neisseriaceae*. In subvolume 3, which also covered the Archaea (archaeobacteria), the *Halobacteriaceae* were included once more.

With the publication of the second edition of *Bergey's Handbook of Systematic Bacteriology*, in five planned volumes of which the first was released in 2001, the second in 2005 and the third in 2009, the goal appears to have been achieved: arrangement of all prokaryotes according to a 16S rRNA sequence based framework, reflecting the current ideas, and made possible by the fact that 16S rRNA sequences of about all type strains of species and genera are now available. To quote from the preface to the first volume (Garrity, 2001):

> *Since publication of the First Edition of the Systematics Manual, we have witnessed a major shift in how we view the relationships among Bacteria and Archaea. While the possibility of a universally applicable natural classification was evident as the First Edition was in preparation, it is only recently that the sequence databases became large enough, and the taxonomic coverage broad enough, to make such an arrangement feasible. We have relied heavily upon these data in organizing the contents of this edition of Bergey's Manual of Systematic Bacteriology, which will follow a phylogenetic framework based on analysis of the nucleotide sequence of the small ribosomal subunit RNA, rather than a phenotypic structure. This departs from the First Edition, as well as the Eighth and the Ninth Editions of the Determinative Manual. While the rationale for presenting the content of this edition in such a manner should be evident to most readers, they should bear in mind that this edition, as have all preceding ones, represents a progress report rather than a final classification of procaryotes.*

Indeed, a final classification will never be possible, and new discoveries in the future may again lead to changes in concepts. But at present the existing framework looks extremely satisfactory.

## Teaching prokaryote evolution: phylogeny and microbiology textbooks

After the above discussions of the most influential articles that have shaped the ideas about microbial phylogeny in the past hundred years, it is also interesting to see how the major textbooks on general microbiology have included these ideas to teach the current models to new generations of scientists.

The first example is from *Bau und Leben der Bakterien* by Wilhelm Benecke, published in 1912:

> *So sind wir denn gezwungen, heuten Tags ein Bakteriensystem dessen wir nicht entraten können, auf recht äußerlichen Merkmalen aufzubauen, und hoffen, daß er nur eine Frage der Zeit sein möchte, daß es gelingt, ein natürliches System auf Grund genauerer Erkenntnis der Zellorganisation der Bakterien aufzustellen, und auch auf Grund der rüstig vorwärtsschreitenden Versuche über Vererbung und Artbildung bei Bakterien, die wie im nächsten Kapitel noch kennen lernen werden. Die Bakteriepaläontologie läßt uns in diesen Fragen fast ganz im Stich.*
>
> *(Thus we are nowadays forced to base a system of the bacteria, with which we cannot dispense, purely on truly superficial properties, and we hope that it only will be a matter of time that it may be possible to design a natural system based on a more exact understanding of the cellular organization of the bacteria, also based on the firmly advancing experiments about genetics and speciation in bacteria, which will be further discussed in the next chapter. About these questions the bacterial paleontology leaves us almost completely in the dark.)*
>
> (Translation A.O.).

No speculations here, no elaborate models to attempt to establish order in the apparent chaos, just a modest recognition that time was not yet ripe to say anything more sensible about bacterial evolution and phylogeny, as the experimental tools for that purpose were not sufficiently developed at the time.

In the first half of the twentieth century there were hardly any good textbooks of general microbiology in the English language. In his memoirs in which he discussed the origins of his own textbook *The Microbial World* (Stanier *et al.*, 1957), Roger Stanier claimed that only one earlier textbook could be considered worthy, and that was Arthur Henrici's *The Biology of Bacteria* (1st edition 1930): 'Ed [Adelberg], Mike [Doudoroff] and I used to grouse intermittently about the very inferior quality of elementary bacteriology textbooks. Only one was intellectually respectable: A.T. Henrici's *The Biology of Bacteria*. This was much out of date ...' (Stanier, 1980). It is therefore of interest to see how the best student text of the time handled the key questions of bacterial evolution. The second edition of Henrici's book (1939) gave the following information under the heading 'Relationships of Bacteria':

> *The ensheathed filamentous Chlamydiobacteriales [encompassing genera such as Leptothrix and Sphaerotilus] show strong resemblances to the filamentous Blue–green Algae, as do also the white sulphur bacteria. The peculiar creeping motions of Beggiatoa imitate exactly those of the alga Oscillatoria. ... More recently, however, the bacteria have been thought to be related to the Fungi rather than to the Algae. Some of the Actinomycetales are so mold-like that certain authorities would include them with the fungi rather than with the bacteria. The Actinomycetales show clear relationships with a number of true bacteria. ... The various similarities of the bacteria to other groups of microbes may be summarized in the following table:*
>
> *Chlamydobacteriales → Cyanophyceae*
> *Thiobacteriales → Cyanophyceae*
> *Caulobacteriales → Unknown*
> *Myxobacteriales → Unknown*
> *Actinomycetales → Fungi*
> *Eubacteriales*
> *Gram-positive species → Fungi*
> *Gram-negative species → Unknown*
> *Spirochaetales → Protozoa*

And in a later chapter on 'Classifiation of Bacteria', Henrici wrote:

> *Further, it is hoped to arrange these divisions so that they will actually represent 'blood' relationships, that they will represent the course of evolution – to place in one genus all those species which have been derived from a common ancestral type, to place in one family all those genera which have had a common origin, and so on.*

Beautiful words indeed, but the whole scheme should have looked highly speculative even at the time the textbook was written. How fungi may have evolved from Gram-positive bacteria and how spirochetes may have given rise to the formation of protozoa was not made clear, and I did not find any support for these statements in the contemporary scientific literature.

The different editions of the beautifully illustrated textbook *The Microbial World*, by Stanier, Doudoroff and Adelberg, (Stanier *et al.*, 1957, 1963, 1970), reflected Roger Stanier's pessimistic views at the time, as discussed above, claiming that all attempts to use a phylogenetic approach in bacterial classification are no more than futile exercises (Stanier, 1970). Thus, the first edition of the book (Stanier *et al.*, 1957) stated:

> *... the construction of the broad outlines of a natural system of bacterial classification involves much guesswork and affords the possibility for endless unprofitable disputes between the holders of different views about bacterial evolution. An eminent contemporary bacteriologist, van Niel, who is noted for his taxonomic studies on several groups of bacteria, has expressed the opinion that it is a waste of time to attempt a natural system of classification for bacteria, and that bacteriologists should concentrate instead on the more humble practical task of devising determinative keys to provide the easiest possible identification of species and genera. This opinion, based on a clear recognition and acceptance of our ignorance concerning bacterial evolution, probably represents the soundest approach to bacterial classification, but it has not gained universal acceptance.*

The second edition of 1963 naturally included the ideas expressed a year earlier in Stanier and van Niel's *The Concept of a Bacterium*:

> *In fact, the basic divergence in cellular structure, which separates the bacteria and blue–green algae from all other cellular organisms, represents the greatest single evolutionary discontinuity to be found in the present-day world.*

> *All these organisms share the distinctive structural properties associated with the prokaryotic cell ..., and we can therefore safely infer a common origin for the whole group in the remote evolutionary past; we can also discern four principal subgroups, blue–green algae, myxobacteria, spirochetes, and eubacteria, which seem to be distinct from one another. ... Beyond this point, however, any systematic attempt to construct a detailed scheme of natural relationships becomes the purest speculation, completely unsupported by any sort of evidence.*
>
> *... the ultimate scientific goal of biological classification cannot be achieved in the case of bacteria.*

I myself was introduced into the basics of microbiology in the spring of 1970, and the textbook used was the then very recently published first edition of Hans Schlegel's *Allgemeine Mikrobiologie*. The following sentences have undoubtedly influenced my own concepts on the bacterial world during my student years. Life on Earth was divided into three kingdoms, animals, plants, and protists, but the differences between prokaryotes and eukaryotes were also stressed as being very basic:

> *Diese Unterschiede sind so tiefgreifend, daß man beide Gruppen als Prokaryonten und Eukaryonten einander gegenüberstellt.*

> *(These differences are so fundamental, that the two groups are opposed as prokaryotes and eukaryotes.)*

> *Es sind zwei Arten von Klassifikationen zu unterscheiden: phylogenetische oder 'naturliche' Klassifikationen einerseits und künstliche Klassifikationen andererseits. Erstere ist das*

*weitgesteckte Ziel der Bakterien-Taxonomie, nämlich verwandte Formen – also solche, die durch gemeinsame Vorfahren verbunden sind – zusammenzuordnen und einen phylogenetischen Stammbaum der Bakterien zu entwickeln. Es besteht keine Zweifel, daß dieses Ziel auf der Grundlage chemischer Merkmale, etwa der Sequenz der Aminosäuren in funktionsgleichen Enzymproteine, einmal erreicht werden wird. Einfache morphologische oder physiologische Merkmale reichen dazu bei Bakterien nicht aus.*

*(One can discriminate between two kinds of classification: phylogenetic or 'natural' classification on one hand and artificial classifications on the other hand. The first is the ultimate goal of bacterial taxonomy, namely to classify together related forms, i.e. those that are linked by a common evolutionary history. There can be no doubt that once this goal will be achieved on the basis of chemical characteristics such as the amino acid sequence of enzyme proteins with similar functions. Simple morphological or physiological characteristics are insufficient for that purpose in the case of the bacteria.)*

(Translation: A.O.)

A very modest and realistic statement at the time; in fact it differs very little from the above-cited text from Wilhelm Benecke's 1912 book. It is clear that very little progress had been made in more than 50 years.

At about the same time in which Schlegel's short microbiology text was first published, the first edition of Thomas Brock's *Biology of Microorganisms* became available. This book, which has regularly been updated and is now in its 12th edition *as Brock Biology of Microorganisms,* can probably be considered as the leading general microbiology textbook currently on the market. The first edition (Brock, 1970) makes a statement that does not greatly differ from the above quotation from Schlegel's book:

*A third and perhaps the most fundamental is the genetic or molecular approach, which aims to ascertain the degree of genetic relatedness of different organisms. The ultimate goal could be to determine the DNA base sequences of the complete genome of the organisms since the better the sequences of two organisms match, the closer is their genetic relatedness. Although this last goal is not yet attainable, several interesting approaches to it can be carried out; ... (DNA base composition, nucleic acid hybridization ...).*

When the third edition was printed in 1978, the three-domain concept Archaebacteria–Eubacteria–Eukaryotes was still too new and not yet sufficiently widely known (and surely not sufficiently widely appreciated!) to be included in the textbook. However, the fourth edition (Brock *et al.*, 1984) already gave a full account of the new ideas. It was probably the first time that the term archaebacteria was introduced in a student textbook, and every word written then is still valid today:

*The picture that is emerging from molecular sequencing studies using ribosomal RNAs that three groups of cellular organisms may have evolved from the ancestral cell, two of these groups being procaryotic, and the remaining one eucaryotic. The two prokaryotic groups have been designated the eubacteria, consisting of the majority of Gram-positive and Gram-negative bacteria plus the cyanobacteria, and the archaebacteria, a group of unusual procaryotes including the methanogenic bacteria, the extremely halophilic bacteria, and certain thermoacidophilic bacteria.*

All subsequent editions have included the latest updates, and other microbiology textbooks have now adopted the three-domain model as well.

In many texts of general biology, high school textbooks as well as university-level texts, Whittaker's five-kingdom system (Whittaker, 1969) is still popular. Therefore a few words should be devoted to this model, based on morphological and ecological considerations rather than on phylogeny. In Whittaker's original concept, four kingdoms were proposed: Protista, Plantae, Fungi and Animalia; the Protista were further divided into Monera (all prokaryotes) and the Eunucleata (the eukaryotic protists) (Whittaker, 1959). The separation of the Fungi from the Plantae was based on ecological considerations, opposing the autotrophic green plants from the heterotrophic world of the fungi. Following Stanier and van Niel's 1962 paper on 'The concept of a bacterium' presenting the prokaryote–eukaryote dichotomy, Whittaker modified his model to establish a five-kingdom with Animalia, Plantae, Fungi, Protista and Monera, and this proposal was widely adopted (see also Mayr, 1990). To justify the change, Whittaker (1969) commented:

> *These contrasts between the procaryotic cells of bacteria and blue–green algae, and the eucaryotic cells of other organisms, define the clearest, most effectively discontinuous separation of levels of organization in the living world ... the difference between procaryotes and eucaryotes remains a line of division deserving recognition in a current system of broad classification.*

## The genomics era and beyond – final comments

> *Old prejudices tend to inhibit, distort, or otherwise shape new ideas, and historical analysis helps to eliminate much of the negative impact of the status quo.*

Thus wrote Carl Woese in 1987 when analysing the great events that have shaped our concepts on microbial phylogeny in the past century. Ideas have changed, sometimes slowly, sometimes more rapidly, as new techniques became available that enabled a better based view of the evolution of microorganisms and especially the evolution of that very heterogeneous group of organisms designated 'procaryotes' in Stanier and van Niel's 1962 classic paper.

The three-domain tree of life, as first proposed by Woese in 1977 on the basis of small subunit rRNA sequence comparisons, was initially slow to become accepted by the scientific community, and, as shown above, this model still comes under attack from time to time long after most biologists have been convinced that it is at least superior to all earlier models. Now complete genome sequences of prokaryotes are becoming available at an ever increasing rate, it has become clear that also the relatively simple three-domain model of evolution cannot explain all the observations. Although the occurrence of horizontal gene transfer was postulated already in the late 1970s from analyses of cytochrome *c* genes in purple bacteria of the *Rhodospirillaceae* (Ambler, 1979a,b), the extent of the phenomenon is only now becoming clear. As lateral gene transfer now appears to be so extensive and so many genes are shared across the bacterial taxonomic spectrum, the issue of whether it is indeed possible to trace bacterial phylogeny re-emerged. One of the sceptics was Ford Doolittle, who drew a tangled net-like 'shrub' rather than a tree to illustrate what may be the true phylogenetic relationships within and among Woese's proposed three domains of life, and he therefore claimed that constructing a universal phylogenetic tree may be altogether impossible (Doolittle, 2000).

It is thus well conceivable that 16S rRNA sequences are good markers for prokaryote genera and higher taxa, but insufficient to delineate species. Staley (2006, 2009) therefore proposed a 'phylogenomic species concept' that relies both on phylogenetic and on genomic approaches. Genomic methods such as multi-locus sequence analysis will help resolve issues of horizontal gene transfer. It is well possible that within a short time our models on microbial, and especially on prokaryotic phylogeny, will have to be modified again.

In view of all it is fitting to end this chapter with yet another quotation from Carl Woese's writings (Woese, 2004):

> *Scientific advance, then, is a succession of newer representations superseding older ones, either because an older one has run its course and is no longer a reliable guide for a field, or because the newer one is more powerful, encompassing, and productive than its predecessor(s).*

## References

Ambler, R.P., Daniel, M., Hermoso, J., Meyer, T.E., Bartsch, R.G., and Kamen, M.D. (1979a). Cytochrome $c_2$ sequence variation among the recognized species of purple nonsulfur photosynthetic bacteria. Nature *278*, 659–660.

Ambler, R.P., Meyer, T.E., and Kamen, M.D. (1979b). Anomalies in amino acid sequences of small cytochrome *c* and cytochrome *c′* from two species of purple photosynthetic bacteria. Nature *278*, 661–662.

Beijerinck, W. (1905). Address made on September 30th, 1905 at the presentation of the Leeuwenhoek Medal of the „Koninklijke Akademie van Wetenschappen te Amsterdam" to Beijerinck. Versl. Kon. Akad. v. Wet. Amsterdam *14*, 203.

Benecke, W. (1912). Bau und Leben der Bakterien (Leipzig: B.G. Teubner).

Bergey, D.H., Harrison, F.C., Breed, R.S., Hammer, B.W., and Huntoon, F.M. (ed.) (1923). Bergey's Manual of Determinative Bacteriology, 1st ed. (Baltimore: The Williams & Wilkins Company).

Bissett, K.A. (1950). Evolution in bacteria and the significance of the bacterial spore. Nature *166*, 431–432.

Bissett, K.A. (1973). Do bacteria have a nuclear membrane? Nature *241*, 45.

Breed, R.S. (1939). In *Bergey's Manual* of Determinative Bacteriology. 5th ed., D.H. Bergey, R.S. Breed, E.G.D. Murray, and A.P. Hitchens, ed. (Baltimore: The Williams & Wilkins Company).

Breed, R.S. (1957). Considerations influencing the classification used in this edition of the Manual. In Bergey's Manual of Determinative Bacteriology. 7th ed., R.S. Breed, E.G.D. Murray, and N.R. Smith, ed. (Baltimore: The Williams & Wilkins Company), pp. 4–14.

Brock, T.D. (1970). Biology of Microorganisms, 1st ed. (Englewood Cliffs: Prentice Hall).

Brock, T.D., Smith, D.W., and Madigan, M.T. (1984). Biology of Microorganisms, 4th ed. (Englewood Cliffs: Prentice Hall).

Broda, E. (1975). The Evolution of the Bioenergetic Processes (Oxford: Pergamon Press).

Buchanan, R.E., and Gibbons, N.E. (1974). Introduction. On using the Manual. In Bergey's Manual of Determinative Bacteriology, 8th ed., R.E. Buchanan, and N.E. Gibbons, ed. (Baltimore: The Williams & Wilkins Company), pp. 1–3.

Burkhardt, F., and Smith, S. (1990). The Correspondence of Charles Darwin, vol. 6, 1856–1856 (Cambridge: Cambridge University Press).

Cavalier-Smith, T. (1992). Bacteria and eukaryotes. Nature 356, 570.

Chatton, E. (1925). *Pansporella perplexa.* Réflexions sur la biologie et la phylogenie des protozoaires. Ann. Sci. Nat. 10e, Ser. vii, 1–84.

Chatton, E. (1938). Titres et Travaux Scientifiques (1906–1937) de Edouard Chatton (Sète: Sottano).

Cohn, F. (1853). Untersuchungen über die Entwicklungsgeschichte der mikroskopischen Algen und Pilze (Bonn: Eduard Weber).

Cohn, F. (1872a). Untersuchungen über Bakterien. I. Beitr. Biol. Pflanz. *1*, 87; 2, 127.

Cohn, F. (1872b). Über die Bakterien, die kleinsten lebenden Wesen. Sammlung Gemeinverständlicher Wissenschaftlicher Vorträge. Berlin.

Cohn, F. (1875). Untersuchungen über Bakterien. II. Beitr. Biol. Pflanz. 3, 141–208.

Copeland, E.B. (1927). What is a plant? Science 65, 388–390.

Crick, F.H.C. (1958). The biological replication of macromolecules. Symp. Soc. Exp. Biol. *12*, 138–163.

Doolittle, W.F. (2000). Uprooting the tree of life. Sci. Am. *282*, 90–95.

Dougherty, E.C. (1955). Comparative evolution and the origin of sexuality. Syst. Zool. *4*, 145–169.

Dougherty, E.C. (1957). Neologism needed for structures of primitive organisms. 1. Types of nuclei. J. Protozool. *4*, 14.

Ehrenberg, C.F. (1838). Die Infusionsthierchen als Volkommene Organismen. Ein Blick in das Tiefere Organische Leben der Natur (Leipzig: L. Voss).

Fox, G.E., Peckman, K.J., and Woese, C.E. (1977a). Comparative cataloging of 16S ribosomal ribonucleic acid: molecular approach to procaryotic systematics. Int. J. Syst. Bacteriol. *27*, 44–57.

Fox, G.E., Magrum, L.J., Balch, W.E., Wolfe, R.S., and Woese, C.R. (1977b). Classification of methanogenic bacteria by 16S ribosomal RNA characterization. Proc. Natl. Acad. Sci. U.S.A. *74*, 4537–4541.

Garrity, G.M. (2001). Preface to the second edition of Bergey's Manual of Systematic Bacteriology. In Bergey's Manual of Systematic Bacteriology, 2nd ed. Vol. 1., D.R. Boone, R.W. Castenholz, and G.M. Garrity, ed. (New York: Springer), p. vii.

Gogarten, J.P., Kibak, H., Dittrich, P., Taiz, L., Bowman, E.J., Bowman, B.J., Manolson, M.F., Poole, R.J., Date, T., and Oshima, T. (1989). Evolution of the vacuolar $H^+$-ATPase: implications for the origin of eukaryotes. Proc. Natl. Acad. Sci. U.S.A. *86*, 6661–6665.

Haeckel, E. (1866). Generelle Morphologie der Organismen (Berlin: Georg Reimer).

Hall, J.B. (1971). Evolution of the prokaryotes. J. Theor. Biol. *30*, 429–454.

Harold, F.M. (2001). The way of the cell. Molecules, organisms and the order of life (Oxford, Oxford University Press).

Henrici, A.T. (1939). The Biology of the Bacteria. An Introduction to General Microbiology. 2nd ed. (Boston, D.C. Heath & Co.).

Horvath, R.S. (1974). Evolution of anaerobic-energy-yielding metabolic pathways of the procaryotes. J. Theor. Biol. *48*, 361–371.

Ingraham, J.L., and Nikaido, H. (1994). The phylogeny of microorganisms. ASM News *60*, 293.

Iwabe, N., Kuma, K., Hasegawa, M., Osawa, S., and Miyata, T. (1989). Evolutionary relationships of Archaebacteria, Eubacteria and Eukaryotes inferred from phylogenetic trees of duplicated genes. Proc. Natl. Acad. Sci. U.S.A. *86*, 9355–9359.

Jensen, R.A. (1985). Biochemical pathways in prokaryotes can be traced backwards through evolutionary time. Mol. Biol. Evol. *2*, 87–120.

Kluyver, A.J., and van Niel, C.B. (1936). Prospects for a natural system of classification of bacteria. Zentralbl. Bakteriol. Parasitenkd. II *94*, 369–403.

Krieg, N.R. (1984). On using the manual. In *Bergey's Manual* of Systematic Bacteriology, 1st ed., Vol. 1., N.R. Krieg and J.G. Holt, ed. (Baltimore: Williams & Wilkins), pp. xix-xxi.

Krieg, N.R., and Holt, J.G. (1984). Preface to first edition of *Bergey's Manual* of Systematic Bacteriology. In *Bergey's Manual* of Systematic Bacteriology, 1st ed., Vol. 1., N.R. Krieg and J.G. Holt, ed. (Baltimore: Williams & Wilkins), pp. xiii-xiv.

Magrum, L.J., Luehrsen, K.R., and Woese, C.R. (1978). Are the extreme halophiles actually "bacteria"? J. Mol. Evol. *11*, 1–8.

Margulis, L. (1970). Origin of Eukaryotic Cells (New Haven: Yale University Press).

Margulis, L., and Guerrero, R. (1991). Kingdoms in turmoil. New Sci. *129*, 46–50.

Martin, W., and Russell, M.J. (2003). On the origins of cells: a hypothesis for the evolutionary transitions from abiotic geochemistry to chemoautotrophic prokaryotes, and from prokaryotes to nucleated cells. Phil. Trans. R. Soc. Lond. B. Biol. Sci. *358*, 59–83.

Mayr, E. (1990). A natural system of organisms. Nature *348*, 491.

Mayr, E. (1998). Two empires or three? Proc. Natl. Acad. Sci. U.S.A. *95*, 9720–9723.

Merezhkowsky, C. (1905). Über Natur und Ursprung der Chromatophoren im Pflanzenreiche. Biol. Centralbl. *25*, 593–604.

Merezhkowsky, C. (1910). Theorie der zwei Plasmaarten als Grundlage der Symbiogenesis, einer neuen Lehre von der Entstehung der Organismen. Biol. Centralbl. *30*, 277–303, 321–347, 353–367.

Merezhkowsky, C. (1920). La plante considérée comme un complexe symbiotique. Bull. Soc. Nat. *6*, 17–98.

Migula, W. (1897, 1900). System der Bakterien. Handbuch der Morphologie, Entwicklungsgeschichte und Systematik der Bakterien. 2 vols. (Jena: G. Fischer).

Müller, O.F. (1786). Animalcula infusoria fluvatilia et marina, quae detextit, systematice descripsit et ad vivum delineari curavit (Hauniæ: Nicolai Mülleri) (online at http://www.archive.org/stream/animalculainfuso00ml#page/n4/mode/1up).

Murray, R.G.E. (1974). A place for bacteria in the living world. In *Bergey's Manual* of Determinative Bacteriology, 8th ed., R.E. Buchanan and N.E. Gibbons, ed. (Baltimore: Williams & Wilkins), pp. 4–9.

Murray, R.G.E., and Holt, J.G. (2001). The history of Bergey's Manual. In Bergey's Manual of Systematic Bacteriology, 2nd ed. Vol. 1., D.R. Boone, R.W. Castenholz, and G.M. Garrity, ed. (New York: Springer), pp. 1–13.

Olsen, G.J., Pace, N.R., Nuell, M., Kaine, B.P., Gupta, R., and Woese, C.R. (1985). Sequence of the 16S rRNA gene from the thermoacidophilic archaebacterium *Sulfolobus solfataricus* and its evolutionary implications. J. Mol. Evol. *22*, 301–307.

Orla-Jensen, S. (1909). Die Hauptlinien des natürlichen Bakteriensystems nebst einer Uebersicht der Gärungsphenomene. Zentralbl. Bakteriol. Parasitenkd. II 22, 305–346.

Portier, P. (1918). Les Symbiotes (Paris: Masson).

Pringsheim, E.G. (1923). Zur Kritik der Bakteriensystematik. Lotos *71*, 357–377.

Pringsheim, E.G. (1949). The relationship between bacteria and myxophyceae. Bacteriol. Rev. *13*, 47–98.

Pringsheim, E.G. (1964). Die verwandtschaftlichen Beziehungen zwischen den Lebewesen mit und ohne Blattgrün. Naturwissenschaften *51*, 154–157.

Sanger, F., Brownlee, G.G., and Barrell, B.G. (1965). A two-dimensional fractionation procedure for radioactive nucleotides. J. Mol. Biol. *13*, 373–398.

Sapp, J. (2005). The prokaryote–eukaryote dichotomy: meanings and mythology. Microbiol. Mol. Biol. Rev. *69*, 292–305.

Sapp, J. (2006). The bacterium's place in nature. In Microbial Phylogeny and Evolution. Concepts and Controversies, J. Sapp, ed. (Oxford: Oxford University Press), pp. 3–52.

Schlegel, H.G. (1969). Allgemeine Mikrobiologie (Stuttgart: Georg Thieme Verlag).

Sehgal, S.N., Kates, M., and Gibbons, N.E. (1962). Lipids of *Halobacterium cutrirubrum*. Can. J. Biochem. Physiol. *40*, 69–81.

Staley, J.T. (2006). The bacterial species dilemma and the genomic-phylogenetic species concept. Phil. Trans. R. Soc. B *361*, 1899–1909.

Staley, J.T. (2009). The phylogenomic species concept for Bacteria and Archaea. Microbe *4*, 361–365.

Staley, J.T., and Krieg, N.R. (1984). Classification of procaryotic organisms: an overview. In *Bergey's Manual* of Systematic Bacteriology, 1st ed., Vol. 1., N.R. Krieg and J.G. Holt, ed. (Baltimore: Williams & Wilkins), pp. 1–4.

Stanier, R.Y. (1970). Some aspects of the biology of cells and their possible evolutionary significance. In Organization and Control in Prokaryotic and Eukaryotic Cells, H.P. Charles and B.C.J.G. Knight, ed. (Cambridge: Cambridge University Press), pp. 1–38.

Stanier, R.Y. (1980). The journey, not the arrival, matters. Ann. Rev. Microbiol. *34*: 1–48.

Stanier, R.Y., and van Niel, C.B. (1941). The main outlines of bacterial classification. J. Bacteriol. *42*, 437–466.

Stanier, R.Y., and van Niel, C.B. (1962). The concept of a bacterium. Arch. f. Mikrobiol. *42*, 17–35.

Stanier, R.Y., Doudoroff, M., and Adelberg, E.A. (1957). The Microbial World. 1st ed. (Englewood Cliffs, NJ.: Prentice Hall).

Stanier, R.Y., Doudoroff, M., and Adelberg, E.A. (1963). The Microbial World. 2nd ed. (Englewood Cliffs, NJ.: Prentice Hall).

Stanier, R.Y., Doudoroff, M., and Adelberg, E.A. (1970). The Microbial World. 3rd ed. (Englewood Cliffs, NJ.: Prentice Hall).

Uzzell, T., and Spolsky, C. (1974). Mitochondria and plastids as endosymbionts: A revival of special creation? Amer. Scient. *62*, 334–343.

van Niel, C.B. (1946). The classification and natural relationships of bacteria. Cold Spring Harbor Symp. Quant. Biol. *11*, 285–301.

van Niel, C.B. (1955). Classification and taxonomy of the bacteria and bluegreen algae. In A Century of Progress in the Natural Sciences 1853–1953, E.L. Kessel, E.L., ed. (San Francisco: California Academy of Sciences), pp. 89–114.

van Niel, C.B. (1956). Evolution as viewed by the microbiologist. In The Microbe's Contribution to Biology, A.J. Kluyver and C.B. van Niel. (Cambridge MA: Harvard University Press), pp. 155–176.

Wallin, I.E. (1927). Symbionticism and the Origin of Species (Baltimore: Willams & Wilkins).

Whittaker, R.H. (1959). On the broad classification of organisms. Quart. Rev. Biol. *34,* 210–228.

Whittaker, R.H. (1969). New concepts of kingdoms of organisms. Science *163,* 150–160.

Winogradsky, S. (1952). Sur la classification des bactéries. Ann. Inst. Pasteur *82,* 125–131.

Winslow, C.E.A. (1914). The characterization and classification of bacterial types. Science *39,* 77–91.

Woese, C.R. (1987). Bacterial evolution. Microbiol. Rev. *51,* 221–271.

Woese, C.R. (1994). There must be a prokaryote somewhere: microbiology's search for itself. Microbiol. Rev. *58,* 1–9.

Woese, C.R. (2004). A new biology for a new century. Microbiol. Mol. Biol. Rev. *68,* 173–186.

Woese, C.R., and Goldenfeld, N. (2009). How the microbial world saved evolution from the Scylla of molecular biology and the Charybdis of the modern synthesis. Microbiol. Mol. Biol. Rev. *73,* 14–21.

Woese, C.R., and Fox, G.E. (1977). Phylogenetic structure of the prokaryotic domain: the primary kingdoms. Proc. Natl. Acad. Sci. U.S.A. *74,* 5088–5090.

Woese, C.R., Kandler, O., and Wheelis, M.L. (1990). Towards a natural system of organisms: proposal for the domains Archaea, Bacteria, and Eucarya. Proc. Natl. Acad. Sci. U.S.A. *87,* 357–366.

Wolfe, R.S. (1991). My kind of biology. Ann. Rev. Microbiol. *45,* 1–35.

Zuckerkandl, E., and Pauling, L. (1965). Molecules as documents of evolutionary history. J. Theor. Biol. *8,* 357–366.

# Methods and Programs for Calculation of Phylogenetic Relationships from Molecular Sequences

# 2

Jongsik Chun and Soon Gyu Hong

Abstract

The purpose of phylogenetic analysis is to understand the past evolutionary path of organisms. Even though we will never know the true phylogeny of any organism for certain, phylogenetic analysis provides best assumptions, thereby providing a framework for various disciplines in microbiology. Due to the technological innovation of modern molecular biology and the rapid advancement in computational science, accurate inference of the phylogeny of a gene or organism seems possible in the near future. There has been a flood of nucleic acid sequence information, bioinformatic tools and phylogenetic inference methods in public domain databases, literature and worldwide web space. Phylogenetic analysis has long played a central role in basic microbiology, for example in taxonomy and ecology. In addition, more recently emerging fields of microbiology, including comparative genomics and phylogenomics, require substantial knowledge and understanding of phylogenetic analysis and computational skills to handle the large-scale data involved. In this chapter we will review major steps in phylogenetic analysis and introduce relevant computer software tools with respect to their accuracy, efficiency and availability.

## Introduction

Phylogeny is the study of the evolutionary history of organisms. Due to the limited number of morphological and physiological characters available in microorganisms, and due to the extensive variation even among closely related taxa, their taxonomy and systematics have increasingly relied on molecular data – typically DNA sequences located on chromosomes. This is particularly true for prokaryotes and fungi, and the framework of their entire taxonomy is now mainly based on phylogenetic trees inferred from ribosomal RNA gene sequence comparison, although chemical and physiological taxonomic markers still play important roles (Tindall *et al.*, 2010).

Microbiologists use phylogenetic analysis to understand the evolutionary history of one or more genes, and increasingly of the organism's entire genome. The latter would be the ultimate goal of many microbiological disciplines including systematics, taxonomy, ecology and epidemiology. Since the first entire bacterial genome was elucidated in 1995 (Fleischmann *et al.*, 1995), more than 1200 prokaryotic and 400 fungal genomes have been sequenced (http://genomesonline.org/). The numerous prokaryotic genomes that have now been sequenced represent most of major phyla known, and have revealed extensive lateral gene transfer between closely related prokaryotes as well as distantly related prokaryotic domains (Brown and Doolittle, 1997; Chun *et al.*, 2009; Wu *et al.*, 2009). Because of this level of lateral gene transfer, it is challenging to infer the exact evolutionary path of microorganisms from such extensively patched genome information.

Another limitation that has recently emerged in the use of phylogenetic analysis to reconstruct the evolutionary history of microorganisms is the computing power available in general biology laboratories. Fortunately, computing cost (i.e. the price of computing hardware) has been decreasing rapidly and exponentially during the last several decades (Moore's Law). However, with the rapid advancement in sequencing technologies, the demands being placed on computing power are ever greater. The new massively parallel DNA sequencing platforms, so-called 'next generation sequencing' (NGS) technology (Shendure and Ji, 2008; Metzker, 2010), generate enormous amounts of DNA sequence data for phylogenetic studies, and this may simply overwhelm the rapid advancement of computing capability. For example, pyrosequencing technology (Ronaghi *et al.*, 1996), a new sequencing method with the lowest data throughput among the currently available NGS platforms, can generate over a million sequencing reads with >400 bp per run, which is more than all the 16S rRNA gene sequences that had accumulated in the public domain database by the end of 2008.

The large number of genome sequences available in the public domain, the development of massively parallel high throughput DNA sequencing technologies, and the introduction of novel phylogenetic inference methods, each provide a challenge as well as an opportunity to microbiologists to study evolution, taxonomy and ecology of microorganisms. The new era of bioinformatics and genomics is unfolding at an unprecedented pace. In this chapter, we will review the currently available methods for phylogenetic inference and related software.

## General procedure for phylogenetic inference

There are four steps in general phylogenetic analysis of molecular sequences: (i) selection of a suitable molecule or molecules (phylogenetic marker), (ii) acquisition of molecular sequences, (iii) multiple sequence alignment (MSA) and (iv) phylogenetic treeing and evaluation.

### Selection of phylogenetic markers

The goal of phylogenetic inference is to reconstruct the evolutionary history of genes or organisms under consideration. The first step of any phylogenetic analysis is to choose a suitable homologous part of the genomes to be compared. Mechanisms of molecular evolution include mutations, duplication of genes, reorganization of genomes, and genetic exchanges such as recombination, reassortment and lateral gene transfer (Vandamme, 2009). Although all of this information can be used to infer phylogenetic relationships of genes or organisms, information on mutations, including substitution, insertion, and deletion, is most frequently used in phylogeny reconstruction. The aim is to infer a correct organismal phylogeny, using orthologous genetic loci, in which common ancestry of two sequences can be traced back to a speciation event. Phylogeny using homologous genetic loci derived by gene duplication (paralogy) or related through lateral gene transfer (xenology), cannot reflect evolutionary history of organisms (Moritz and Hillis, 1996). Although using *a priori* information for paralogy and xenology is the best approach to select orthologous genetic loci for phylogenetic studies, this information is unavailable in many cases until phylogeny of the gene is analysed. Genes in multiple copies, and non-essential genes (such as genes related to secondary metabolism), tend to experience duplication, loss, and acquisition events more frequently, leading to more paralogous and xenologous relationships, compared to essential single copy genes.

The next important criterion in choosing phylogenetic markers is the rate of evolutionary change of the given genetic loci. Highly conserved genetic loci will not provide any information for the closely related taxa. In contrast, highly variable genetic loci accumulate mutations at the same nucleotide or amino acid sites and variation of genetic information becomes noise rather than information. Therefore, it is crucial to choose an appropriate phylogenetic marker by considering the range of phylogenetic distances among the taxa under consideration. For example, the 16S rRNA gene is a highly conserved genetic locus and frequently used to build the tree encompassing bacterial phyla (Woese, 1987), but it is too conserved to dissect phylogenetic relationships of isolates belonging to a species or closely related species. In such a case, more rapidly evolving genes should be used. Most popular phylogenetic markers are genes coding for house-keeping proteins, including *rpoB*, *gyrB*, and *groEL* (Yamamoto and Harayama, 1995; Adekambi and Drancourt, 2004; Glazunova *et al.*, 2009). However, given the extensive lateral gene transfer and gene duplication found in microbial genomes, special attention must be paid to this step.

Even though complete genome sequences of hundreds of microbial species are available, in all except a few special cases just one or few genes have been chosen for large scale sequencing surveys to unravel biodiversity in nature. A typical example is the 16S rRNA coding region of the prokaryotic genome, which has been examined for most known microbial species (Chun *et al.*, 2007). It is well known that building molecular phylogenies for distantly related species is a difficult task, since the inferred evolutionary paths have varied according to the choice of phylogenetic methods and the genes selected.

In cases where multiple genes are incorporated in the phylogenetic inference, each set of homologous genes is independently aligned by computer programs for multiple sequence alignment (MSA), can be followed by concatenation of all resulting alignments. This approach has been used successfully for elucidating the overall phylogenetic structure among prokaryotes using complete genome sequences (Wu *et al.*, 2009), species within a genus (Guo *et al.*, 2008), and strains within a species (Chun *et al.*, 2009). Using this procedure, a large MSA, up to the scale of megabases, can be generated from genome sequences.

There are ways to infer phylogenetic trees in the absence of MSA, such as through overlapping genes (Jiang *et al.*, 2008; http://bioalgorithm.life.nctu.edu.tw/OGtree/) and conservation of local gene order (Wolf *et al.*, 2001). The process for gene finding, annotation and detection of homologues can be completely omitted in rapid calculation of phylogenetic trees using Lempel-Ziv complexity (Otu and Sayood, 2003) and oligonucleotide frequency (Takahashi *et al.*, 2009).

Since genome sequences from most major prokaryotic phyla are available from the public domain, any gene, set of genes (e.g. set of conserved ribosomal proteins) or even complete genomes can be subjected to phylogenetic analysis, though the computing cost for genome scale calculation will be high.

### *Acquisition of molecular sequences*

Nucleotide sequences for phylogenetic analyses have usually been obtained from DNA fragments from PCR amplification, but sequence information from whole genome sequencing and metagenome sequencing has recently been widely applied for phylogenetic analyses to understand genome evolution or organismal evolution. Now, through the development of NGS technology, a huge amount of sequence information for either phylogenetic marker genes or the whole genome or metagenome can be obtained very rapidly and at low cost.

## Multiple sequence alignment

### *General principle*

Once DNA sequence data are generated, they are subjected to a MSA process. This involves finding homologous sites, that is, positions derived from the same ancestral organism in the molecules under study. A set of sequences can be aligned with another by introducing 'alignment gaps' (known in brief as 'gaps). In general, MSA starts by aligning a pair of sequences (pairwise alignment), and is then expanded to multiple sequences using various algorithms (Lipman *et al.*, 1989; Notredame *et al.*, 2000; Thompson *et al.*, 2002; Edgar, 2004b). An approximate phylogenetic tree is produced to guide the order of addition of each sequence to the final MSA. In general, this technique may not guarantee the most optimal alignment among the almost infinite number of possible multiple alignments due to limited computing cost. The general rule underpinning computer-generated MSA is that the increase in sequence similarity due to the introduction of gaps must be greater than would be expected based on random insertion of gaps. Judicious insertion of gaps is inadvisable. By allowing an infinite number of gaps, even two very different sequences can be forced to be matched perfectly. In most multiple alignment algorithms, gaps are inserted by balancing penalties for introducing a gap and allowing a mismatch. Even though computer programs for multiple alignment provide default values for alignment parameters such as gap penalty and gap extension penalty, users must not consider these to be the only correct solution for every case of MSA. Often, applying suitable alignment parameters allows a correct outcome in special circumstances. For example, by applying small gap extension penalty, one can identify insertion of a very large non-homologous DNA sequence in one sequence within a multiple alignment. Weight or scoring matrices containing costs for each mismatched pair of nucleotide (for DNA or RNA) are used to calculate various mismatch penalties, and it is important to choose adequate weight matrices depending on the level of similarity among subject sequences.

This MSA procedure provides the backbone of general comparative sequence analysis. Accurate DNA sequence alignment is a prerequisite to various methods for downstream data processing including phylogenetic tree construction. It has been shown that there is direct correlation between error rates of alignment and phylogenetic tree inference in a simulation study (Kumar and Filipski, 2007). In other words, alignments that have failed to find correct homologous sites will lead to inaccurate inference of phylogenetic trees. In some phylogenetic tree-making methods, the order of sequences in MSA may affect the outcome, that is, the resulting phylogenetic trees.

*Computer programs for multiple sequence alignment*

Many algorithms and computer programs have been developed in the last few decades for multiple sequence alignment (Table 2.1), but the original Clustal series programs (Chenna *et al.*, 2003) are still most widely used and produce reasonably good quality MSA for small data sets. ClustalW is a command line program, whereas ClustalX provides a graphical user interface. A parallel version of Clustal, called ClustalW-MPI (Li, 2003), can be used on multiple-processor environment for large datasets. Like Clustal, T-Coffee program (Notredame *et al.*, 2000) also uses a progressive algorithm, but its accuracy was improved by pre-processing all pairwise alignments between the sequences in a dataset. For a large dataset, such as massive pyrosequencing reads, the MUSCLE program (Edgar, 2004a) can generate good compromise between accuracy and speed. This program employs oligonucleotide counting, instead of full-length pairwise sequence alignment, for the initial distance estimation. The MAFFT program (Katoh and Toh, 2008) utilizes several different algorithmic approaches and can be used for either small or very large datasets. There are also other computer programs developed for general multiple sequence alignment, but the above three have been most popular and are routinely used in publications in various microbiological disciplines. Wallace *et al.* (2006) developed the M-Coffee program to combine multiple sequence alignments generated from different multiple alignment programs.

*Editing and viewing computer-generated alignments*

Unfortunately, there are no general rules for choosing suitable multiple sequence alignment algorithms (computer programs) and alignment parameters, such as the gap-open and gap-extension penalties. It is therefore important to test different computer programs and parameters and evaluate the resultant alignments using a viewer program. Manual editing of the computer generated MSA is often needed to improve the final result.

The ClustalX program provides a MSA viewer that can display nucleotides or amino acids in colours based on different functional categories. The SeaView program (http://pbil.univ-lyon1.fr/software/seaview.html) also provides viewing and editing functions as well as several popular multiple alignment and phylogenetic treeing algorithms for both Microsoft (MS) Windows and Linux platforms. The BioEdit program is popular among MS Windows users, and also contains various MSA and phylogenetic treeing functions (http://www.mbio.ncsu.edu/BioEdit/bioedit.html). The Jalview is a JAVA-based alignment editor which also incorporates structural biological component (Waterhouse *et al.*, 2009; http://www.jalview.org/). These computer programs/suites were initially developed as a sequence alignment viewer/editor, then evolved to include many popular programs for MSA and treeing, so now serve as workbench where users can do most processes necessary for phylogenetic analysis. The MEGA (Kumar *et al.*, 2004) is a specialized program for phylogenetic inference, though it has an excellent user interface for sequence editing. All of steps in phylogenetic inferences can be readily carried out using the MEGA program alone, even though it has not yet implemented maximum likelihood method.

**Table 2.1** Most widely used computer programs for multiple sequence alignment

| Program | Available platforms[a] | Throughput | Availability | References |
|---|---|---|---|---|
| ClustalW[b] | W/L/M/N | Low | http://www.clustal.org/ | Thompson *et al.* (2002) |
| ClustalX | W/L/M/N | Low | http://www.clustal.org/ | Thompson *et al.* (2002) |
| Kalign | L/N | High | http://msa.sbc.su.se/ | Lassmann and Sonnhammer (2005) |
| MAFFT | W/L/M/N | High | http://align.bmr.kyushu-u.ac.jp/mafft/software/ | Katoh *et al.* (2005) |
| MUSCLE | W/L/N | High | http://www.drive5.com/muscle/ | Edgar (2004a) |
| T-Coffee | L/M/N | Low | http://www.tcoffee.org/ | Notredame *et al.* (2000) |

[a]Operating system: D, Microsoft Windows; L, Linux; M, Mac OS; N, web service is available.

[b]Also a parallel processing implementation is available at http://www.bii.a-star.edu.sg/achievements/applications/clustalw/.

*Consideration of methods for rRNA sequence alignment*

It is evident that ribosomal RNA gene sequences are a centrepiece of microbiology, especially taxonomy, systematics and ecology. Unlike protein sequence alignment, rRNA gene alignment can benefit from additional information such as knowledge of the RNA secondary structure. Proper secondary and tertiary structures of the rRNA molecules are essential for their function, and this structure information can be used to identify the homologous sites/residues even when the nucleotide sequence similarity is low. In Fig. 2.1 we demonstrate the use of secondary structure in rRNA gene alignment. In this example, multiple alignment from a region (positions between 451 and 482 according to *Escherichia coli* reference sequence) in 16S rRNA genes of three related bacteria is shown (Fig. 2.1A). Fig. 2.1B shows a MSA generated by the ClustalX program, and the loop part of this hairpin structure (Fig. 2.1A) could not be identified in this computer-assisted process. However, by manually editing this alignment based on the secondary structure model given in Fig. 2.1C, the correct alignment for the four nucleotide residues (indicated by asterisks) in the loop region was readily figured out, even though the stem area (indicated by < and >) still cannot be aligned with high confidence.

Fortunately, secondary structure models of the small subunit rRNA of the major microbial taxa have been extensively studied by Gutell and his colleagues (Cannone *et al.*, 2002), and can be downloaded from http://www.rna.ccbb.utexas.edu/. Employing this functional constraint into MSA, wherever possible, should offer better accuracy to the subsequent phylogenetic analyses. However, applying this concept in practice may not be straight forward without using specialty computer software tools.

There are two computer programs that offer manual editing function while rRNA secondary structural information being displayed (Fig. 2.2). The ARB software is a graphically oriented package comprising various tools for sequence database handling and data analysis (Ludwig *et al.*, 2004; http://www.arb-silva.de/). It is one of the most sophisticated software packages to date, including not only MSA and phylogenetic treeing capabilities but also highly sophisticated functions such as designing oligonucleotide probes and primers from thousands of sequences. However, it only runs on Unix/Linux and compatible operating systems which may not be readily available or familiar to all microbiology laboratories. The jPHYDIT program (Jeon *et al.*, 2005; http://chunlab.snu.ac.kr/jphydit/) is a JAVA-based nucleotide sequence editor program that is a direct successor of the PHYDIT program

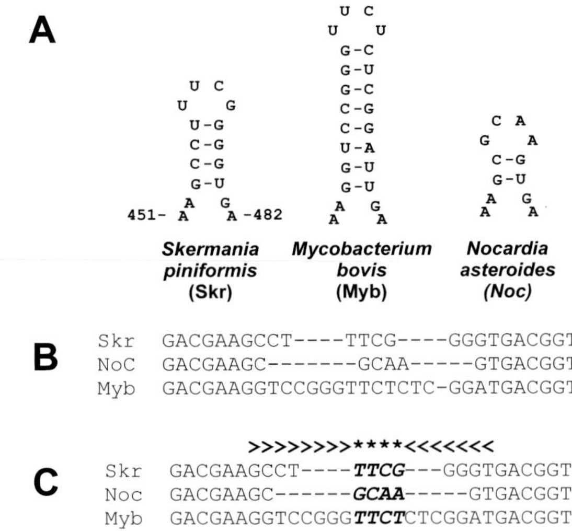


**Figure 2.1** Multiple sequence alignment using rRNA secondary structure. (A) Secondary structures of three related bacteria species are depicted for the region representing positions between 451 and 481 of the *Escherichia coli* numbering system. (B) Corresponding multiple sequence alignment generated by the ClustalW program. (C) Corresponding multiple sequence alignment that was manually aligned on the basis of secondary structure information using the jPHYDIT program.

(http://plaza.snu.ac.kr/~jchun/phydit/). Multiple sequence alignment based on secondary structure greatly improves the quality of phylogenetic analysis and has been highly recommended to be used for taxonomic purpose (Tindall *et al.*, 2010).

After the introduction of the Frederick Sanger's sequencing method over three decades ago, substantial amounts of sequence information has accumulated in the public domain. In case of 16S rRNA gene sequences, more than one million sequence entries were already registered in the public domain database. Aligning this large sized data set in each MSA is enormous task as well as impractical. A solution to this problem is to generate a MSA of all available sequences and then add new sequences to this pre-existing alignment. This approach has been adopted in the Ribosomal Database Project (Cole *et al.*, 2009) and the Silva database project (Pruesse *et al.*, 2007), and is widely used among microbiologists working in various disciplines.

*Considerations for protein sequence alignment*

Even though protein sequences were determined before the first DNA or RNA sequences became available, most of the protein sequences in the public database are derived from translation of DNA or RNA sequences that code protein. In this case, both nucleotide and its translated amino acid sequences can be used for MSA and phylogenetic treeing, and they sometimes may generate contradictory results.

Generally, amino acid sequences of an evolutionary closely related pair of proteins may not show substantial difference due to conservation of amino acids via purifying selection even when nucleotide sequences have substantial variation on the redundant codon positions. On the contrary, nucleotide sequences among distantly related proteins may contain little similarity or are even unalignable at all. For such a case, aligning amino acid sequences should be used. Since protein sequence is maintained as codons, i.e. three nucleotides, it is possible to align nucleotide sequences with respect to their amino acid sequence alignments. This can be achieved by carrying out MSA at the amino acid level and applying it to insert gaps in the corresponding nucleotide sequence alignment. Several computer programs offer this codon-based MSA. PROTAL2DNA is a simple PERL language script that generates nucleotide

A

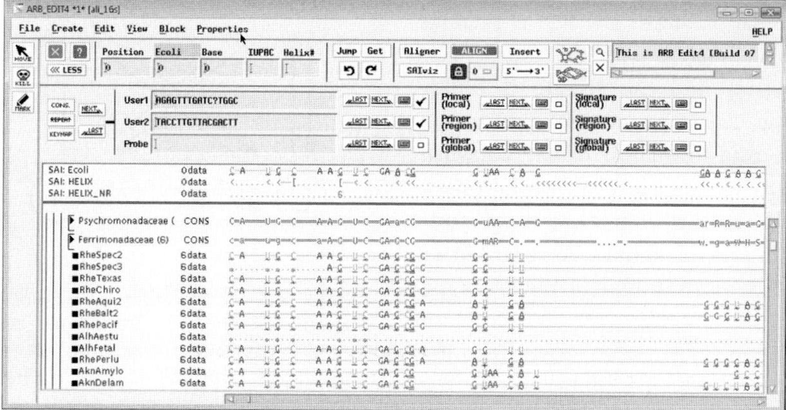


B

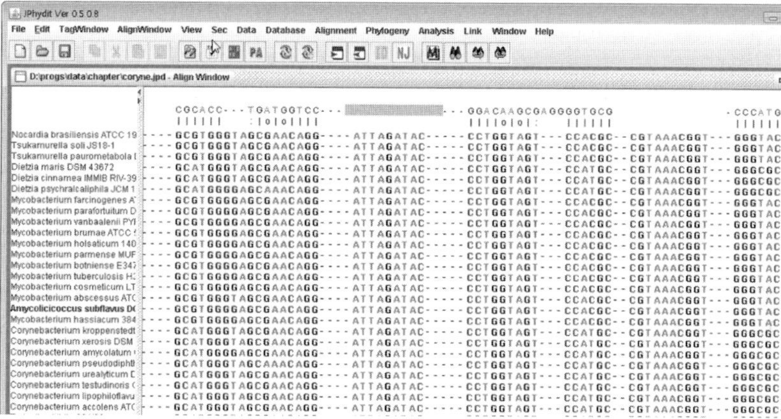


**Figure 2.2** Multiple sequence alignment assisted by rRNA secondary structure using (A) the Arb and (B) the jPHYDIT package.

MSA using amino acid as a seed (ftp://ftp.pasteur.fr/pub/gensoft/unix/alignment/). The DAMBE (Xia and Xie, 2001) and PHYDIT programs (http://plaza.snu.ac.kr/~jchun/phydit/) offer a graphical interface for this type of MSA, and are recommended for aligning the protein coding part of genome sequences.

## Reconstruction of phylogenetic trees

A phylogenetic tree is a graphical representation of the evolutionary relationship among four of more genes or organisms, with phylogenetic analysis as its final goal. Once homologous residues (nucleotides or amino acids) are aligned, phylogenetic trees can be calculated. Numerous methods and algorithms have been developed and incorporated into various software packages and suites. They are generally categorized into four groups: distance, maximum parsimony (MP), maximum likelihood (ML) and Bayesian methods. Distance methods try to generate a tree that fits to a matrix of pairwise evolutionary distances; thereby they do not use sequence information directly (non-character-based). In contrast, MP, ML and Bayesian methods utilize direct sequence information (character-based), while only the latter two are strictly based on an explicit evolutionary model. All these methods are widely used and considered here.

*Distance methods*

Distance methods for phylogenetic construction involve two calculation steps. The first is calculation of a distance matrix with an appropriate evolutionary model, and the second is building phylogenetic trees by suitable clustering algorithms. The transformation of the pairwise alignment of two sequences to an evolutionary distance value requires a sophisticated substitution model. This process should be regarded as important as MSA and subsequent phylogenetic treeing steps.

If the difference between two aligned sequences is very small, it will be simply the count of different nucleotides. However, due to multiple mutational events over time, in most cases exact numbers of substitutions that have occurred should be inferred from the observed sequence differences on the basis of a given evolutionary model. Jukes and Cantor assumed that the frequencies of each nucleotide are 25% each and that any nucleotide has the same probability to be substituted by any other, which resulted in a simple model (Jukes, 1969). There are six ways for substitutions to occur among four different nucleotides (Fig. 2.3), and Jukes and Cantor's correction considered that their rates are all same. Later, Kimura (1980) developed a two-parameter model that took into account the different substitution rates of transition and transversion. Hasegawa *et al.* (1985) proposed a model similar to Kimura's two parameter model in that it treats transition and transversion differently, but considers unequal equilibrium frequencies of each nucleotide. The general reversible model (Rodriguez *et al.*, 1990) is the most 'general' one, allowing independent rates of all possible six way substitutions and unequal base frequencies. It is now well known that the rate of nucleotide substitution can vary for different positions over a sequence. A typical example is the third position in codons as the mutation in this position often does not cause change in amino acid sequence. Another example is differential sequence variations across the stretch of a sequence, typically stem and loop regions, in 16S rRNA sequences. To take heterogeneity of substitution rates over sites into consideration, a suitable model is required that can be applied to different categories of sites (Yang *et al.*, 1994).

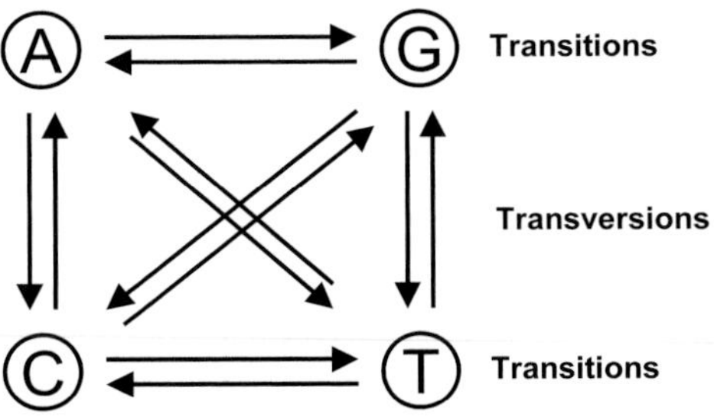


**Figure 2.3** All possible substitutions among four nucleotides in DNA. Transitions are interchanges between two-ringed purines (A, G) or between one-ringed pyrimidines (C, T). Transversions are interchanges between a purine and a pyrimidine base. In general, transversions occur less frequently than transition, though the opposite is often observed.

In case of protein coding genes, both nucleotide and amino acid sequences can be used for phylogenetic analyses. The decision between the two choices can be made depending on the degree of sequence divergence. Amino acid sequences of well conserved genes in closely related taxa may contain very little information, thus informative homologous sites. In contrast, highly diverged DNA sequences are usually hard to align with confidence and the redundant codon positions, i.e. third nucleotide position, likely experienced multiple substitution events. Eukaryotic coding regions are frequently discontinued by insertion of introns. Due to the different selection constraints on the exon and intron domains, introns evolve much faster than exon. Although intron sequences may provide valuable phylogenetic information in some cases, they are generally omitted for the study of protein gene evolution. This can be accomplished by identifying intron sequences through comparative sequence analysis with intron-missing related sequences and recognition of conserved sequences of exon-intron junction sites. Evolutionary distance matrix of aligned amino acid sequences can be calculated based on models of amino acid substitutions by Jones–Taylor–Thornton model (Jones *et al.*, 1992), PMB model (Veerassamy *et al.*, 2003) DCMut model (Kosiol and Goldman, 2005) or Kimura's distance (Kimura, 1983). Several computer programs are available for calculation of amino acid sequence-based evolutionary distances, including CLUSTAL series programs, PROTDIST program of PHYLIP package, and TREEPUZZLE program (http://www.tree-puzzle.de/).

Once a distance matrix has been calculated under an appropriate evolutionary model, a phylogenetic tree can be generated using several clustering algorithms. These methods simply agglomerate sequences in a one by one manner, and never go back to check the accuracy of the resulting tree. The Unweighted Pair Group Method with Arithmetic mean (UPGMA), also known as average linkage method, is widely used to generate dendrograms from phenotypic data in numerical taxonomic studies (Goodfellow, 1971), but is rarely employed in phylogenetic research. It is important to bear in mind that ultrametric distances are assumed to construct the UPGMA tree, where ultrametric distances imply that the distance between any two taxa is equal to the sum of the branches joining them (additive), and that the tree can be rooted so that all taxa are equidistant from the root. When the evolutionary rates among taxa or lineages are not same, which is true for most cases, the tree built by UPGMA often does not represent the true phylogeny.

The neighbour-joining (NJ) method (Saitou and Nei, 1987) is also a kind of clustering algorithm, but does not assume equal amount of divergence from the common, albeit hypothetical, ancestral sequence/organism (non-additive). Gascuel developed an improved version of NJ method, called BIONJ (Gascuel, 1997; http://www.atgc-montpellier.fr/bionj/), which uses a simple first-order model of the variances and covariances of evolutionary distance estimates. Similarly, Bruno and his colleagues developed the weighted neighbour-joining algorithm and the corresponding computer program, called Weighbor (Bruno *et al.*, 2000; http://www.t6.lanl.gov/billb/weighbor/). It gives significantly less weight to the longer distances in the distance matrix to overcome the long branch attraction, a well known problem caused by the maximum parsimony method.

The distance methods have the advantage that they are easy to implement in a computer program that provides very fast building of phylogenetic trees. In addition, these algorithms always produce a single tree even though it may not represent the true phylogeny. This combination of speed, low computing cost and apparently unambiguous outcome is naturally appealing and makes this kind of clustering methods very popular. However, clustering methods do not allow the evaluation of competing hypotheses, that is, phylogenetic trees. Nonetheless, distance methods can be very useful in large-scale phylogenetic studies if they are used as complementary means assisting more elaborate methods such as MP and ML methods.

*Maximum parsimony*

While the distance method is an algorithmic method, the MP and ML are methods to choose best trees among alternative competing tree topologies by optimality criteria. For this reason, these methods require a significantly higher computing cost than distance methods.

In the MP method, trees with minimum step number, number of changes on the tree, are used as an optimality criterion to choose the best tree. Nucleotide sites that are conserved for all taxa or sites with only autoapomorphic change, changes at terminal are not used in the MP approach, because all possible tree topologies have the same step numbers for those sites. These are called 'uninformative sites' and removed prior to the tree building by computer programs. Therefore, unlike distances, the MP method does not utilize all information available in the input MSA.

Generally there are several algorithms available to search for the best tree (most parsimonious tree) which meets the given optimality criterion. The exhaustive search, also called brute-force search, is a general algorithm to evaluate all types of possible phylogenetic tree topologies. This method guarantees finding the most parsimonious solution (tree), but is rarely practical in a real situation because of the enormous number of trees under consideration. The branch-and-bound algorithm is also a general algorithm for finding optimal solutions. Like the exhaustive algorithm, it guarantees that no more parsimonious trees are missed during the search process. Even though this method is several orders of magnitude faster than the exhaustive search, it is still impractical when a relatively large number of sequences (e.g. >30 sequences with 1 kb long alignment) is involved.

The heuristic approach is a collection of algorithms that help to find optimal tree(s) in many different ways. One of the most popular algorithms is the branch swapping of trees constructed by stepwise addition of taxa. At each step of stepwise addition of taxa, trees with the minimum step number remain, the next taxon is added on the selected tree, and it continues until all taxa are added. The first tree to be swapped can also be built by other tree-building method such as the UPGMA and NJ method. Heuristic searches work best if the starting tree is a good approximation of the most parsimonious tree. It should be noted that heuristic algorithms do not guarantee finding of the most parsimonious tree, so the resultant tree may be a suboptimal one. Unlike distance methods, MP method tends to produce multiple most parsimonious trees, and interpretation and presentation of these trees pose an extra challenge to researchers.

The MP method has been well known for a problem called 'long branch attraction' (Bergsten, 2005). This phenomenon happens when rates of evolution show considerable variation among highly divergent sequences. This can be easily detected by applying other phylogenetic methods that are not affected by this situation.

Several computer programs were developed for MP analysis. The PAUP* (Phylogenetic Analysis Using Parsimony; http://paup.csit.fsu.edu/) program developed by David Swofford offers most diverse and sophisticated MP algorithms, including the exhaustive, branch-and-bound and heuristic approaches. Three branch swapping algorithms, namely the nearest-neighbour interchanges, subtree pruning–regrafting, and tree bisection–reconnection, are available in the PAUP* software. After branch swapping of a saved tree, the step number is evaluated and compared with other saved best trees. When the swapped new tree has better score, it will replace the saved trees. New branch swapping begins with the newly saved tree topologies. However, branch swapping does not guarantee to test all of the possible tree topologies and choose the best tree for the given dataset. The most serious problem of a heuristic search is that the best tree cannot be obtained when the tree topology can be produced by more than two tree swapping steps and the intermediate tree does not have a good score. Because of this problem, random sequence addition option with many replications of stepwise sequence addition is highly recommended.

Similarly, a variety of MP algorithms are available in the Joe Felsenstein's PHYLIP package (http://evolution.genetics.washington.edu/phylip.html). This suite contains the largest collection of phylogenetic inference methods including the distance, MP and ML methods, though it is provided as command line version without graphical user interface. However, the MEGA program is based on an integrated environment with graphical user interface while offering the distance and MP methods. For building trees using MP methods, selecting a right search algorithm is most important as the outcome is totally dependent on this choice.

*Maximum likelihood*

The maximum likelihood method for inferring trees from DNA or RNA sequences was first developed by Felsenstein (1981). Like the MP method, the ML is a method of finding the best tree(s) with an optimality criterion. In contrast to the simple optimality criterion (minimum number of substitutions in given tree) employed in the MP method, the ML method provides a purely statistical way of building phylogenetic trees. In this approach, probabilities are considered for every individual residue in the MSA for any given tree, so it does utilize information at each nucleotide or amino acid position.

Like the MP method, appropriate search strategies are required in search for the best tree, in this case, a tree with the maximum likelihood. For the calculation of probabilities of a given tree, an evolutionary model, such as Jukes and Cater or Kimura's two parameter models, should be given to the computer programs. The PAUP* version 4 can be used for ML inference on Linux and Macintosh platforms. The Modeltest program (Posada and Crandall, 1998; http://darwin.uvigo.es/software/modeltest.html) is an accessory program to PAUP* and is generally used to select the model of nucleotide substitution, from 56 different models, that best fits the data. The PHYLIP package provides several different programs for ML analysis. The DNAML program generates ML phylogenetic tree from DNA sequences allowing for variable frequencies of the four nucleotides, for unequal rates of transitions and transversions, and for different rates of substitution in different categories of sites. DNAMLK works similarly to DNAML, but assumes a molecular clock and generates an ultrametric tree. For protein sequences, PHYLIP provides PROTML and PROMLK programs where the latter also assumes a molecular clock. Gary Olsen and his colleagues wrote an improved version of Felsenstein's DNAML that can deal with larger dataset (Olsen *et al.*, 1994).

Even though ML is most the powerful and statistically sound tree building method to date, its use is generally limited by the enormous time required for computers available in general laboratories. Several computer programs were developed to overcome this problem. PHYML (Guindon and Gascuel, 2003; http://atgc.lirmm.fr/phyml/) attempts to find the tree with the maximum likelihood value using a simple hill-climbing algorithm that adjusts tree topology and branch lengths simultaneously. This program can be run under MS Windows and Linux operating systems, and also a web service is provided at http://www.atgc-montpellier.fr/phyml/. More recently, the RAxML (Randomized Axelerated Maximum Likelihood; Stamatakis *et al.*, 2005) is becoming popular as it generates a ML tree from large dataset within a reasonable time. Both PHYML and RAxML are known to generate good quality ML trees, though only the latter provides a variety of versions for parallel processing, giving a huge advantage when it is used for dataset containing >10,000 sequences (Stamatakis, 2006). Given the availability of faster computing hardware, the ML method is highly recommended as it is, to date, based on the most statistically sound basis.

*Bayesian phylogenetic analysis*

Like the MP and ML methods, Bayesian analysis deals with each character/residue of nucleotide or amino acid. Similar to ML, it is a statistical analysis involving estimation of the probability that a given model has occurred, based on prior assumptions. The assumed model can be an evolutionary model and tree topology. Incorporating prior information to a tree-building process can be an advantage as well as disadvantage, depending on the availability and accuracy of such information. It is generally considered that the Bayesian phylogenetic method, together with the ML method, is one of the most accurate methods for constructing phylogenetic trees. The most popular implementation of the Bayesian analysis is MrBayes program (Ronquist and Huelsenbeck, 2003; http://mrbayes.csit.fsu.edu/) which uses a simulation technique called Markov chain Monte Carlo (MCMC) to approximate the posterior probabilities of trees. This program is also available in parallel processor version.

*Rooting phylogenetic trees*

All phylogenetic inference methods, except for the UPGMA algorithm, produce 'unrooted' trees which have no evolutionary polarity. To interpret a phylogenetic tree for taxonomical and other microbiological purposes, it is often necessary to identify the position of root. As pointed out by Nixon and Carpenter (1993), this task is not a trivial one and should be treated as such. Even though the root position in phylogenetic tree can be purely computationally decided, the general practice is the use of an outgroup. An outgroup is any sequence or organism which is not descended from the nearest common ancestor of the sequences, called ingroup, under consideration. Often extrinsic information such as morphological and physiological properties of organisms serves as the primary criterion for selecting the outgroup, though this is not always possible in the microbial world where molecular phylogeny is the only relevant information available.

A proper choice of the outgroup is important, but this process is a subjective one, so it cannot be easily automated. Outgroups become invalid for rooting if they are too distantly related to the ingroup. In contrast, if outgroups are too close to the ingroup, there is a chance that outgroups may be erroneously placed as members of the ingroup. A difficult problem to tackle occurs when different outgroups, either single or multiple, indicate different root positions. Fig. 2.4 demonstrates the inconsistency in finding a root in the phylogenetic tree of the suborder Corynebacterineae; this taxon is well defined by chemotaxonomy and rRNA-based phylogeny (Stackebrandt *et al.*, 1997). In this case, the different outgroup(s)

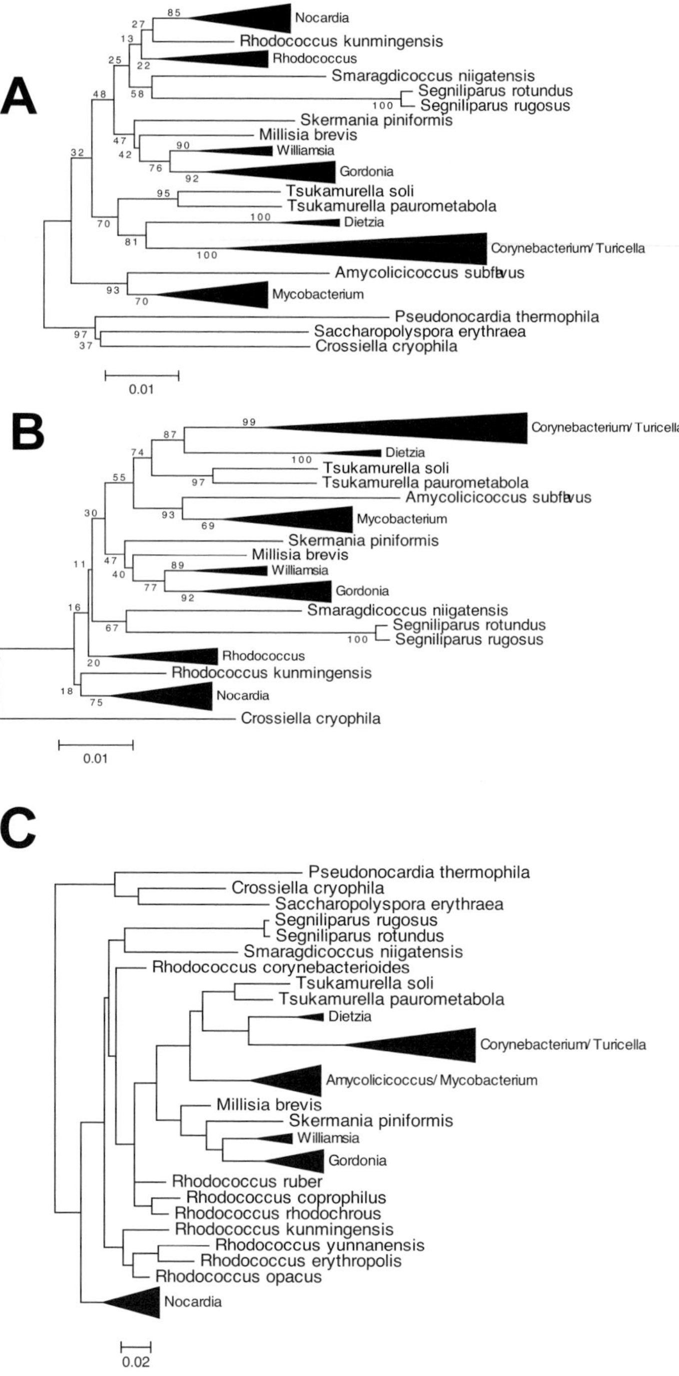


**Figure 2.4** Phylogenetic trees showing relationship among members of the suborder Corynebacterineae. 16S rRNA gene sequences of representative taxa belonging to the suborder Corynebacterineae were aligned using the jPHYDIT program on the basis of secondary structure model. (A) Neighbour-joining tree including three outgroups (*Crossiella cryophila*, *Pseudonocardia thermophila*, *Saccharopolyspora erythraea*). Maximum likelihood composite distances were calculated by the MEGA version 4 program. (B) Neighbour-joining tree generated using only one outgroup (*Crossiella cryophila*). (C) Maximum likelihood tree for the dataset used in the panel A. The tree was generated by the RAxML program. A similar tree was also recovered by the PHyML web service (http://www.atgc-montpellier.fr/phyml/) using the default setting. Note that the positions of the root in this dataset differ depending on outgroup(s) and phylogenetic inference methods.

and tree-building methods pinpoint different root positions, which poses a serious problem for classification of families within this suborder. Therefore, rooting based on outgroup analysis should be carried out with a variety of combination of outgroups and phylogenetic inference methods, and the result should be interpreted with a great caution.

*Evaluation of phylogenetic trees*

Several procedures have been developed and are available for the evaluation of phylogenetic analysis, that is to assess the robustness of phylogenetic tree topology. The most generalized method is bootstrap analysis, which can be used with any tree-building method described above. This approach is a sampling technique for estimating statistical error in situations where the underlying sampling distribution is unknown. Since Felsenstein (1985) introduced bootstrap analysis into phylogenetic methods to estimate confidence levels, it has been most widely used and successfully applied.

Bootstrapping process simply resamples a new MSA from the original MSA by randomly choosing alignment columns while permitting multiple sampling of any given column. This process is carried out in multiple times, resulting multiple MSAs which are similar to each other or to the original MSA, but rarely identical. Then, phylogenetic treeing is performed on each resampled MSA (replicate) to produce multiple bootstrapped trees. From these trees, the statistical confidence of each branch, representing grouping of particular set of sequences/organisms, can be estimated as the proportion of correct recovery out of the total number of trees examined. This value, often called bootstrap support, is generally given as percentages. Several computer programs, including PAUP*, PHYLIP and MEGA, can be used to analyse and display bootstrapped phylogenetic trees. The consensus phylogenetic tree, produced as consensus of the resultant bootstrapped trees, may not be identical to the original tree. The general practice for presenting bootstrapped phylogenetic trees is to display bootstrap values, as percentages, at each branch of the original tree. The MEGA program provides this function.

Even though this method provides a generally applicable way of evaluating phylogenetic trees, it can only be used for relatively less time consuming methods, such as distance methods. At least 100 and usually 200–2000 resamplings are recommended. Under normal circumstances, bootstrap support of 70% or higher is regarded as a good indication of reliable grouping/clustering in a given phylogenetic tree (Van de Peer, 2003). It is worth noting that as the number of sequences increases, bootstrap support values tend to decrease.

## Working with large-scale sequence data

Given the recent innovations in DNA sequencing technology and the demand for accurate phylogenetic analysis in a variety of microbiological disciplines, scientists often face the task of generating a phylogenetic tree from many sequences within a reasonable time (computational cost), without compromising its accuracy. In such a situation, pre-existing alignment can play a crucial role. The Ribosomal Database Project (RDP) provides over 1.2 million 16S rRNA sequences of Bacteria and Archaea as a pre-aligned format. Similarly, the Silva database provides the data files of the Arb program for both small and large subunit RNA gene sequences, where all sequences are already aligned with each other based on rRNA secondary structure. Both databases also offer high-throughput sequence alignment functions. For example, RDP database provides the Pyrosequencing Aligner (http://pyro.cme.msu.edu/spring/align.spr) that can align up to 500,000 short sequences generated by the next generation sequencing technology through a web-based service. Greengenes (http://greengenes.lbl.gov/) is also a versatile web application providing access to the comprehensive 16S rRNA gene sequence alignment for browsing, blasting, probing, and downloading. It employed the NAST (Nearest Alignment Space Termination) algorithm (DeSantis *et al.*, 2006) for generating large MSA. A similar strategy of pre-aligned database can be applied to protein sequence database, perhaps using codon-based sequence alignment to generate template reference alignments.

### General recommendations

Given the plethora of DNA and translated protein sequences in the public domain and the invention of massively parallel high throughput DNA sequencing technologies, phylogenetics has come of age as a discipline in recent years. Fortunately, the availability of better computing hardware now allows application of more sophisticated phylogenetic methods on larger datasets. For example, an ordinary personal computer can be used for the bootstrap analysis of the ML method, one of the most complex methods known, if the number of sequences is less than 50. There are also a flood of computer programs and algorithms that claim better solutions to known problems in phylogenetic analysis. A user who wishes to carry out phylogenetic tree must navigate this sea of information with right map, i.e. theoretical knowledge and suitable computer programs. Several multipurpose software packages are conveniently available for one stop analysis of MSA and phylogenetic tree construction (Table 2.2).

We recommend the following for general purpose phylogenetic analysis:

1 Use functional constraint in the multiple sequence alignment whenever possible. For example, secondary structure should be used to improve alignment quality for ribosomal RNA sequences, and other RNA genes that form secondary and tertiary structures.
2 The columns of the multiple sequence alignment should be carefully chosen for the subsequent phylogenetic treeing. This process is often referred 'masking' or 'filtering'. Visual inspection and manual selection of columns for the further analyses are highly recommended. This can be achieved using graphically oriented sequence editors (Arb, MEGA, PHYDIT and jPHYDIT). Nonetheless, this process is subjective one which requires substantial experience.
3 Always apply multiple phylogenetic inference methods whose theoretical bases are different. Perhaps the best indication of finding true tree/evolutionary path derives when such a relationship is recovered from different tree-building methods. This has been proved from a simulation-based study (Kim, 1993), and is also highly recommended for the description of new taxa (Tindall *et al.*, 2010).
4 If a rooted tree is produced, always apply different combinations of outgroups and tree-building methods to check the stability of the root position. This is often neglected, but has a fundamental consequence when a group/clade becomes a formal taxon.

### Further reading

For readers who are interested in the principles of phylogenetic analyses, book chapters by Hillis *et al.* (1996) and Swofford *et al.* (1996), and the books by Nei and Kumar (2000) and Graur and Li (2000) will be very helpful. The manual of PAUP* version 3.1 (http://paup.csit.fsu.edu/) is good source of the basic terminology and principles of maximum parsimony analysis. Step-by-step procedures using MEGA, PHYML, and MrBayes can be found in the introductory book by Hall (2008). A book edited by Lemey *et al.*(2009) offers theory and practical guidelines and can be used as a detailed reference for phylogenetic analyses.

## Concluding remarks and future perspectives

Microbiologists often seem to regard phylogenetic analyses as a 'black box' integrated into computer software. This leads to fundamental mistakes in publications in which algorithms and methods are not explicitly presented for later validation. This is now especially problematic as most phylogenetic packages contain multiple evolutionary models and treeing algorithms. For example, a phrase like 'the phylogenetic tree was generated by the MEGA program' cannot be scientifically valid as this program contains several different tree-building methods.

Phylogenetic analysis is merely inferring, not identifying, processes in the past evolutionary path of organisms, and the results should be interpreted as carefully as possible. This is particularly true when a taxonomic proposal is involved (Tindall *et al.*, 2010).

Owing to the presence of laterally transferred genes and genomic islands in prokaryotic genomes, more innovative concepts may be needed for addressing evolution of genomes rather than genes. Traditional phylogenetic trees represented

**Table 2.2** Most widely used computer programs/packages for building phylogenetic trees[a]

| Program | Available platforms[b] | Graphical user interface | Editing function | Distance method | Maximum parsimony method | Maximum likelihood method | Bayesian method | Drawing tree | Available at | References |
|---|---|---|---|---|---|---|---|---|---|---|
| Arb | L/M | O | O [c] | O/L | L | L | X | O | http://www.arb-home.de/ | Ludwig *et al.* (2004) |
| Bioedit | W | O | O [c] | L | L | L | X | X | http://www.mbio.ncsu.edu/BioEdit/bioedit.html | |
| jPHYDIT | W/L/M | O | O | O/L | L | L | L | X | http://chunlab.snu.ac.kr/jphydit/ | Jeon *et al.* (2005) |
| MEGA | W | O | O | O | O | X | X | O | http://www.megasoftware.net/ | Kumar *et al.* (2004) |
| MrBayes | W/L/M/B | X | X | X | X | X | O | X | http://mrbayes.csit.fsu.edu/ | Ronquist and Huelsenbeck (2003) |
| PHYLIP | W/L/M | X | X | O | O | O | X | O | http://evolution.genetics.washington.edu/phylip.html | Felsenstein (1993) |
| PAUP* | W/L/M | O[d] | X | O | O | O | O | O | http://paup.csit.fsu.edu/[e] | |
| PhyML | L/B | X | X | X | X | O | X | X | http://atgc.lirmm.fr/phyml/ | Guindon and Gascuel (2003) |
| RaxML | L/B | X | X | X | X | O | X | X | http://icwww.epfl.ch/~stamatak/index-Dateien/Page443.htm | Stamatakis *et al.* (2005) |

[a]O, function is available; L, function is available as a link to other programs; X, function is not available.

[b]Operating system: W, Microsoft Windows; L, Linux; M, Mac OS; B, web service is available. Emulated environments are not included.

[c]Ribosomal RNA secondary structure is displayed for viewer function.

[d]Graphical user interface is only available in Mac Classic version (not available in OS_10.0 or later.

[e]The current version (4.0) is not free for academic institutions.

by recursively bifurcating branches are suitable for representing evolution of eukaryotes which lack genetic recombination in the horizontal way. Therefore, a network approach is proposed for interpreting the mosaic evolutionary history of prokaryotic genomes (Dagan *et al.*, 2008; Bapteste *et al.*, 2009). The use of this approach in microbial taxonomy and ecology has not yet been considered, though it has a potential to incorporate this pan-genomic view of evolution into the more practical microbiological fields.

Another tide of change coming to general microbiology laboratories is revolution of DNA sequencing technology, exemplified by NGS. The currently available platforms, such as the Illumina Genetic Analyzer (formerly Solexa), can produce DNA sequence data equivalent to 100-fold of *Escherichia coli* genome with a cost of a couple of thousands of dollars. It is evident that the cost of DNA sequencing will soon be reduced, perhaps dramatically, and the availability of whole genome sequence information will be increased for routine microbiological research, such as identification, classification and phylogenetics. Given this dynamic environment with rapidly evolving DNA sequencing technology, molecular biology and computer science, phylogenetic analysis will continue to play a central role in various scientific disciplines including taxonomy, systematics, evolutionary microbiology and ecology.

## Acknowledgement

We are grateful to Professor Jonathan Adams for critically reviewing the manuscript.

## References

Adekambi, T., and Drancourt, M. (2004). Dissection of phylogenetic relationships among 19 rapidly growing Mycobacterium species by 16S rRNA, hsp65, sodA, recA and rpoB gene sequencing. Int. J. Syst. Evol. Microbiol. *54*, 2095–2105.

Bapteste, E., O'Malley, M.A., Beiko, R.G., Ereshefsky, M., Gogarten, J.P., Franklin-Hall, L., Lapointe, F.J., Dupre, J., Dagan, T., Boucher, Y., and Martin, W. (2009). Prokaryotic evolution and the tree of life are two different things. Biol. Direct *4*, 34.

Bergsten, J. (2005). A review of long-branch attraction. Cladistics *21*, 163–193.

Brown, J.R., and Doolittle, W.F. (1997). Archaea and the prokaryote-to-eukaryote transition. Microbiol. Mol. Biol. Rev. *61*, 456–502.

Bruno, W.J., Socci, N.D., and Halpern, A.L. (2000). Weighted neighbor joining: a likelihood-based approach to distance-based phylogeny reconstruction. Mol. Biol. Evol. *17*, 189–197.

Cannone, J.J., Subramanian, S., Schnare, M.N., Collett, J.R., D'Souza, L.M., Du, Y., Feng, B., Lin, N., Madabusi, L.V., Müller, K.M., Pande, N., Shang, Z., Yu, N., and Gutell, R.R. (2002). The comparative RNA web (CRW) site: an online database of comparative sequence and structure information for ribosomal, intron, and other RNAs. BMC Bioinformatics *3*, 2.

Chenna, R., Sugawara, H., Koike, T., Lopez, R., Gibson, T.J., Higgins, D.G., and Thompson, J.D. (2003). Multiple sequence alignment with the Clustal series of programs. Nucl. Acids Res. *31*, 3497–3500.

Chun, J., Grim, C.J., Hasan, N.A., Lee, J.H., Choi, S.Y., Haley, B.J., Taviani, E., Jeon, Y.S., Kim, D.W., Brettin, T.S., Bruce D.C., Challacombe J.F., Detter J.C., Han C.S., Munk A.C., Chertkov O., Meincke L., Saunders E., Walters R.A., Huq A., Nair G.B., and Colwell R.R. (2009). Comparative genomics reveals mechanism for short-term and long-term clonal transitions in pandemic *Vibrio cholerae*. Proc. Natl. Acad. Sci. U.S.A. *106*, 15442–15447.

Chun, J., Lee, J.H., Jung, Y., Kim, M., Kim, S., Kim, B.K., and Lim, Y.W. (2007). EzTaxon: a web-based tool for the identification of prokaryotes based on 16S ribosomal RNA gene sequences. Int. J. Syst. Evol. Microbiol. *57*, 2259–2261.

Cole, J.R., Wang, Q., Cardenas, E., Fish, J., Chai, B., Farris, R.J., Kulam-Syed-Mohideen, A.S., McGarrell, D.M., Marsh, T., Garrity, G.M., and Tiedje, J.M. (2009). The Ribosomal Database Project: improved alignments and new tools for rRNA analysis. Nucl. Acids Res. *37*, D141–145.

Dagan, T., Artzy-Randrup, Y., and Martin, W. (2008). Modular networks and cumulative impact of lateral transfer in prokaryote genome evolution. Proc. Natl. Acad. Sci. U.S.A. *105*, 10039–10044.

DeSantis, T.Z., Jr., Hugenholtz, P., Keller, K., Brodie, E.L., Larsen, N., Piceno, Y.M., Phan, R., and Andersen, G.L. (2006). NAST: a multiple sequence alignment server for comparative analysis of 16S rRNA genes. Nucl. Acids Res. *34*, W394–399.

Edgar, R.C. (2004a). MUSCLE: a multiple sequence alignment method with reduced time and space complexity. BMC Bioinformatics *5*, 113.

Edgar, R.C. (2004b). MUSCLE: multiple sequence alignment with high accuracy and high throughput. Nucl. Acids Res. *32*, 1792–1797.

Felsenstein, J. (1981). Evolutionary trees from DNA sequences: a maximum likelihood approach. J. Mol. Evol. *17*, 368–376.

Felsenstein, J. (1985). Confidence limits on phylogenies: an approach using the bootstrap. Evolution *39*, 783–791.

Felsenstein, J. (1993). PHYLIP (phylogenetic inference package) version 3.5c (Seattle: University of Washington).

Fleischmann, R.D., Adams, M.D., White, O., Clayton, R.A., Kirkness, E.F., Kerlavage, A.R., Bult, C.J., Tomb, J.F., Dougherty, B.A., Merrick, J.M., McKenney, K., Sutton, G., Fitzhugh, W., Fields, C., Gocayne, J.D., Scott, J., Shirley, R., Liu, L.-I., Glodek, A., Kelly, J.M.,

Weidman, J.F., Phillips, C.A., Spriggs, T., Hedblom, E., Cotton, M.D., Utterback, T.R., Hanna, M.C., Nguyen, D.T., Saudek, D.M., Brandon, R.C., Fine, L.D., Fritchman, J.L., Fuhrmann, J.L., Geoghagen, N.S.M., Gnehm, C.L., McDonald, L.A., Small, K.V., Fraser, C.M., Smith, H.O., and Venter, J.C. (1995). Whole-genome random sequencing and assembly of *Haemophilus influenzae* Rd. Science *269*, 496–512.

Gascuel, O. (1997). BIONJ: an improved version of the NJ algorithm based on a simple model of sequence data. Mol. Biol. Evol. *14*, 685–695.

Glazunova, O.O., Raoult, D., and Roux, V. (2009). Partial sequence comparison of the rpoB, sodA, groEL and gyrB genes within the genus *Streptococcus*. Int. J. Syst. Evol. Microbiol. *59*, 2317–2322.

Goodfellow, M. (1971). Numerical taxonomy of some nocardioform bacteria. J. Gen. Microbiol. *69*, 33–80.

Graur, D., and Li, W.-H. (2000). Fundamentals of Molecular Evolution, 2nd ed. edn (Sunderland, Massachusetts, Sinauer Associates).

Guindon, S., and Gascuel, O. (2003). A simple, fast, and accurate algorithm to estimate large phylogenies by maximum likelihood. Syst. Biol. *52*, 696–704.

Guo, Y., Zheng, W., Rong, X., and Huang, Y. (2008). A multilocus phylogeny of the *Streptomyces griseus* 16S rRNA gene clade: use of multilocus sequence analysis for streptomycete systematics. Int. J. Syst. Evol. Microbiol. *58*, 149–159.

Hall, B.G. (2008). Phylogenetic Trees Made Easy: A How-to Manual., 3rd ed. edn (Sunderland, Massachusetts: Sinauer Associates).

Hasegawa, M., Kishino, H., and Yano, T. (1985). Dating of the human-ape splitting by a molecular clock of mitochondrial DNA. J. Mol. Evol. *22*, 160–174.

Hillis, D.M., Mable, B.K., and Moritz, C. (1996). Phylogenetic inference. In Molecular Systematics, 2nd ed. (Sunderland, Massachusetts: Sinauer Associates, Inc.).

Jeon, Y.S., Chung, H., Park, S., Hur, I., Lee, J.H., and Chun, J. (2005). jPHYDIT: a JAVA-based integrated environment for molecular phylogeny of ribosomal RNA sequences. Bioinformatics *21*, 3171–3173.

Jiang, L.W., Lin, K.L., and Lu, C.L. (2008). OGtree: a tool for creating genome trees of prokaryotes based on overlapping genes. Nucl. Acids Res. *36*, W475–480.

Jones, D.T., Taylor, W.R., and Thornton, J.M. (1992). The rapid generation of mutation data matrices from protein sequences. Comput. Appl. Biosci. *8*, 275–282.

Jukes, T.H., and Cantor, C.R. (1969). Evolution of protein molecules. In Mammalian Protein Metabolism, H.N. Munro, ed. (New York: Academic Press).

Katoh, K., and Toh, H. (2008). Recent developments in the MAFFT multiple sequence alignment program. Brief Bioinform. *9*, 286–298.

Katoh, K., Kuma, K., Toh, H., and Miyata, T. (2005). MAFFT version 5: improvement in accuracy of multiple sequence alignment. Nucl. Acids Res. *33*, 511–518.

Kim, J. (1993). Improving the accuracy of phylogenetic estimation by combining different methods. Syst. Biol. *42*, 331–340.

Kimura, M. (1980). A simple method for estimating evolutionary rate of base substitutions through comparative studies of nucleotide sequences. J. Mol. Evol. *16*, 111–120.

Kimura, M. (1983). The Neutral Theory of Molecular Evolution (Cambridge: Cambridge University Press).

Kosiol, C., and Goldman, N. (2005). Different versions of the Dayhoff rate matrix. Mol. Biol. Evol. *22*, 193–199.

Kumar, S., and Filipski, A. (2007). Multiple sequence alignment: in pursuit of homologous DNA positions. Genome Res. *17*, 127–135.

Kumar, S., Tamura, K., and Nei, M. (2004). MEGA3: Integrated software for Molecular Evolutionary Genetics Analysis and sequence alignment. Brief Bioinform. *5*, 150–163.

Lassmann, T., and Sonnhammer, E.L. (2005). Kalign-an accurate and fast multiple sequence alignment algorithm. BMC Bioinformatics *6*, 298.

Lemey, P., Salemi, M.S., and Vandamme, A.-M. (2009). The Phylogenetic Handbook: A Practical Approach to Phylogenetic Analysis and Hypothesis Testing, 2nd ed. (Cambridge: Cambridge University Press).

Li, K.B. (2003). ClustalW-MPI: ClustalW analysis using distributed and parallel computing. Bioinformatics *19*, 1585–1586.

Lipman, D.J., Altschul, S.F., and Kececioglu, J.D. (1989). A tool for multiple sequence alignment. Proc. Natl. Acad. Sci. U.S.A. *86*, 4412–4415.

Ludwig, W., Strunk, O., Westram, R., Richter, L., Meier, H., Yadhukumar, Buchner, A., Lai, T., Steppi, S., Jobb, G., Forster, W., Brettske, I., Gerber, S., Ginhart, A.W., Gross, O., Grumann, S., Hermann, S., Jost, R., König, A., Liss, T., Lussmann, R., May, M., Nonhoff, B., Reichel, B., Strehlow, R., Stamatakis, A., Stuckmann, N., Vilbig, A., Lenke, M., Ludwig, T., Bode, A., and Schleifer, K.H. (2004). ARB: a software environment for sequence data. Nucl. Acids Res. *32*, 1363–1371.

Metzker, M.L. (2010). Sequencing technologies – the next generation. Nature Rev. Genet. *11*, 31–46.

Moritz, C., and Hillis, D.M. (1996). Molecular systematics: context and controversies. In Molecular Systematisc, D.M. Hillis, C. Moritz, and B.K. Mable, eds. (Sunderland, Massachusetts: Sinauer Associates), pp. 1–13.

Nei, M., and Kumar, S. (2000). Molecular Evolution and Phylogenetics (New York: Oxford University Press).

Nixon, K.C., and Carpenter, J.M. (1993). On outgroups. Cladistics *9*, 413–4126.

Notredame, C., Higgins, D.G., and Heringa, J. (2000). T-Coffee: A novel method for fast and accurate multiple sequence alignment. J. Mol. Biol. *302*, 205–217.

Olsen, G.J., Matsuda, H., Hagstrom, R., and Overbeek, R. (1994). fastDNAmL: a tool for construction of phylogenetic trees of DNA sequences using maximum likelihood. Comput. Appl. Biosci. *10*, 41–48.

Otu, H.H., and Sayood, K. (2003). A new sequence distance measure for phylogenetic tree construction. Bioinformatics *19*, 2122–2130.

Posada, D., and Crandall, K.A. (1998). MODELTEST: testing the model of DNA substitution. Bioinformatics *14*, 817–818.

Pruesse, E., Quast, C., Knittel, K., Fuchs, B.M., Ludwig, W., Peplies, J., and Glockner, F.O. (2007). SILVA: a comprehensive online resource for quality checked and aligned ribosomal RNA sequence data compatible with ARB. Nucl. Acids Res. *35*, 7188–7196.

Rodriguez, F., Oliver, J.L., Marin, A., and Medina, J.R. (1990). The general stochastic model of nucleotide substitution. J. Theor. Biol. *142*, 485–501.

Ronaghi, M., Karamohamed, S., Pettersson, B., Uhlen, M., and Nyren, P. (1996). Real-time DNA sequencing using detection of pyrophosphate release. Anal. Biochem. *242*, 84–89.

Ronquist, F., and Huelsenbeck, J.P. (2003). MrBayes 3: Bayesian phylogenetic inference under mixed models. Bioinformatics *19*, 1572–1574.

Saitou, N., and Nei, M. (1987). The neighbor-joining method: a new method for reconstructing phylogenetic trees. Mol. Biol. Evol. *4*, 406–425.

Shendure, J., and Ji, H. (2008). Next-generation DNA sequencing. Nature Biotechnol. *26*, 1135–1145.

Stackebrandt, E., Rainey, F.A., and Ward-Rainey, N.L. (1997). Proposal for a new hierarchic classification system, Actinobacteria classis nov. Int. J. Syst. Bacteriol. *47*, 479–491.

Stamatakis, A. (2006). RAxML-VI-HPC: maximum likelihood-based phylogenetic analyses with thousands of taxa and mixed models. Bioinformatics *22*, 2688–2690.

Stamatakis, A., Ludwig, T., and Meier, H. (2005). RAxML-III: a fast program for maximum likelihood-based inference of large phylogenetic trees. Bioinformatics *21*, 456–463.

Swofford, D.L., Olsen, G.J., Waddell, P.J., and Hillis, D.M. (1996). Phylogenetic inference. In Molecular systematis, D.M. Hillis, C. Moritz, and B.K. Mable, eds. (Sunderland, Massachusetts: Sinauer Associates), pp. 407–514.

Takahashi, M., Kryukov, K., and Saitou, N. (2009). Estimation of bacterial species phylogeny through oligonucleotide frequency distances. Genomics *93*, 525–533.

Thompson, J.D., Gibson, T.J., and Higgins, D.G. (2002). Multiple sequence alignment using ClustalW and ClustalX. Curr. Protoc. Bioinform., Chapter 2, Unit 2 3.

Tindall, B.J., Rosselló-Mora, R., Busse, H.J., Ludwig, W., and Kämpfer, P. (2010). Notes on the characterization of prokaryote strains for taxonomic purposes. Int. J. Syst. Evol. Microbiol. *60*, 249–266.

Vandamme, A.M. (2009). Basic concepts of molecular evolution. In The Phylogenetic Handbook: A Practical Approach to Phylogenetic Analysis and Hypothesis Testing, P. Lemey, M. Salemi, and A.M. Vandamme, eds. (Cambridge: Cambridge University Press), pp. 3–29.

Van de Peer, Y. (2003). Phylogeny inference based on distance methods. Theory. In The Phylogenetic Handbook, M. Salemi, and A.M. Vandamme, eds. (Cambridge: Cambridge University Press), pp. 101–119.

Veerassamy, S., Smith, A., and Tillier, E.R. (2003). A transition probability model for amino acid substitutions from blocks. J. Comput. Biol. *10*, 997–1010.

Wallace, I.M., O'Sullivan, O., Higgins, D.G., and Notredame, C. (2006). M-Coffee: combining multiple sequence alignment methods with T-Coffee. Nucl. Acids Res. *34*, 1692–1699.

Waterhouse, A.M., Procter, J.B., Martin, D.M., Clamp, M., and Barton, G.J. (2009). Jalview Version 2 – a multiple sequence alignment editor and analysis workbench. Bioinformatics *25*, 1189–1191.

Woese, C.R. (1987). Bacterial evolution. Microbiol Rev *51*, 221–271.

Wolf, Y.I., Rogozin, I.B., Grishin, N.V., Tatusov, R.L., and Koonin, E.V. (2001). Genome trees constructed using five different approaches suggest new major bacterial clades. BMC Evol. Biol. *1*, 8.

Wu, D., Hugenholtz, P., Mavromatis, K., Pukall, R., Dalin, E., Ivanova, N.N., Kunin, V., Goodwin, L., Wu, M., Tindall, B.J., Hooper, S.D., Pati, A., Lykidis, A., Spring, S., Anderson, I.J., D'haeseleer, P., Zemla, A., Singer, M., Lapidus, A., Nolan, M., Copeland, A., Han, C., Chen, F., Cheng, J.F., Lucas, S., Kerfeld, C., Lang, E., Gronow, S., Chain, P., Bruce, D., Rubin, E.M., Kyrpides, N.C., Klenk, H.P., and Eisen, J.A. (2009). A phylogeny-driven genomic encyclopaedia of Bacteria and Archaea. Nature *462*, 1056–1060.

Xia, X., and Xie, Z. (2001). DAMBE: software package for data analysis in molecular biology and evolution. J. Hered. *92*, 371–373.

Yamamoto, S., and Harayama, S. (1995). PCR Amplification and direct sequencing of gyrB genes with universal primers and their application to the detection and taxonomic analysis of *Pseudomonas putida* strains. Appl. Environ. Microbiol. *61*, 1104–1109.

Yang, Z., Goldman, N., and Friday, A. (1994). Comparison of models for nucleotide substitution used in maximum-likelihood phylogenetic estimation. Mol. Biol. Evol. *11*, 316–324.

# Multilocus Sequence Analysis and Bacterial Species Phylogeny Estimation

3

Pablo Vinuesa

Abstract

This chapter presents a review of critical factors that have to be considered and evaluated in multilocus sequence analysis (MLSA) in order to make robust estimates of bacterial species phylogenies. The theoretical arguments in favour of the conditional data combination will be presented. I will briefly review criteria for marker selection, and will provide practical advice on the computational aspects, potential pitfalls, and software choices available for each step in a MLSA. For this purpose, a detailed case study using *atpD*, *glnII*, *recA* and *rpoB* sequences of symbiotic root nodule bacteria of the genus *Bradyrhizobium* will be presented. I will discuss and illustrate the use of phylogenetic congruence analysis, and a strategy to evaluate the additivity of the phylogenetic signals in the different partitions. The importance of using multiple isolates per species/lineage, proper model selection and thorough tree searches to get a good estimate of a multispecies phylogeny will be emphasized. Maximum likelihood and Bayesian phylogeny estimation using supermatrices are thoroughly discussed and critically compared with the new, concatenation-independent, Bayesian Estimation of Species Trees (BEST) algorithm, which is based on the multispecies coalescent.

## Introduction

### Multilocus sequence analysis in the broader context of molecular systematics

Multilocus sequence analysis (MLSA) represents the novel standard in microbial molecular systematics (Gevers *et al.*, 2005). In this context, MLSA is implemented in a relatively straightforward way, consisting essentially in the concatenation of several sequence partitions for the same set of organisms, resulting in a 'supermatrix' which is used to infer a phylogeny by means of distance-matrix or optimality criterion-based methods (Devulder *et al.*, 2005; Vinuesa *et al.*, 2005b, 2008; Wertz *et al.*, 2003). This approach is expected to have an increased resolving power due to the large number of characters analysed, and a lower sensitivity to the impact of conflicting signals (i.e. phylogenetic incongruence) that result from eventual horizontal gene transfer events (Escobar-Páramo *et al.*, 2004; Lerat *et al.*, 2003; Rokas *et al.*, 2003).

Although claimed as a new standard in microbial systematics, MLSA is by no means new in the broader context of molecular systematics, as reflected by the rich literature available on how to treat partitioned data. The strategies used to deal with multiple partitions can be grouped in three broad categories: the total evidence, separate analysis and combination approaches (Huelsenbeck *et al.*, 1996). The concatenation approach that dominates MLSAs in the microbial molecular systematics literature is known to systematists working with plants and animals as the 'total molecular evidence' approach, and has been used to solve difficult phylogenetic questions such as the relationships among the major groups of cetaceans (Hasegawa *et al.*, 1997), that of microsporidia and fungi (Hirt *et al.*, 1999), or the phylogeny of major plant lineages (Soltis *et al.*, 1999), to mention a few classic examples.

The total molecular evidence approach has been criticized because by directly concatenating all available sequence alignments, the evidence of conflicting phylogenetic signals in the different data partitions is lost along with the possibility to uncover the evolutionary processes that gave rise to such contradictory signals (Bull, 1993). The nature of these conflicts is varied, but in the microbial world the strongest conflicting signals often derive from the existence of horizontal gene transfer (HGT) events in the dataset (Boucher *et al.*, 2003; Gogarten and Townsend, 2005; Ochman *et al.*, 2005). If the individuals containing xenologous loci are not identified and removed from the supermatrix prior to phylogeny inference, the resulting hypothesis may be strongly distorted, since standard treeing methods assume a single underlying evolutionary history (Posada and Crandall, 2002; Schierup and Hein, 2000; Vinuesa *et al.*, 2005c). Based on these arguments, the conditional data combination strategy is to be generally preferred in bacterial MLSA (Vinuesa *et al.*, 2005c, 2008).

### Gene trees versus species trees

Owing to the stochasticity of gene lineage sorting during the speciation process, no single gene tree is likely to adequately reflect the species phylogeny (Degnan and Rosenberg, 2006; Maddison and Knowles, 2006; Nichols, 2001; Rosenberg, 2002; Rosenberg and Nordborg, 2002). Therefore, single gene trees (i.e. 16S rRNA-based phylogenies) should not be considered equal to species trees (Maddison, 1997; Nichols, 2001; Vinuesa *et al.*, 2005a, 2005b). Theoretically, the best strategy to get a reasonable estimate of the species tree is to consider multiple genealogies inferred from unlinked loci, and to use multiple individuals per species (Rosenberg and Nordborg, 2002; Vinuesa *et al.*, 2008). These important theoretical considerations have been largely neglected in the bulk of the microbial molecular systematics and taxonomic literature (Vinuesa *et al.*, 2005c, 2008). This fact is clearly reflected by the dominant practice of proposing novel bacterial species based on the analysis of a single isolate, the species' type strain. This specimen is compared with other such closely related type strains by DNA–DNA hybridization and phenotypic traits, providing phylogenetic evidence for the relationships among the 'species' under consideration based on a single gene sequence (almost invariably the SSU rRNA gene). This practice forms the so-called polyphasic approach to taxonomy that dominates the field (Stackebrandt *et al.*, 2002; Stackebrandt and Goebel, 1994; Vandamme *et al.*, 1996). In addition to the above mentioned limitations of phylogenies estimated from single strains per species, such phylogenies are of little use in a truly molecular systematics framework, since a minimum of two strains per species would be required for them to form a clade (i.e. a monophyletic group). Ideally, multiple individuals from a single species sampled from different demes (populations) should form monophyletic lineages on the estimated species tree. To have multiple strains per species is also very valuable to detect potential HGT events, and a minimum of 8–12 strains per species or population is required to make population genetic analyses in order to demarcate the species boundaries using evolutionary criteria, and to make inferences about trait evolution, population genetic structures, population subdivision or biogeographic distribution patterns (Didelot and Falush, 2007; Falush *et al.*, 2003; Ivars-Martinez *et al.*, 2008; Martiny *et al.*, 2006; Miragaia *et al.*, 2007; Papke *et al.*, 2003, 2007; Perez-Losada *et al.*, 2006; Ramette and Tiedje, 2007; Vinuesa *et al.*, 2005c, 2008; Whitaker *et al.*, 2003). Owing to space constraints, I will not discuss the use of multilocus data in bacterial population genetics or ecological contexts, referring the interested reader to the references provided above.

## What are the basic attributes of good phylogenetic markers for MLSA?

### Core versus accessory genes

Many bacterial species have so-called open pangenomes (Tettelin *et al.*, 2005, 2008), with a small fraction of conserved 'core' genes and a large fraction of strain-specific 'accessory' genes found only in some individuals of the species. This was first illustrated by the work of Welch and collaborators in a comparative study of three *Escherichia coli* genomes. They found that only 39% of the single copy genes were shared by the commensal, uropathogenic and enterohaemorrhagic strains compared in that study (Welch *et al.*, 2002). They also showed that most of the accessory genes were clustered as genomic islands interspersed between syntenic orthologous genes and operons, a finding that has been corroborated in more recent studies, showing that the genomic diversity of *E. coli* represents an open pangenome model containing a reservoir of more than 13,000 genes, many of which may be uncharacterized but important virulence factors (Rasko *et al.*, 2008).

Only the orthologous core loci are useful for MLSA-based inference of multispecies phylogenies and for within-species population genetic analyses (Vinuesa *et al.*, 2005c, 2008). Accessory loci often encode interesting functions for ecological specialization such as virulence, resistance or symbiosis genes which are very informative about ecological attributes of the strains that harbour them. However, they frequently have a contrasting evolutionary history, mode and tempo of evolution when compared with those of the core genes. Therefore, accessory genes are not suitable molecular markers to estimate species phylogenies (Enright *et al.*, 2002; Ogura *et al.*, 2009; Reid *et al.*, 2000; Silva *et al.*, 2005; Vinuesa *et al.*, 2005c; Wirth *et al.*, 2006; Young *et al.*, 2006). This highlights the mosaic nature of many bacterial genomes, which allows rapid genomic adaptation to changing environments (Ogura *et al.*, 2007; Schubert *et al.*, 2009; Sullivan and Ronson, 1998; Touchon *et al.*, 2009; Young *et al.*, 2006).

## Ubiquitous core genes, primer design and marker selection for MLSA

A limitation in MLSA and MLST studies is the lack of definition of a 'universal' set of loci to target as molecular markers. With 'universal' I don't mean universal primers that would amplify the corresponding loci from whatever organism one may encounter, but conserved orthologous loci across higher taxonomic ranks on which to base the group-specific primer formulations. The lack of such a definition has made it difficult to evaluate the relative merits of different markers and PCR primers used in different studies, often precluding the comparative analysis of evolutionary processes and patterns among closely related species and genera.

Recent estimates of the conserved core genome of all bacteria indicate that it is most likely composed of < 50 genes, of which 34 are ubiquitous (Charlebois and Doolittle, 2004), a figure not much larger than the universal set of single copy orthologous genes shared by the three domains of life, which has been estimated to be 31 genes (Ciccarelli *et al.*, 2006). These figures are very low, and don't increase much if the ubiquity criterion is relaxed to presence in 90% or 80% of members in each prokaryotic phylum, reaching 60 and 71 genes, respectively (Charlebois and Doolittle, 2004). In a recent phylogenomic analysis of 41 fully sequenced genomes of the *Rhizobiales* we found 62 of the relaxed set of 71 ubiquitous genes mentioned above (Contreras-Moreira and Vinuesa, in preparation). These genes are potentially useful to develop a broadly applicable MLSA scheme for bacteria in this order. However, 19 of them are < 600 nucleotides long (Table 3.1) and therefore most likely of little value as targets to develop informative molecular markers.

We have recently published the primers4-clades web server, which implements an easy to use and highly customizable analysis pipeline to design lineage-specific degenerate PCR primer pairs in an automatic fashion, based on multiple protein sequence alignments using an extended CODEHOP algorithm (Contreras-Moreira *et al.*, 2009). The server does not only provide the user with primer formulations, sorted according to their theoretical quality (measured on the basis of a comprehensive set of thermodynamic criteria), but also estimates the phylogenetic information content of the expected amplicons, as will be explained in detail later. These features greatly help the user in making an informed choice when confronted with alternative primer-pair formulations for a single gene, a situation often encountered with large genes such as *ileS* or *rpoB* (Table 3.1). We are currently developing and evaluating primer-pair formulations for MLSA studies in the *Rhizobiales* and other bacterial taxa (*Bacillales*, *Enterobacteriales*, *Mycobacteriaceae*,

**Table 3.1** Median lengths of a set of 62 highly conserved universal orthologous genes found among 41 *Rhizobiales* genomes, sorted by decreasing length (L.), corresponding to the set of 71 'relaxed' universal orthologous genes shared by *Archaea* and *Bacteria*, as defined by Charlebois and Doolittle (2004)

| Gene | Median L. | Gene | Median L. |
|---|---|---|---|
| *rpoC* | 4203 | *trpS* | 1065 |
| *rpoB* | 4140 | *rpsB* | 996 |
| *ileS* | 3006 | *trxB* | 966 |
| *valS* | 2859 | *ksgA* | 861 |
| *topA* | 2736 | *rplB* | 837 |
| *alaS* | 2676 | *map* | 825 |
| *infB* | 2661 | *uppS* | 759 |
| *leuS_v2* | 2631 | *rplC* | 726 |
| *pheT* | 2424 | *pyrH* | 723 |
| *fus* | 2073 | *rpsC* | 708 |
| *thrS* | 2007 | *rplA* | 693 |
| *dnaG* | 2004 | *rpsD* | 618 |
| *ftsH* | 1923 | *rpsE* | 573 |
| *dnaX* | 1875 | *efp* | 567 |
| *argS_v2* | 1794 | *rplE* | 558 |
| *pyrG* | 1632 | *nusG* | 531 |
| *nusA* | 1617 | *rpsI* | 477 |
| *guaA* | 1608 | *rpsG* | 471 |
| *metG* | 1551 | *rplM* | 465 |
| *atpD* | 1431 | *rplK* | 429 |
| *gltX* | 1428 | *rpsH* | 399 |
| *lpdA* | 1407 | *rpsK* | 390 |
| *cysS* | 1401 | *rplV* | 390 |
| *secY* | 1332 | *rpsL* | 372 |
| *proS* | 1320 | *rpsM* | 369 |
| *serS* | 1305 | *rplN* | 369 |
| *glyA* | 1302 | *rplX* | 315 |
| *eno* | 1284 | *rpsJ* | 309 |
| *recA* | 1092 | *rpsN* | 306 |
| *gcp* | 1089 | *rpsS* | 279 |
| *pheS* | 1083 | *rpmC* | 207 |

*Pseudomonadales* and *Vibrionales*) using our extended CODEHOP algorithm and the genes listed in Table 3.1.

## A multilocus sequence analysis case study: estimating a species tree for the genus *Bradyrhizobium* using the concatenation approach

In this section I will present a tutorial on how to perform a MLSA. In doing so, I will provide detailed information on what factors to consider in order to design a powerful MLSA for multispecies phylogeny estimation, along with recommendations on software choices available for each step in the analysis for different computing platforms. The data that will be used here are a subset of the *atpD, glnII, recA* and *rpoB* sequences used in a recently published MLSA-based study of the biogeography and evolutionary genetics of four *Bradyrhizobium* species (Vinuesa *et al.*, 2008). That study analysed the sequence data of 80 soybean nodule isolates obtained from India, Myanmar, Nepal and Vietnam. Thirty three reference strains from a previous study (Vinuesa *et al.*, 2005c) were included in the combined phylogenetic and population genetic approaches used for species demarcation and to estimate the magnitude of evolutionary forces acting within lineages. The aim of the study was to assess the power and practical utility of the multilocus sequence analysis approach (Gevers *et al.*, 2005) for *Bradyrhizobium* molecular systematics and ecological inference. This bacterial genus is considered as a 'taxonomically difficult' group of organisms due to their highly conserved *rrs* sequences and poor correlation between the groupings formed on the basis of genotypic and phenotypic traits, raising questions about the suitability of the polyphasic taxonomic approach to *Bradyrhizobium* systematics (So *et al.*, 1994; van Rossum *et al.*, 1995).

### Multiple sequence alignment, saturation analysis and model selection

A first critical aspect for making good estimates of a gene phylogeny is to obtain a reliable alignment. MLSA is typically based on protein-coding gene fragments. This is of great advantage to ensure a high quality of the multiple sequence alignment because protein alignments are generally much more reliable than nucleotide alignments since the former have 5 times as many character states as the latter. Therefore, if relatively divergent protein-coding sequences are to be aligned, the CDSs should be first translated in order to align them at the protein level. The resulting alignment is then used to generate the underlying codon alignment. There are many programs and algorithms available that generate multiple sequence alignments. Probably the most commonly used programs are those of the Clustal family (Chenna *et al.*, 2003; Larkin *et al.*, 2007). However, newer, faster and more accurate algorithms are implemented in software such as Mafft (Katoh and Toh, 2008) and Muscle (Edgar, 2004), available for all major computing platforms. There are several Windows programs like BioEdit (Hall, 1999) and DAMBE (Xia and Xie, 2001) that allow the translation of CDSs to their encoded protein sequences. Both packages run ClustalW to perform multiple sequence alignments. DAMBE even implements a function that allows the user to align CDSs to the translation products in order to generate the underlying codon alignment. Other options to perform this operation are the RevTrans server or the corresponding Python script that can be freely downloaded (Wernersson and Pedersen, 2003), the PAL2NAL server (Suyama *et al.*, 2006), or simple custom Perl scripts that import BioPerl (Stajich *et al.*, 2002) modules and classes will help in getting this critically important job done on any platform.

If the sequences are relatively divergent, as could be the case when taxa from different genera or families are to be compared, it may be necessary to use the protein alignments instead of the codon alignments as character sources for phylogeny estimation, because DNA alignments get saturated much faster (Page and Holmes, 1998). Software packages like DAMBE (Xia and Xie, 2001) make it easy to get saturation plots for any codon positions in the alignment. If many gaps are found in a multiple sequence alignment, it may be useful to filter out poorly aligned residues by using either GBlocks (Castresana, 2000) or trimAl (Capella-Gutierrez *et al.*, 2009), although this is rarely necessary in typical MLSAs that focus on closely related species or genera.

Once reliable multiple sequence alignments are available for the different gene partitions it is important to select a proper model of sequence evolution (Posada and Buckley, 2004; Posada and Crandall, 2001). One of the most popular programs for automating nucleotide model selection is ModelTest (Posada and Crandall, 1998). Its recent successor jModelTest (Posada, 2008) has the virtue of using the freely available PhyML phylogeny inference software (Guindon and Gascuel, 2003) to calculate the likelihood scores and ML parameter value estimates of each competing model. It is also a cross-platform program since it is written in Java. Furthermore, it implements 5 different model selection strategies, including 'hierarchical and dynamical likelihood ratio tests', the 'Akaike information criterion', the 'Bayesian information criterion', and a 'decision-theoretic performance-based' approach. For protein alignments, the equivalent software for selecting the best fitting substitution matrix is ProtTest (Abascal *et al.*, 2005). If you want it all in one program, use ModelGenerator (Keane *et al.*, 2006), a model selection program written in Java, that selects optimal amino acid and nucleotide substitution models from FASTA or Phylip alignments.

If 16S rRNA gene sequences are used, multiple sequence alignments should be performed taking secondary structure criteria into account. There are several good options to do so. ARB (Ludwig *et al.*, 2004) is a dedicated software environment for UNIX/Linux systems that uses curated, pre-aligned *rrs* sequences to align user-provided sequences (Pruesse *et al.*, 2007). The ARB package comprises additional tools for data import and export, profile and filter calculation, phylogenetic analyses, specific hybridization probe design and evaluation and other components for data analysis. It is therefore the platform of choice for serious work with *rrs* sequences. It should be noted that although it was initially designed for the analysis of ribosomal RNA data, it can be used for any nucleic and amino acid sequence data as well. Other excellent alternatives for proper *rrs* sequence alignment are offered by the GreenGenes (DeSantis *et al.*, 2006) and RDP-II (Cole *et al.*, 2007; Maidak *et al.*, 2001) web services. I don't have space here for discussing the many useful features offered by these web services, but the interested reader can check my online tutorial on using these services, which is accessible from http://www.ccg.unam.mx/~vinuesa/Using_the_GreenGenes_and_RDPII_servers.html.

## Phylogenetic exploration of sequence partitions using distance-matrix reconstruction methods

A convenient way to classify phylogeny inference methods is based on two criteria: (i) the type of data they use to reconstruct the tree(s) (i.e. distance matrices vs. discrete characters) and (ii) the reconstruction strategy (algorithmic vs. tree searches using an optimality criterion) (Felsenstein, 2004a; Swofford *et al.*, 1996). Distance-matrix methods require that the discrete data matrices (i.e. the multiple sequence alignments) are converted to distance matrices. Different nucleotide substitution models or protein substitution matrices can be used to get corrected estimates of the pair-wise genetic distances between all sequences considered (Felsenstein, 2004a; Swofford *et al.*, 1996). A short online tutorial on the use and selection of nucleotide and protein evolutionary models in phylogenetics can be found at the author's web site (http://www.ccg.unam.mx/~vinuesa/Model_fitting_in_phylogenetics.html).

The big strength of distance-matrix reconstruction methods such as the neighbour-joining (NJ) algorithm (Saitou and Nei, 1987) lies in their speed. Distance matrices for several hundred OTUs can be clustered in a few seconds on a standard desktop or laptop computer. This time scales linearly with the number of bootstrap replicates one may want to analyse in order to get an idea of the impact of sampling error on different clades of the phylogeny (Felsenstein, 1985). These attributes make NJ analyses very suitable for preliminary phylogenetic data exploration. I would recommend that NJ trees are reconstructed for each individual sequence partition, along with a bootstrap analysis, and the resulting trees visually inspected. These analyses can be easily performed on Windows systems using the popular Molecular Evolutionary Genetic Analysis (MEGA4) software package (Tamura *et al.*, 2007). It implements several distance-matrix methods under a restricted set of nucleotide substitution models, including the new Maximum

Composite Likelihood (MCL) method for estimating evolutionary distances between all pairs of sequences simultaneously, with and without incorporating rate variation among sites and substitution pattern heterogeneities among lineages. The MCL method can also be used to estimate transition/transversion bias and nucleotide substitution pattern without knowledge of the phylogenetic tree. It further offers editing of DNA sequence data from autosequencers, mining web databases, performing automatic and manual sequence alignment based on ClustalW, definition of sequence partitions and taxon groups, and performing diverse population genetics analyses and evolutionary hypothesis tests. A powerful tree editor is also integrated. MEGA4 and other native Windows programs can be run in a Linux desktop environment (via the Wine compatibility layer), and on Intel-based Macintosh computers under the Parallels program. Those users that like the command line and automating analyses on UNIX/Linux platforms via simple Shell, Perl or Python scripts will probably favour Joe Felsenstein's PHYLogeny Inference Package (PHYLIP) (Felsenstein, 2004b), or David Swofford's PAUP* package (Swofford, 2002), although the latter is not freely available (it is actually the only non-freely available software mentioned in this review). Distance trees reconstructed with any of these packages can be conveniently visualized and edited with tree editors such as Rod Page's TreeView (Page, 2002) or Andrew Rambaut's FigTree (Rambaut, 2009), both freely available for all major computing platforms.

Once the NJ trees are available, the user should compare the sequence/taxon composition of well-supported monophyletic lineages (clades) across the different phylogenies in order to identify obvious incongruent clustering of particular strains across them. This is a first and strong indication of potential HGT events. Large bacterial multilocus datasets often contain individuals that are recipients of xenologous alleles at a particular locus (Silva *et al.*, 2005; Vinuesa *et al.*, 2005c, 2008). Otherwise, single long branches that emerge either from a compact and homogeneous clade, or have an isolated position on the tree, with very low bootstrap support values, may be indicative of intragenic sequence mosaics. Such suspicious sequences should be further analysed, searching for evidence of sequence mosaicism (Vinuesa *et al.*, 2005b, 2005c), using algorithms such as the bootscan (Martin *et al.*, 2005) or RDP (Martin and Rybicki, 2000), implemented in the Windows package RDP3 (Martin, 2009).

For the inference of a robust species phylogeny using standard treeing methods and concatenated alignments it is critical that individuals (strains) showing clear evidence of being recipients of xenologous alleles, or harbouring gene mosaics, are removed from the supermatrix. The reason for it is that standard tree reconstruction methods assume a single underlying evolutionary history, reflected in the dichotomously branching pattern of the inferred phylogenetic hypothesis. Organisms (multilocus haplotypes) containing xenologous or chimaeric sequences violate this basic assumption, seriously compromising the accuracy of the phylogenetic estimate (Posada and Crandall, 2002; Schierup and Hein, 2000; Vinuesa *et al.*, 2005b,c). Network graphing methods such as split-decomposition or neighbour-net are useful for visualizing parallel evolutionary pathways, an important alternative to classical treeing methods when recombination or HGT have been at play (Huson and Bryant, 2006; Huson *et al.*, 2007; Kloepper and Huson, 2008; Vinuesa *et al.*, 2005c).

## Phylogeny inference under the maximum likelihood criterion using PhyML v3.0

The maximum likelihood (ML) criterion can be defined as the conditional probability of observing the data (D), given a hypothesis or model (H), that is: $L = \Pr(D|H)$. In phylogenetics this equation translates into finding the topology ($\Gamma$), set of branch lengths ($\gamma$) and substitution model parameters ($\sigma$) that maximize the probability of observing the data, i.e. the character states in the multiple sequence alignment. Therefore, we could re-write the expression given above as $L = \Pr(D|\Gamma, \gamma, \sigma)$. Phylogenetic inference under the ML criterion is therefore an optimization problem, although a difficult one. The difficulty derives from the nature of the phylogenetic model itself. Branch lengths and parameters of the substitution model are continuous variables, whereas the topology is a discrete parameter. Furthermore, as more taxa/sequences are added, the number of unique, unrooted and strictly bifurcating topologies increases factorially, as indicated by the following expression: $(2n-5)!/(n-3)^{2n-3}$. Therefore, for four taxa we have three distinct bifurcating topologies, for eight taxa this number increases to 10,395, and for only 22 taxa it is somewhere around $3\times10^{23}$, like 'a mol of trees'! Searching for the ML tree therefore relies on sophisticated, heuristic methods that combine both discrete and continuous optimization procedures to explore a likelihood function of high dimensionality, which for most biological datasets defines a rugged landscape with multiple peaks. In practical terms this means that for large datasets tree searches often get stuck in local maxima. Since the search algorithms are heuristic, there is no guarantee of finding the global (best) ML phylogeny (Felsenstein, 2004a; Swofford *et al.*, 1996). Therefore it is important to make a thorough search of the tree space, which can be most efficiently performed by launching multiple searches starting from different seed trees, generated for example by a sequential random addition algorithm (Swofford, 2002; Swofford *et al.*, 1996). These seed trees are used to explore the astronomically large space of tree topologies that exists for large datasets by doing topological rearrangements of the seed trees by either nearest-neighbour interchange (NNI), subtree pruning and regrafting (SPR) or tree bisection and reconnection (TBR) moves or algorithms (Swofford *et al.*, 1996). Each of these moves explores alternatives of the tree space, but the neighbourhood defined by the TBR moves is much larger than that defined by SPR moves, which in turn defines a much larger neighbourhood than NNI rearrangements of the original topology on which they operate. Hence the computing times required by these algorithms are as follows: TBR > SPR > NNI. Alternative topologies found by any of these moves that improve the score of the optimality criterion (e.g. parsimony or likelihood) are retained and used as the seed for the next round of branch swappings, until no further increase in the likelihood function is attained.

The most efficient algorithms are those that provide the best trade-off between their ability to maximize the likelihood function (escaping as many local maxima as possible) and the processor time used to achieve this. PhyML is one of the most efficient modern ML tree search algorithms currently available (Guindon and Gascuel, 2003). The last version of the software is PhyML v3.0 (Guindon *et al.*, 2009), which implements three branch swapping methods (NNI, NNI + SPR and SPR only), while v2 only implemented NNI moves, and was therefore much more prone to get stuck in local maxima than the new version run with NNI+SPR or pure SPR moves (Guindon *et al.*, 2009; Hordijk and Gascuel, 2005). By default, both PhyML v2 and v3 initiate the search from a BioNJ distance tree (Gascuel, 1997). To perform thorough searches of tree space, it is possible to use random trees in addition to the default BioNJ tree to seed the searches from multiple distinct points in tree space, a strategy that almost invariably finds better trees than those found in searches starting from the BioNJ topology. We have recently demonstrated this in a MLSA analysis of *Bradyrhizobium* strains that involved 62 multilocus haplotypes, for which there are $\sim1.945514\times10^{181}$ distinct unrooted bifurcating topologies (Vinuesa *et al.*, 2008).

Another advantage of PhyML v3 is that it implements fast approximate likelihood ratio tests (aLRTs) of branch support (Anisimova and Gascuel, 2006; Guindon *et al.*, 2009), which is an efficient alternative to the traditional and time consuming bootstrap analysis (Vinuesa *et al.*, 2008). In the latter publication we proposed an aLRT-based measure of the phylogenetic information content of different sequence partitions, which I will introduce in the next section. PhyML also implements classical bootstrap analyses. Finally, it works with both protein and DNA sequences, implementing many protein distance matrices and the full range of 203 models of nucleotide substitution of the GTR + I + G family. Therefore, PhyML v3 is an excellent, fast and accurate, all purpose tree search program under the maximum likelihood criterion, which can be run on all major computing platforms. The casual user can run PhyML searches remotely over the Internet using different web servers (Dereeper *et al.*, 2008; Guindon *et al.*, 2005). Another very useful, multiplatform and free multipurpose package that runs PhyML using a graphical user interface is SeaView4 (Gouy *et al.*, 2010), which

I will introduce in a section below on alignment concatenation.

Looking again at the expression of the likelihood function [$L = \Pr(D| \Gamma, \gamma, \sigma)$] makes it clear that in addition to the critical issue of doing a thorough search of tree space (more and more important as the number of taxa to analyse increases), it is also important to select the best fit substitution model for the data at hand. As mentioned in a previous section, this can be conveniently done with programs such as jModelTest (Posada, 2008) or ProtTest (Abascal *et al.*, 2005). These programs do not only select for the best approximating model, but also provide ML estimates of the corresponding parameters (i.e. transition/transversion rate ratio, alpha value of the gamma distribution to model among-site rate heterogeneity, etc). When using PhyML, we only need to provide the selected model (i.e. its parameterization) to the program, and it will optimize the parameter values along with the topology and branch lengths in multiple iterations of topology, branch length and substitution model parameter optimizations.

## Computing phylogenetic congruence and signal content of sequence partitions under the maximum likelihood criterion

As mentioned in the introduction, gene trees are expected to differ from each other and from the species tree because of the stochastic nature of the gene lineage sorting process during speciation. Lineage sorting occurs much earlier than speciation, and this difference in timing of the cladogenesis process varies from gene to gene, resulting not only in different gene tree topologies but also notable differences in branch lengths (Maddison, 1997). It is therefore important to gather the evidence from multiple loci to get a better estimate of the species phylogeny in which the gene trees are embedded (Edwards, 2009). There is the potential, however, that the subset of loci selected for a MLSA has strongly conflicting signals (Papke *et al.*, 2007). The evolutionary process that generated them is often difficult to identify, as it could be the result of a mixture of forces or processes such cross-species HGT or hybridization, mutational saturation, hidden paralogy or retention of ancestral polymorphisms due to incomplete lineage sorting (Degnan and Rosenberg, 2006; Maddison, 1997; Maddison and Knowles, 2006). This may result in a poorly resolved species tree if it is estimated from a concatenated supermatrix (Jeffroy *et al.*, 2006; Kubatko and Degnan, 2007; Vinuesa *et al.*, 2008), as generally done in microbial MLSA studies (Devulder *et al.*, 2005; Lerat *et al.*, 2003; Rokas *et al.*, 2003; Vinuesa *et al.*, 2008). If the signals from the different sequence partitions are strongly conflicting, then the estimated gene trees will not only be incongruent with each other, which is actually expected based on coalescence theory (Degnan and Rosenberg, 2006; Rosenberg, 2002; Rosenberg and Nordborg, 2002), but also with different subsets of concatenated partitions. In the worst case, the conflicting signals in the different partitions don't contribute to a net increase in overall phylogenetic resolving power after concatenation and the inferred phylogeny won't be a good estimate of the underlying species tree (Jeffroy *et al.*, 2006; Kubatko and Degnan, 2007; Vinuesa *et al.*, 2008). Given these potential problems, it is convenient to statistically evaluate the degree of conflict or 'cooperativeness' of the signals in the different sequence partitions to be used for the construction of a supermatrix. This information can then be used to make an informed choice of the partitions to use for the construction of the supermatrix (Jeffroy *et al.*, 2006).

Once good estimates of the individual gene trees have been obtained, there are several methods to evaluate the significance of the phylogenetic congruence between them (Felsenstein, 2004a). PAUP* (Swofford, 2002) implements parsimony-specific methods such as the 'incongruence length difference' or ILD test (Farris *et al.*, 1994), which has been strongly criticized (Darlu and Lecointre, 2002; Dolphin *et al.*, 2000; Dowton and Austin, 2002), and the so-called 'paired sites tests', such as the *t*-test, Wilcoxon signed-rank test, and the RELL or Kishino–Hasegawa (KH) test (Felsenstein, 2004a). The basic idea behind these tests is that two trees can be compared for either their parsimony or likelihood scores. The expected log-likelihood of a tree is the average log-likelihood we would get per site as the number of sites grows without limit. Assuming independence of character evolution, if two trees have equal expected log-likelihood or

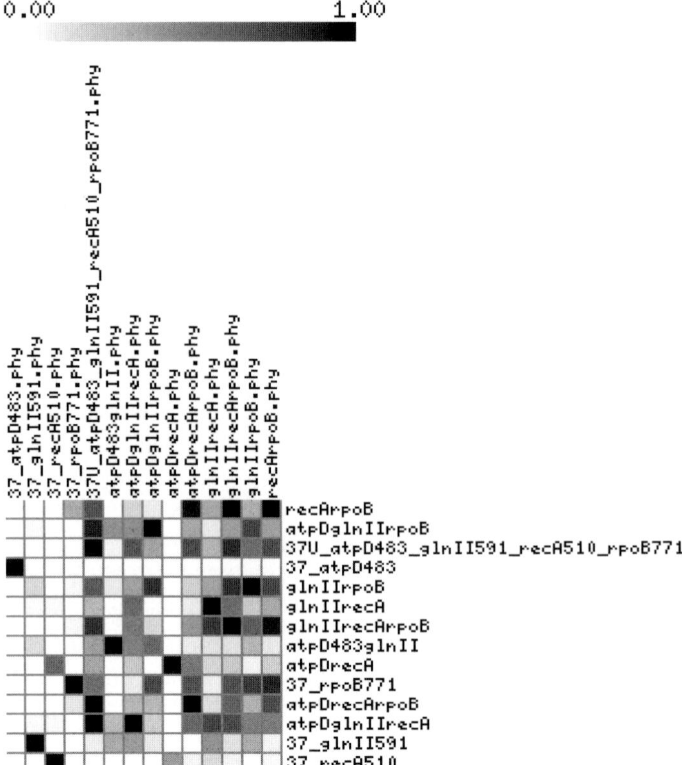


**Figure 3.1** Matrix showing *P* values of Shimodaira–Hasegawa congruence tests among all pairs of single and combined sequence partitions, as indicated. White corresponds to $P = 0$ and black to $P = 1$, meaning completely incongruent (highly significant rejection of the null hypothesis of congruence) and perfectly congruent trees, respectively.

parsimony scores, then the differences in scores at each site will be drawn independently from some distribution whose expectation is zero. Statistical tests can be performed to test whether the mean of these differences is zero, therefore testing the hypothesis that both topologies under comparison are phylogenetically congruent, i.e. equally good estimates. If this null hypothesis is rejected, the topology with the best global parsimony or likelihood score is significantly better than the competing topology. These tests, in their one-sided version, are only adequate for pair-wise comparison of topologies between the most parsimonious or maximum likelihood trees and an alternative tree (e.g. a constrained topology or one found by a different reconstruction method). They are not suitable, however, to perform multiple tree comparisons, not even if some sort of correction for multiple tests such as the simple Bonferroni correction is applied (Goldman *et al.*, 2000; Shimodaira and Hasegawa, 1999). For multiple tree comparisons the user should rather use the Shimodaira–Hasegawa test (SHT) (Shimodaira and Hasegawa, 1999), which implements a resampling method that approximately corrects for testing multiple trees. However, this test loses power if there are many highly unlikely trees in the set of trees to compare. Shimodaira and Hasegawa (1999) warn against including too many trees in the analysis, as it will dilute the power of discriminating among the plausible trees. A more powerful test is the 'approximately unbiased test' (Shimodaira, 2002), implemented in CONSEL under the maximum likelihood criterion, along with the classical KH and SH tests (Shimodaira and Hasegawa, 2001). The SH, KH and a expected likelihood weights (ELWT) tests (Strimmer and Rambaut, 2002) are also implemented in TREE-PUZZLE (Schmidt *et al.*, 2002), a powerful quartet-based maximum likelihood phylogenetic inference and hypothesis testing program, freely available for all platforms, that works with both protein and DNA sequences alike.

Fig. 3.1 shows the results of a phylogenetic congruence analysis performed on the four *Bradyrhizobium* spp. sequence partitions and all combinations of them used in this review. The individual gene tree phylogenies were estimated under best-fitting substitution models selected with jModelTest (Posada, 2008) using PhyML v3 (Guindon *et al.*, 2010) under the 'best' search method (evaluates both NNI and SPR moves, selecting the best of them). These phylogenies were fed as 'user-trees' into TREE-PUZZLE (Schmidt *et al.*, 2002) to let it perform the SHTs. As can be seen on Fig. 3.1, all single gene trees are incongruent with each other, but with the exception of the *atpD* locus, overall congruence increases as more partitions are concatenated. This is a first piece of statistical evidence suggesting that partition concatenation may be adequate as a strategy to combine the signals from different loci to estimate the species phylogeny.

A second and independent piece of evidence that indicates the suitability of data concatenation derives from the analysis that we call 'signal additivity' (Vinuesa *et al.*, 2008). As shown in Table 3.2, the overall tree resolution increases as we add more partitions to the analysis. The global tree resolution level was computed as the mean or median bipartition SH-like support values calculated by the newer versions of PhyML (Anisimova and Gascuel, 2006; Guindon *et al.*, 2009) with the help of a custom Perl script. Taken together, the independent sources of evidences presented in Fig. 3.1 and Table 3.2 suggest that alignment concatenation is suitable for the partitions at hand for downstream phylogenetic analysis.

## Concatenation of individual sequence alignments and supermatrix filtering

After the generation of reliable multiple sequence alignments and the performance of saturation, model selection, phylogenetic congruence and signal additivity analyses, the user is ready to concatenate selected alignments into a 'supermatrix' or 'superalignment', the dominant strategy used by microbial systematists to infer species trees. This step can be easily done with Windows programs such as BioEdit (Hall, 1999) or DAMBE (Xia and Xie, 2001). This operation can be performed on multiple platforms with the excellent and versatile SeaView4 package (Gouy *et al.*, 2010), which, among other goodies, integrates multiple sequence alignment generation and manipulation, with phylogeny inference under distance, parsimony and maximum likelihood criteria. It also implements simple but convenient functions for tree visualization and edition. Alternatively, you could use web services such as FaBox (Villesen, 2007), or write your own simple Perl scripts, eventually making use of the 'cat' method of the Bio::Align::Utilities module from the BioPerl suite (Stajich *et al.*, 2002). An important note to make at this point is that you should of course have the sequences equally sorted in all partitions before concatenating them, otherwise you will be generating '*in silico* sequence chimaeras' that will seriously compromise the accuracy of the species tree estimation.

Once concatenated, and before starting the proper estimation of the species phyhlogeny, the sequences should be collapsed to haplotypes, that is, repeated multilocus sequences should be dereplicated. Again, Windows programs such as DAMBE (Xia and Xie, 2001) will do this automatically. A convenient cross-platform multiple sequence alignment editor and analysis workbench is Jalview2 (Waterhouse *et al.*, 2009), which can also collapse repeated sequences to haplotypes. The FaBox web server can also do this task (Villesen, 2007). Writing *ad hoc* Perl code for this job will allow you to perform this operation on as many files as needed in an automatic fashion.

Some phylogeny programs such as MrBayes (Ronquist and Huelsenbeck, 2003), RAxML (Stamatakis, 2006) or PAUP* (Swofford, 2002) allow for partitioned models. In order to make use of such models, which provide a much better fit (i.e. are more realistic) than selecting a single best-approximating model for the entire supermatrix (Nylander *et al.*, 2004; Posada and Crandall, 2001; Vinuesa *et al.*, 2005b,c), the user needs to keep track of the coordinates of the individual partitions within the supermatrix.

**Table 3.2** Relative performance of individual molecular markers and some of their combinations assessed by computing descriptive statistics of Shimodaira-Hasegawa-like *P* values of bipartition support values for the corresponding maximum likelihood phylogenies inferred with PhyML

| Partition | mean[a] | median[a] | std_dev[a] | variance[a] | %_NRB[b] | %_PRB[c] | %_MRB[d] | %_WRB[e] | %_HRB[f] |
|---|---|---|---|---|---|---|---|---|---|
| atpD483 | 0.74 | 0.82 | 0.24 | 0.0600 | 32.35 | 32.35 | 23.53 | 5.88 | 5.88 |
| rpoB771 | 0.73 | 0.83 | 0.27 | 0.0711 | 26.47 | 35.29 | 20.59 | 11.76 | 5.88 |
| glnII591 | 0.83 | 0.86 | 0.19 | 0.0355 | 11.76 | 38.24 | 17.65 | 20.59 | 11.76 |
| recA510 | 0.73 | 0.88 | 0.30 | 0.0930 | 26.47 | 14.71 | 41.18 | 14.71 | 2.94 |
| glnIIrecA | 0.85 | 0.94 | 0.23 | 0.0538 | 14.71 | 8.82 | 29.41 | 11.76 | 35.29 |
| glnIIrecArpoB | 0.84 | 0.95 | 0.23 | 0.0507 | 17.65 | 14.71 | 17.65 | 11.76 | 38.24 |
| atpDglnIIrecA | 0.85 | 0.96 | 0.26 | 0.0674 | 14.71 | 5.88 | 26.47 | 23.53 | 29.41 |
| atpDglnIIrpoB | 0.85 | 0.96 | 0.23 | 0.0550 | 14.71 | 11.76 | 17.65 | 35.29 | 20.59 |
| atpDrecArpoB | 0.86 | 0.96 | 0.20 | 0.0410 | 14.71 | 14.71 | 17.65 | 26.47 | 26.47 |
| atpDglnIIrecArpoB | 0.87 | 0.97 | 0.20 | 0.0406 | 20.59 | 0.00 | 20.59 | 26.47 | 32.35 |

[a]The mean, median, standard deviation and variance of SH-like P values were computed on ML trees inferred with PhyML v3 under the 'best' branch swapping method and best fitting substitution models selected with jModelTest.

[b]Percentage of non resolved bipartitions, with $P < 0.70$.

[c]Percentage of poorly resolved bipartitions, with $0.70 \leq P < 0.85$.

[d]Percentage of moderately resolved bipartitions, with $0.85 \leq P < 0.95$.

[e]Percentage of well resolved bipartitions, with $0.95 \leq P < 0.99$.

[f]Percentage of highly resolved bipartitions, with $P \geq 0.99$.

### Phylogenetic inference using partitioned models under ML and Bayesian optimality criteria

A current limitation of PhyML v3.0 is that it does not implement partitioned models (Nylander *et al.*, 2004; Vinuesa *et al.*, 2005b,c), that is, it will not allow defining a best-fitting model for each gene partition. TreeFinder (Jobb *et al.*, 2004) is an easy-to-use analysis environment for molecular phylogenetics under the maximum likelihood criterion written in ANSI C and Java, that implements fast heuristics for tree searching and evaluation of phylogenetic hypotheses using large datasets. In addition, it provides a user-friendly graphical interface and a phylogenetic programming language. Among its many features, it implements partitioned models and a broad range of phylogenetic congruence tests. Another powerful, versatile and very fast maximum likelihood tree search algorithm that allows for partitioned models is implemented in RAxML (randomized accelerated maximum likelihood for high performance computing). It is a sequential and parallel program for inference of large phylogenies under the maximum likelihood criterion (Stamatakis, 2006). Low-level technical optimizations such as efficient memory use, and the use of the GTR+CAT approximation as replacement for GTR+Gamma yield a program that is extremely fast and accurate, allowing the inference of ML trees using multiple sequence partitions and hundreds or even thousands of protein or nucleotide sequences. It also runs on two publicly accessible web-servers (Stamatakis *et al.*, 2008). All these programs are freely available.

An alternative to maximum likelihood within the realm of statistical phylogenetics is the Bayesian framework (Holder and Lewis, 2003; Huelsenbeck *et al.*, 2001, 2002). MrBayes is currently the most popular software implementation for Bayesian estimation of phylogenies (Ronquist and Huelsenbeck, 2003). It is freely available for all major platforms from http://mrbayes.csit.fsu.edu/. Excellent tutorial materials are available from that site. MrBayes implements a Metropolis–Hastings coupled Markov-chain Monte Carlo (MCMCMC or $MC^3$) tree sampler that permits the combination of information from different partitions evolving under different stochastic evolutionary models. This allows the user to analyse heterogeneous data sets consisting of different data types – e.g. morphological, nucleotide, and protein – and to explore a wide variety of structured models, mixing partition-unique and shared parameters (Nylander *et al.*, 2004; Vinuesa *et al.*, 2005b, 2005c). In addition to standard tree searches, MrBayes also implements constrained searches and the analysis of selection at the molecular level using codon-based substitution models. The program employs MPI to parallelize Metropolis coupling on Macintosh or UNIX clusters, making it suitable for the analysis of large datasets.

Fig. 3.2 shows a Bayesian species tree inferred from our multilocus sequence data set using MrBayes and the command block shown below:

```
BEGIN mrbayes;
   log start filename=37U_atpD_glnII_
      recA_rpoB_bygene_RUN1.log1 append;
   outgroup Rho_pal_1;
   charset atpD = 1-483; charset glnII =
      484-1074;
   charset recA = 1075-1584; charset
      rpoB = 1585-2355;
   partition by_gene = 4: atpD, glnII,
      recA, rpoB;

   set autoclose=yes nowarnings=yes; set
      partition = by_gene;

   Prset ratepr = variable;
   Lset applyto =(all) nst=6
      rates=invgamma Ngammacat=4;
   unlink shape=(all) pinvar=(all)
      statefreq=(all) revmat=(all);

   mcmc ngen=3000000 printfreq=10000
      samplefreq=100 nchains=4 nruns=2
      temp=0.18;
```

This block specifies four sequence partitions, each evolving at its own rate, under the best-fitting GTR+I+G model selected for each partition by MrModelTest 2.3 (Nylander, 2004). Notice that by unlinking the state (nucleotide) frequencies, substitution rates (revmat), shape parameter of the gamma distribution (shape), and proportion of invariant sites (pinvar), the values for each of these model parameters will be optimized independently for each of the four partitions. The $MC^3$ chain will be run for $3 \times 10^6$ generations,

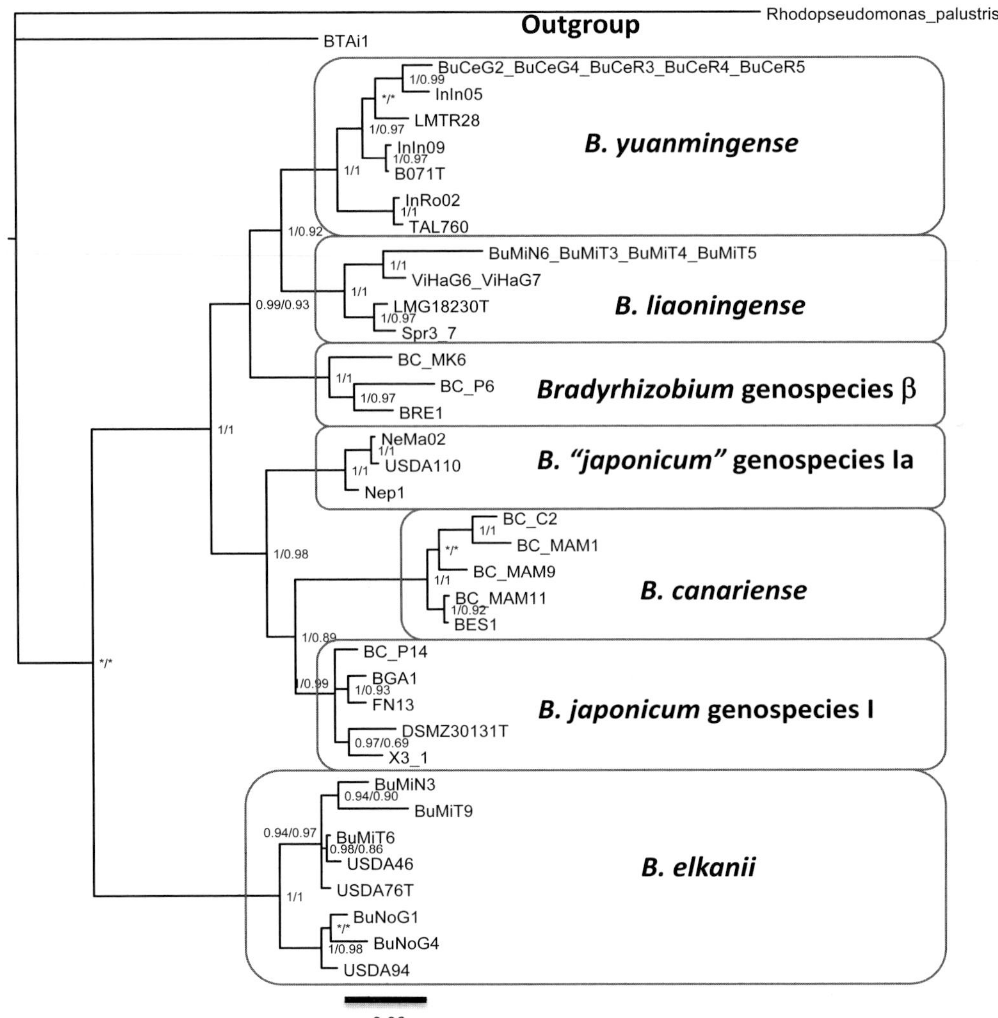


**Figure 3.2** Bayesian species phylogeny estimated under gene-partitioned, best fitting models, based on a supermatrix of 2355 characters for a selection of 37 unique *atpD+glnII+recA+rpoB* sequences (multilocus haplotypes) of representative strains of 6 *Bradyrhizobium* species previously studied by Vinuesa *et al.* (2008). The numbers on the nodes denote the Bayesian posterior probabilities(PP)/Shimodaira–Hasegawa-like (SH) bipartition support values. Asterisks indicate any of these values when < 0.90. The Bayesian PP values were computed from 10,000 post-burnin trees pooled from two independent, convergent, MC$^3$ simulations run for 3x10$^6$ generations. The MrBayes block used to run the analysis is presented in the main text. The SH-like support values were obtained with PhyML v3, run with the 'best' branch swapping mode (see text for the details). The bar scale represents the expected number of substitutions per site, under the mixed model.

sampling parameters from the chain every 100th in order to avoid excessive autocorrelation between successive samples. The MCMC command settings actually indicate that two replicate runs will be executed, each with three heated chains (heat parameter set to 0.18) and a cold chain. These same settings were used for two additional runs, so that we ended up with data from six independent chains (three runs, each with two replicates, executed on three processors).

One of the critical issues in Bayesian MCMC-based analyses is to obtain convincing evidence that the Markov chain has been run long enough so that the collection of trees and parameters sampled from it are a good representation of the posterior probability distribution (Huelsenbeck *et al.*, 2002). This evidence can be obtained indirectly from different sources, but none is truly conclusive. The key consideration here is to run several replicate MC$^3$ chains, each starting from independent random trees (the default setting in MrBayes). The first check that should be performed is to examine the so-called trace or generation plots for all parameters, in which the log probability of the data given the parameter values are plotted against the number of generations (MCMC cycles), as shown in Fig. 3.3. In the beginning of the run the log likelihood of the cold chain typically increases rapidly. This phase of the run is referred to as the burnin, and the samples taken from the MCMC chain during these early steps of the search phase are typically discarded. Ideally, however, the chain should be run long enough, so that the burnin phase has no noticeable effect on the results. Once the likelihood of the cold chain stops to increase and starts to randomly fluctuate within a more or less stable range, the run may have reached stationarity. At this stage, the cold and heated chains should ideally be exchanging states frequently. As a rough rule of thumb, an efficient Metropolis–Hastings MCMC sampler will have acceptance rates somewhere in the range of 10–70%. If they are much lower, a first possibility to improve this misbehaviour of the chains is to set a lower temperature value, which is set to `temp=0.2` by default. Fig. 3.3 reveals that two of the chains (labelled as run1_rep1 and run2_rep1) converged at a relatively late point of the runs (generations > 2,250,000), while chains run1_rep2 and run3_rep2 did not find the region of higher posterior probabilities hit by the other four chains. In these analyses

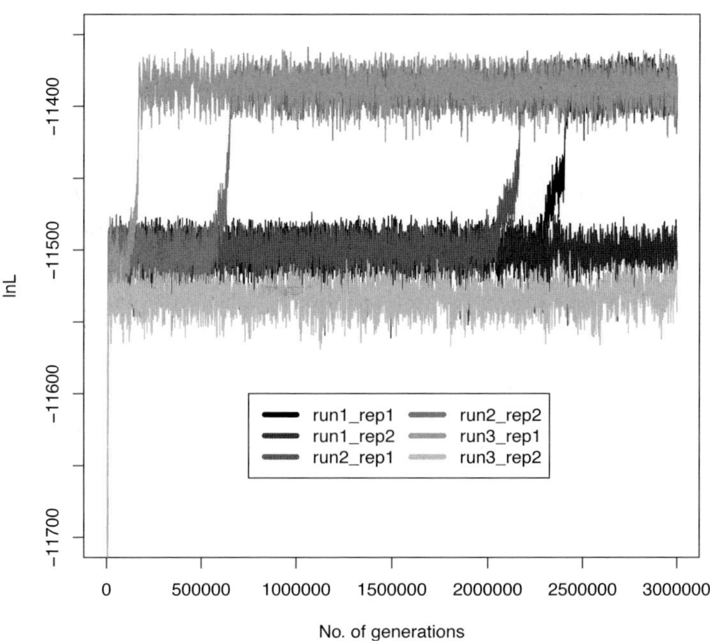


**Figure 3.3** Log likelihood traces (generation plots) of six MC$^3$ chains, each run with three heated chains and a cold chain, as indicated in the MrBayes block shown in the text. Runs labelled as run2_rep2 and run3_rep1 were used for the Bayesian species phylogeny estimation presented in Figure 3.2. These plots were generated with R (R Development Core Team, 2009) Note the good mixing of the chains, as revealed by the fast random fluctuation of the ln*L* values within a more or less stable range, indicating that the cold and heated chains are exchanging states frequently. Runs labelled as run1_rep2 and run3_rep2 got stuck in a suboptimal region of parameter space. The y-axis was severely shortened in order to present a better resolution of the chains in the region of higher posterior probabilities.

the heating parameter was set to `temp=0.18`, but should probably have been lowered further to ~0.15. This would allow the heated chains to explore peaks in parameter space that are 'further away', that is, separated from currently sampled local maxima by 'deeper valleys'. Tracer (Rambaut and Drummond, 2007) is a very convenient tool for examining multiple trace files (the *.p parameter files generated by MrBayes) at once, providing convenient summary statistics for each run, along with marginal density, joint marginal and trace plots for each parameter. It calculates the effective sample size (ESS), that is, the number of effectively independent draws from the posterior distribution that the Markov chain is equivalent to. This parameter should be well over 100, the higher the better (ESS for the global ln*L* of run3_rep1 is 993.6577). Tracer is written in Java and is therefore cross-platform.

What we can learn from this analysis is that when using relatively large datasets and complex partitioned models, such as the one used in this analysis (GTR+I+G model, unlinked for each partition), long MCMC runs have to be performed and replicated as many times as possible. MrBayes' default values of $1 \times 10^6$ generations (`mcmc ngen=1000000`) and two replicates (`nruns = 2`) are clearly too low for such analyses and should at least be triplicated, as done herein.

Based on the trace plots shown in Fig. 3.3 we should limit our downstream analysis to the data sampled from chains run3_rep1 and run2_rep2, discarding at least the first 7000 samples of the latter run as burnin. MrBayes provides the important and convenient `sump` and `sumt` functions to summarize the model parameters and trees sampled from selected chains, respectively. The `sump` function can be executed for single runs using the following line: '`sump filename=file_basenam_run2_rep2 burnin=7001 nruns=1`'. Notice that by default MrBayes will execute the `sump` command on all replicate runs issued from a single MrBayes command block. However, only the two replicates of run 2 found the region of highest posterior probability. For this particular run and replicates we could have issued the command '`sump filename=file_basename_run2 burnin=22501 nruns=2`' in order to summarize the parameters from both replicates, although discarding the first 22501 suboptimal samples of run2_rep2 (see Fig. 3.3). The `sump` output includes a rough generation plot and a table with summary statistics for the model parameters for each partition, including their 95% credibility intervals. The `sumt` command has a similar syntax, allowing the user to get a summary of the sampled trees ('`sumt filename = file_basename_run3_rep1 burnin=3001 nruns=1`'). The latter command would write a consensus tree to disk with the *.con extension based on 27000 sampled trees ($3 \times 10^6$ generations sampled every 100th) from run3_rep1. This is the file that the user should open with a tree editor such as TreeView or FigTree to visualize the estimated Bayesian phylogeny with branch support values indicated as posterior probabilities.

Further convergence diagnostics can be generated by MrBayes from selected pairs of independent runs by issuing the `comparetree` command using the following syntax: '`comparetree filename1=file_basename_run3_rep1.t file_basename_run2_rep2.t burnin=7501`'. This command will display a rough bivariate plot for a random subsample of clade probability values found in filename1 and those for the same clades found in filename2. It will also display rough generation plots for different tree-to-tree distance measures. Fig. 3.4 shows the bivariate plot of clade probabilities for run3_rep1 vs. run2_rep2, according to the code line shown above. This plot and the Pearson correlation analysis shown (Fig. 3.4) were generated with R (R Development Core Team, 2009) using the *.comp.dist file written to disk by MrBayes as the source data. This analysis shows a very strong and significant correlation of the support values for the same clades in both runs, suggesting that the corresponding trees were sampled from the same posterior distribution.

The AWTY (Are We There Yet?) online convergence diagnostics tool can further aid in deciding if the chains have been run long enough and have converged (Nylander *et al.*, 2008). It provides a graphical representation of additional parameters such as convergence rates of posterior split probabilities and branch lengths. AWTY therefore complements the diagnostics described and available in Tracer and MrBayes.

As mentioned before, convergence diagnostics can never prove that the globally best phylogeny

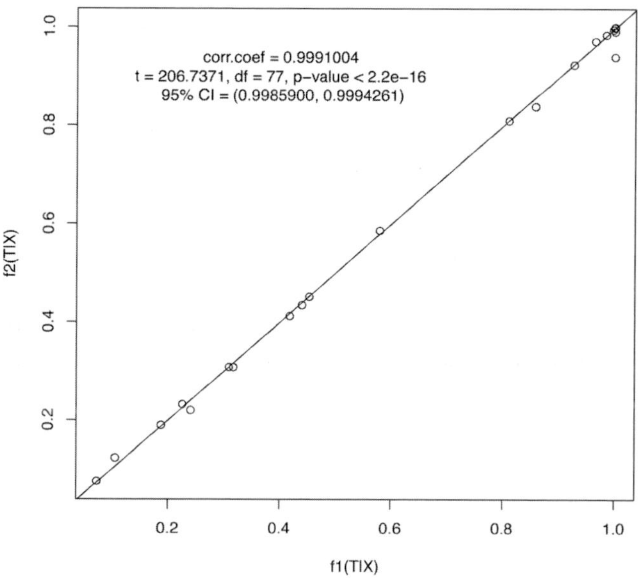


**Figure 3.4** Bivariate plot of clade probabilities for run3_rep1 vs. run2_rep2 (see Figure 3.3) computed by the 'comparetree' command of MrBayes, according to the command shown in the text. This plot and the Pearson correlation analysis were generated with R (R Development Core Team, 2009). The *f*(T|X) labels on the axes denote the clade posterior probabilities of the two runs compared.

has been found. The possibility exists that there are still better tree islands to be found, but it should be clear at this point that long runs, well selected substitution models, clean data matrices and careful selection of priors and proposals are key conditions for making robust Bayesian inferences of phylogeny. However, we can be quite confident in our estimate (Fig. 3.2), because in addition to the evidence gathered from diverse convergence diagnostics, we did also make the inference under the ML criterion using PhyML v3. Both trees are nearly identical, the Robinson–Foulds (R–F) symmetric difference between them being 2, as computed with the Treedist program from the Phylip package (Felsenstein, 2004b). That is, they differ in only a single branch from each other at the base (root) of the tree. This difference is irrelevant, since MrBayes places the outgroup as a basal polytomy, whereas PhyML does not. Furthermore, the SH-like branch support values estimated with PhyML were mapped on the Bayesian tree shown in Fig. 3.2. The first number on the bipartitions of the phylogeny corresponds to Bayesian posterior probabilities, whereas the second number corresponds to the SH-like support values. Bayesian posterior probabilities (PP) of clade support have been criticized as providing 'overcredibility' (Alfaro *et al.*, 2003; Buckley, 2002; Douady *et al.*, 2003; Erixon *et al.*, 2003), whereas the SH-like tests are known to be very conservative (Shimodaira and Hasegawa, 1999), as mentioned in the section on phylogenetic congruence tests. Therefore we could take the PP and SH-like support values as upper and lower credibility bounds of branch support, respectively. Since both values are highly significant for most bipartitions, we can safely conclude that the species phylogeny estimate presented in Fig. 3.2 is robust and well resolved.

## Prospect: estimating species trees from multilocus data without concatenation – an example using BEST

Recent empirical studies have demonstrated that concatenation of sequences from multiple genes prior to phylogenetic analysis often results in inference of a single, well-supported phylogeny (Devulder *et al.*, 2005; Lerat *et al.*, 2003; Rokas *et al.*, 2003; Vinuesa *et al.*, 2008). Theoretical work, however, has shown that the coalescent can produce substantial variation in single-gene histories, reflected in a high diversity of topologies and branch lengths across gene trees (Degnan and Rosenberg, 2006; Degnan and Salter, 2005; Rosenberg, 2002). Using simulation, these ideas were recently tested by Laura Kubatko and James Degnan to examine the performance of the concatenation approach under conditions in which the coalescent produces a high level of discord among individual gene trees and showed that it leads to statistically inconsistent estimation in this setting (Kubatko and Degnan, 2007). Furthermore, they showed in that study that the use of the bootstrap to measure support for the inferred phylogeny can result in moderate to strong support for an incorrect tree under such conditions. These results highlight the importance of incorporating variation in gene histories into multilocus phylogenetics (Degnan and Rosenberg, 2009; Edwards, 2009). This variation in genealogies (topologies and branch lengths) cannot be taken into account using the concatenation strategy, not even when partitioned models are used with program such as MrBayes, since they impose a common topology and set of branch lengths on the estimate made from each partition. The concatenation approaches estimate a phylogeny that reflects some average of gene trees but do not explicitly estimate species trees. The latter are the trees that really matter to systematists (Edwards, 2009; Edwards *et al.*, 2007). This shortcoming of standard treeing methods to infer species trees based on supermatrices has been recently addressed by several groups, who have developed and released first versions of software that will provide estimates of the species trees from multilocus sequence data, without concatenation.

Liang Liu and Dennis Pearl developed BEST (Bayesian Estimation of Species Trees) (Edwards *et al.*, 2007; Liu, 2008; Liu and Pearl, 2007; Liu *et al.*, 2008). It is based on a modified MrBayes code base that implements a Bayesian hierarchical model to jointly estimate gene trees and the species tree from multilocus sequences. The technique of simulated annealing is adopted along with Metropolis coupling as performed in MrBayes to improve the convergence rate of the Markov Chain Monte Carlo algorithm. BEST takes advantage of the information contained in multiple gene trees to perform a hierarchical Bayesian analysis to estimate the topology of the species tree, divergence times in coalescent units and ancestral population sizes. The BEST model is based on two conditional probability distributions: the probability distribution $f(D|G)$, which is the previously discussed likelihood function used to estimate gene trees, and the probability distribution $f\ (G|S)$, which is the likelihood of gene trees (G) conditional on the species tree (S), as derived from the multispecies coalescent (Degnan and Rosenberg, 2009). The algorithm samples from the joint posterior distribution over a set of gene trees and the species tree. Therefore the model involves a larger number of parameters to be estimated from the data than in a standard Bayesian or maximum likelihood gene tree search. This additional computational burden can make it very challenging for the BEST algorithm to converge if the datasets are large and 'misbehaved'. In particular, the BEST model (as all other currently available species tree inference algorithms) assumes that there is no HGT, hybridization, gene flow or intragenic recombination in the data, but free recombination between loci (Liu, 2008). It is likely that many microbial multilocus datasets violate some of these assumptions. It is therefore critical that the data sets are carefully chosen to minimize the impact of systematic bias and to improve convergence rate among replicate runs, as exemplified below.

Fig. 3.5A shows the species tree estimate obtained with BEST using the type strains of each taxon in a replicate run using only the *glnII*, *recA* and *rpoB* partitions, default prior values, the HKY85 + G substitution model for each partition, $10 \times 10^6$ $MC^3$ generations with two chains (`temp=0.17`), sampling every 100th

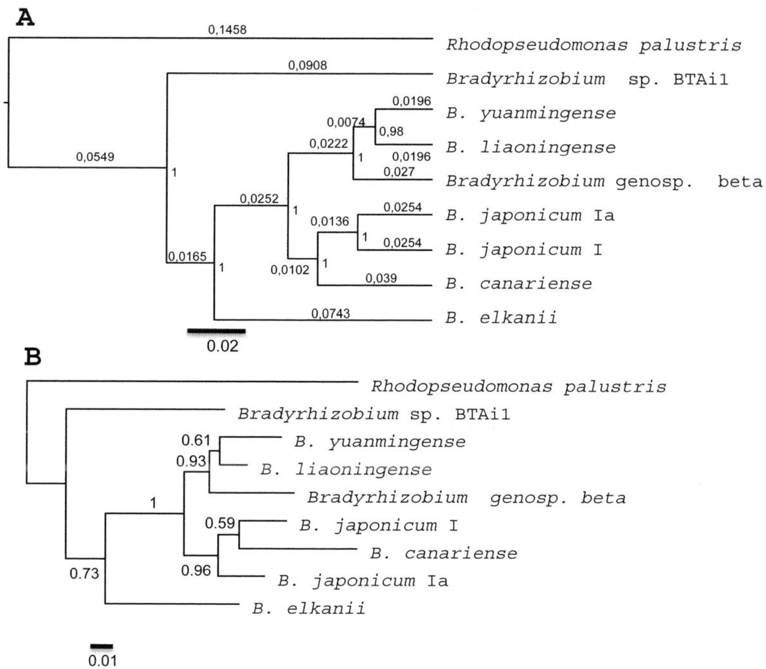


**Figure 3.5** Comparison of the species trees estimated by the BEST (A) and PhyML (B) programs from the *glnII*, *recA* and *rpoB* gene partitions, as explained in the text. The BEST algorithm does not depend on concatenation of the sequence data, whereas PhyML needs the supermatrix to integrate the information from the three loci. Notice the different topologies yielded by both methods. Numbers on the nodes of the tree depicted in panel A represent Bayesian posterior probabilities, whereas those on the branches indicate divergence time in coalescent units. The number depicted on the nodes of tree shown in (B) represent Shimodaria–Hasegawa-like *P* values.

and discarding 25% of the samples as burnin. Exclusion of the *atpD* locus was critical to achieve convergence. This and previous studies revealed that this sequence partition is conflictive in several respects, such as being particularly prone to HGT and intragenic recombination (Vinuesa *et al.*, 2005c, 2008), yielding the gene tree with the strongest topological discordance among those analysed (see Fig. 3.1).

The species tree shown in Fig. 3.5A is highly resolved, all partitions having a posterior probability ≥ 0.98. When compared with the ML tree shown in Fig. 3.5B, estimated with PhyML v3.0 for the corresponding supermatrix under the HKY+G model using a BioNJ and 5 random starting trees and the 'best' search mode (NNI+SPR), a single difference is apparent, namely, that *B. japonicum* I and Ia are sister species in the former but not in the latter tree. This difference may seem trivial, but highlights the difference between estimating some sort of an average gene tree from a concatenated alignment and estimating a species tree that takes into account topological and branch length heterogeneity in the gene trees underlying the species tree. Pairwise Robinson–Foulds symmetric distances between the BEST species tree, and the PhyML trees for the *glnII* + *recA* + *rpoB* supermatrix and the *glnII*, *recA* and *rpoB* partitions were 2, 4, 0 and 6, respectively, illustrating the diversity of topologies that can be found even within small datasets such as this case study.

It will be interesting to evaluate the impact of adding more loci and strains (alleles) per species on the estimation of the species tree using the traditional concatenation approach and the more rigorous species tree inference using *ad hoc* algorithms such as BEST (Liu, 2008). A maximum likelihood alternative called STEM (Species Tree Estimation using Maximum likelihood) has been recently released (Kubatko *et al.*, 2009). STEM provides a capability for searching the space of species trees for a collection of *k* species trees with high likelihood, where *k* is set by the user. These and other new software packages and methods represent promising and exciting avenues for future MLSA-based research. Extensive empirical and simulation work is now required to identify potential pitfalls of these methods and aspects of the current software implementations that need to be improved to make them broadly useful for practicing microbial systematists and evolutionary biologists.

## Acknowledgements

I would like to thank the editors for their invitation to write this chapter, as well as for their suggestions and criticism on a first draft of the manuscript. I gratefully acknowledge funding from the Autonomous University of Mexico through grant DGAPA/PAPIIT-UNAM IN201806–2, and CONACyT-Mexico through grant P1–60071.

## References


Abascal, F., Zardoya, R., and Posada, D. (2005). ProtTest: selection of best-fit models of protein evolution. Bioinformatics *21*, 2104–2105.

Alfaro, M.E., Zoller, S., and Lutzoni, F. (2003). Bayes or bootstrap? A simulation study comparing the performance of Bayesian Markov chain Monte Carlo sampling and bootstrapping in assessing phylogenetic confidence. Mol. Biol. Evol. *20*, 255–266.

Anisimova, M., and Gascuel, O. (2006). Approximate likelihood-ratio test for branches: A fast, accurate, and powerful alternative. Syst. Biol. *55*, 539–352.

Boucher, Y., Douady, C.J., Papke, R.T., Walsh, D.A., Boudreau, M.E., Nesbø, C.L., Case, R.J., and Doolittle, W.F. (2003). Lateral gene transfer and the origins of prokaryotic groups. Annu. Rev. Genet. *37*, 283–328.

Buckley, T.R. (2002). Model misspecification and probabilistic tests of topology: evidence from empirical data sets. Syst. Biol. *51*, 509–523.

Bull, J.J. (1993). Partitioning and combining data in phylogenetic analysis. Syst. Biol. 384–397.

Capella-Gutierrez, S., Silla-Martinez, J.M., and Gabaldon, T. (2009). trimAl: a tool for automated alignment trimming in large-scale phylogenetic analyses. Bioinformatics *25*, 1972–1973.

Castresana, J. (2000). Selection of conserved blocks from multiple alignments for their use in phylogenetic analysis. Mol. Biol. Evol. *17*, 540–552.

Ciccarelli, F.D., Doerks, T., von Mering, C., Creevey, C.J., Snel, B., and Bork, P. (2006). Toward automatic reconstruction of a highly resolved tree of life. Science *311*, 1283–1287.

Cole, J.R., Chai, B., Farris, R.J., Wang, Q., Kulam-Syed-Mohideen, A.S., McGarrell, D.M., Bandela, A.M., Cardenas, E., Garrity, G.M., and Tiedje, J.M. (2007). The ribosomal database project (RDP-II): introducing myRDP space and quality controlled public data. Nucleic Acids Res. *35*, 169–172.

Contreras-Moreira, B., Sachman-Ruiz, B., Figueroa-Palacios, I., and Vinuesa, P. (2009). primers4- clades: a web server that uses phylogenetic trees to design lineage-specific PCR primers for metagenomic and diversity studies. Nucleic Acids Res. *37*, W95-W100.

Charlebois, R.L., and Doolittle, W.F. (2004). Computing prokaryotic gene ubiquity: Rescuing the core from extinction. Genome Res. *14*, 2469–2477.

Chenna, R., Sugawara, H., Koike, T., Lopez, R., Gibson, T.J., Higgins, D.G., and Thompson, J.D. (2003). Multiple sequence alignment with the Clustal series of programs. Nucleic Acids Res. *31*, 3497–3500.

Darlu, P., and Lecointre, G. (2002). When does the incongruence length difference test fail? Mol. Biol. Evol. *19*, 432–437.

Degnan, J.H., and Rosenberg, N.A. (2006). Discordance of species trees with their most likely gene trees. PLoS Genet. *2*, e68.

Degnan, J.H., and Rosenberg, N.A. (2009). Gene tree discordance, phylogenetic inference and the multispecies coalescent. Trends Ecol. Evol. *24*, 332–340.

Degnan, J.H., and Salter, L.A. (2005). Gene tree distributions under the coalescent process. Evolution *59*, 24–37.

Dereeper, A., Guignon, V., Blanc, G., Audic, S., Buffet, S., Chevenet, F., Dufayard, J.F., Guindon, S., Lefort, V., Lescot, M., Claverie, J.M., and Gascuel, O. (2008). Phylogeny.fr: robust phylogenetic analysis for the non-specialist. Nucleic Acids Res. *36*, W465–469.

DeSantis, T.Z., Hugenholtz, P., Larsen, N., Rojas, M., Brodie, E.L., Keller, K., Huber, T., Dalevi, D., Hu, P., and Andersen, G.L. (2006). Greengenes, a chimera-checked 16S rRNA gene database and workbench compatible with ARB. Appl. Environ. Microbiol. *72*, 5069–5072.

Devulder, G., Perouse de Montclos, M., and Flandrois, J.P. (2005). A multigene approach to phylogenetic analysis using the genus *Mycobacterium* as a model. Int. J. Syst. Evol. Microbiol. *55*, 293–302.

Didelot, X., and Falush, D. (2007). Inference of bacterial microevolution using multilocus sequence data. Genetics *175*, 1251–1266.

Dolphin, K., Belshaw, R., Orme, C.D., and Quicke, D.L. (2000). Noise and incongruence: interpreting results of the incongruence length difference test. Mol. Phylogenet. Evol. *17*, 401–406.

Douady, C.J., Delsuc, F., Boucher, Y., Doolittle, W.F., and Douzery, E.J. (2003). Comparison of Bayesian and maximum likelihood bootstrap measures of phylogenetic reliability. Mol. Biol. Evol. *20*, 248–254.

Dowton, M., and Austin, A.D. (2002). Increased congruence does not necessarily indicate increased phylogenetic accuracy--the behavior of the incongruence length difference test in mixed-model analyses. Syst. Biol. *51*, 19–31.

Edgar, R.C. (2004). MUSCLE: multiple sequence alignment with high accuracy and high throughput. Nucleic Acids Res. *32*, 1792–1797.

Edwards, S.V. (2009). Is a new and general theory of molecular systematics emerging? Evolution *63*, 1–19.

Edwards, S.V., Liu, L., and Pearl, D.K. (2007). High-resolution species trees without concatenation. Proc. Natl. Acad. Sci. U.S.A. *104*, 5936–5941.

Enright, M.C., Robinson, D.A., Randle, G., Feil, E.J., Grundmann, H., and Spratt, B.G. (2002). The evolutionary history of methicillin-resistant *Staphylococcus aureus* (MRSA). Proc. Natl. Acad. Sci. U.S.A. *99*, 7687–7692.

Erixon, P., Svennblad, B., Britton, T., and Oxelman, B. (2003). Reliability of Bayesian posterior probabilities and bootstrap frequencies in phylogenetics. Syst. Biol. *52*, 665–673.

Escobar-Páramo, P., Sabbagh, A., Darlu, P., Pradillon, O., Vaury, C., Denamur, E., and Lecointre, G. (2004). Decreasing the effects of horizontal gene transfer on bacterial phylogeny: the *Escherichia coli* case study. Mol. Phylogenet. Evol. *30*, 243–250.

Falush, D., Stephens, M., and Pritchard, J.K. (2003). Inference of population structure using multilocus genotype data. Linked loci and correlated allele frequencies. Genetics *164*, 1567–1587.

Farris, J.S., Källersjö, J., Kluge, A.G., and Bult, C. (1994). Testing significance of congruence. Cladistics *10*, 315–319.

Felsenstein, J. (1985). Confidence limits on phylogenies: An approach using the bootstrap. Evolution *39*, 783–791.

Felsenstein, J. (2004a). Inferring phylogenies (Sunderland, MA: Sinauer Associates, INC.).

Felsenstein, J. (2004b). PHYLIP (Phylogeny Inference Package) (Seattle, Distributed by the author. Department of Genetics, University of Washington).

Gascuel, O. (1997). BIONJ: an improved version of the NJ algorithm based on a simple model of sequence data. Mol. Biol. Evol. *14*, 685–695.

Gevers, D., Cohan, F.M., Lawrence, J.G., Spratt, B.G., Coenye, T., Feil, E.J., Stackebrandt, E., Van de Peer, Y., Vandamme, P., Thompson, F.L., and Swings, J. (2005). Opinion: Re-evaluating prokaryotic species. Nature Rev. Microbiol. 3, 733–739.

Gogarten, J.P., and Townsend, J.P. (2005). Horizontal gene transfer, genome innovation and evolution. Nature Rev. Microbiol. *3*, 679–687.

Goldman, N., Anderson, J.P., and Rodrigo, A.G. (2000). Likelihood-based tests of topologies in phylogenetics. Syst. Biol. *49*, 652–670.

Gouy, M., Guindon, S., and Gascuel, O. (2010). SeaView version 4: a multiplatform graphical user interface for sequence alignment and phylogenetic tree building. Mol. Biol. Evol. *27*, 221–224.

Guindon, S., and Gascuel, O. (2003). A simple, fast, and accurate algorithm to estimate large phylogenies by maximum likelihood. Syst. Biol. *52*, 696–704.

Guindon, S., Lethiec, F., Duroux, P., and Gascuel, O. (2005). PHYML Online – a web server for fast maximum likelihood-based phylogenetic inference. Nucleic Acids Res. *33*, W557–559.

Guindon, S., Delsuc, F., Dufayard, J.F., and Gascuel, O. (2009). Estimating maximum likelihood phylogenies with PhyML. Meth. Mol. Biol. *537*, 113–137.

Hall, T.A. (1999). BioEdit: a user-friendly biological sequence alignment editor and analysis program for Windows 95/98/NT. Nucl. Acids. Symp. Ser. *41*, 95–98.

Hasegawa, M., Adachi, J., and Milinkovitch, M.C. (1997). Novel phylogeny of whales supported by total molecular evidence. J. Mol. Evol. *44 Suppl 1*, 117–120.

Hirt, R.P., Logsdon, J.M., Jr., Healy, B., Dorey, M.W., Doolittle, W.F., and Embley, T.M. (1999). Microsporidia are related to Fungi: evidence from the largest subunit of RNA polymerase II and other proteins. Proc. Natl. Acad. Sci. U.S.A. *96*, 580–585.

Holder, M., and Lewis, P.O. (2003). Phylogeny estimation: traditional and Bayesian approaches. Nature Rev. Genet. *4*, 275–284.

Hordijk, W., and Gascuel, O. (2005). Improving the efficiency of SPR moves in phylogenetic tree search methods based on maximum likelihood. Bioinformatics *21*, 4338–4347.

Huelsenbeck, J.P., Bull, J.J., and Cunningham, C.W. (1996). Combining data in phylogenetic analysis. Trends Ecol. Evol. *11*, 152–157.

Huelsenbeck, J.P., Larget, B., Miller, R.E., and Ronquist, F. (2002). Potential applications and pitfalls of Bayesian inference of phylogeny. Syst. Biol. *51*, 673–688.

Huelsenbeck, J.P., Ronquist, F., Nielsen, R., and Bollback, J.P. (2001). Bayesian inference of phylogeny and its impact on evolutionary biology. Science *294*, 2310–2314.

Huson, D.H., and Bryant, D. (2006). Application of phylogenetic networks in evolutionary studies. Mol. Biol. Evol. *23*, 254–267.

Huson, D.H., Richter, D.C., Rausch, C., Dezulian, T., Franz, M., and Rupp, R. (2007). Dendroscope: An interactive viewer for large phylogenetic trees. BMC Bioinformatics *8*, 460.

Ivars-Martinez, E., D'Auria, G., Rodriguez-Valera, F., Sanchez-Porro, C., Ventosa, A., Joint, I., and Muhling, M. (2008). Biogeography of the ubiquitous marine bacterium *Alteromonas macleodii* determined by multilocus sequence analysis. Mol. Ecol. *17*, 4092–4106.

Jeffroy, O., Brinkmann, H., Delsuc, F., and Philippe, H. (2006). Phylogenomics: the beginning of incongruence? Trends Genet. *22*, 225–231.

Jobb, G., von Haeseler, A., and Strimmer, K. (2004). TREEFINDER: a powerful graphical analysis environment for molecular phylogenetics. BMC Evol. Biol. *4*, 18.

Katoh, K., and Toh, H. (2008). Recent developments in the MAFFT multiple sequence alignment program. Brief Bioinform. *9*, 286–298.

Keane, T.M., Creevey, C.J., Pentony, M.M., Naughton, T.J., and McInerney, J.O. (2006). Assessment of methods for amino acid matrix selection and their use on empirical data shows that ad hoc assumptions for choice of matrix are not justified. BMC Evol. Biol. *6*, 29.

Kloepper, T.H., and Huson, D.H. (2008). Drawing explicit phylogenetic networks and their integration into SplitsTree. BMC Evol. Biol. *8*, 22.

Kubatko, L.S., Carstens, B.C., and Knowles, L.L. (2009). STEM: species tree estimation using maximum likelihood for gene trees under coalescence. Bioinformatics *25*, 971–973.

Kubatko, L.S., and Degnan, J.H. (2007). Inconsistency of phylogenetic estimates from concatenated data under coalescence. Syst. Biol. *56*, 17–24.

Larkin, M.A., Blackshields, G., Brown, N.P., Chenna, R., McGettigan, P.A., McWilliam, H., Valentin, F., Wallace, I.M., Wilm, A., Lopez, R., Thompson, J.D., Gibson, T.J., and Higgins, D.G. (2007). Clustal W and Clustal X version 2.0. Bioinformatics *23*, 2947–2948.

Lerat, E., Daubin, V., and Moran, N.A. (2003). From gene trees to organismal phylogeny in prokaryotes: the case of the gamma-Proteobacteria. PLoS Biol. *1*, E19.

Liu, L. (2008). BEST: Bayesian estimation of species trees under the coalescent model. Bioinformatics *24*, 2542–2543.

Liu, L., and Pearl, D.K. (2007). Species trees from gene trees: reconstructing Bayesian posterior distributions of a species phylogeny using estimated gene tree distributions. Syst. Biol. *56*, 504–514.

Liu, L., Pearl, D.K., Brumfield, R.T., and Edwards, S.V. (2008). Estimating species trees using multiple-allele DNA sequence data. Evolution *62*, 2080–2091.

Ludwig, W., Strunk, O., Westram, R., Richter, L., Meier, H., Yadhukumar, Buchner, A., Lai, T., Steppi, S., Jobb, G., Forster, W., Brettske, I., Gerber, S., Ginhart, A.W., Gross, O., Grumann, S., Hermann, S., Jost, R., König, A., Liss, T., Lussmann, R., May, M., Nonhoff, B., Reichel, B., Strehlow, R., Stamatakis, A., Stuckmann, N., Vilbig, A., Lenke, M., Ludwig, T., Bode, A., and Schleifer, K.H. (2004). ARB: a software environment for sequence data. Nucleic Acids Res. *32*, 1363–1371.

Maddison, W.P. (1997). Gene trees in species trees. Syst. Biol. *46*, 523–536.

Maddison, W.P., and Knowles, L.L. (2006). Inferring phylogeny despite incomplete lineage sorting. Syst. Biol. *55*, 21–30.

Maidak, B.L., Cole, J.R., Lilburn, T.G., Parker, C.T., Jr., Saxman, P.R., Farris, R.J., Garrity, G.M., Olsen, G.J., Schmidt, T.M., and Tiedje, J.M. (2001). The RDP-II (Ribosomal Database Project). Nucleic Acids Res. *29*, 173–174.

Martin, D., and Rybicki, E. (2000). RDP: detection of recombination amongst aligned sequences. Bioinformatics *16*, 562–563.

Martin, D.P. (2009). Recombination detection and analysis using RDP3. Meth. Mol. Biol. *537*, 185–205.

Martin, D.P., Posada, D., Crandall, K.A., and Williamson, C. (2005). A modified bootscan algorithm for automated identification of recombinant sequences and recombination breakpoints. AIDS Res. Hum. Retroviruses *21*, 98–102.

Martiny, J.B., Bohannan, B.J., Brown, J.H., Colwell, R.K., Fuhrman, J.A., Green, J.L., Horner-Devine, M.C., Kane, M., Krumins, J.A., Kuske, C.R., Morin, P.J., Naeem, S., Øvreås, L., Reysenbach, A.L., Smith, V.H., and Staley, J.T. (2006). Microbial biogeography: putting microorganisms on the map. Nature Rev. Microbiol. *4*, 102–112.

Miragaia, M., Thomas, J.C., Couto, I., Enright, M.C., and de Lencastre, H. (2007). Inferring a population structure for *Staphylococcus epidermidis* from multilocus sequence typing data. J. Bacteriol. *189*, 2540–2552.

Nichols, R. (2001). Gene trees and species trees are not the same. Trends Ecol. Evol. *16*, 358–364.

Nylander, J.A., Ronquist, F., Huelsenbeck, J.P., and Nieves-Aldrey, J.L. (2004). Bayesian phylogenetic analysis of combined data. Syst. Biol. *53*, 47–67.

Nylander, J.A., Wilgenbusch, J.C., Warren, D.L., and Swofford, D.L. (2008). AWTY (are we there yet?): a system for graphical exploration of MCMC convergence in Bayesian phylogenetics. Bioinformatics *24*, 581–583.

Nylander, J.A.A. (2004). MrModeltest v2. Program distributed by the author. Evolutionary Biology Centre, Uppsala University.

Ochman, H., Lerat, E., and Daubin, V. (2005). Examining bacterial species under the specter of gene transfer and exchange. Proc. Natl. Acad. Sci. U.S.A. *102 Suppl 1*, 6595–6599.

Ogura, Y., Ooka, T., Asadulghani, Terajima, J., Nougayrede, J.P., Kurokawa, K., Tashiro, K., Tobe, T., Nakayama, K., Kuhara, S., Oswald, E., Watanabe, H., and Hayashi, T. (2007). Extensive genomic diversity and selective conservation of virulence-determinants in enterohemorrhagic *Escherichia coli* strains of O157 and non-O157 serotypes. Genome Biol. *8*, R138.

Ogura, Y., Ooka, T., Iguchi, A., Toh, H., Asadulghani, M., Oshima, K., Kodama, T., Abe, H., Nakayama, K., Kurokawa, K., Tobe, T., Hattori, M., and Hayashi, T. (2009). Comparative genomics reveal the mechanism of the parallel evolution of O157 and non-O157 enterohemorrhagic *Escherichia coli*. Proc. Natl. Acad. Sci. U.S.A. *106*, 17939–17944.

Page, R.D. (2002). Visualizing phylogenetic trees using TreeView. Curr. Protoc. Bioinformatics Chapter 6, Unit 6 2.

Page, R.D.M., and Holmes, E.C. (1998). Molecular Evolution – A Phylogenetic Approach (Oxford: Blackwell Science Ltd.).

Papke, R.T., Ramsing, N.B., Bateson, M.M., and Ward, D.M. (2003). Geographical isolation in hot spring cyanobacteria. Environ. Microbiol. *5*, 650–659.

Papke, R.T., Zhaxybayeva, O., Feil, E.J., Sommerfeld, K., Muise, D., and Doolittle, W.F. (2007). Searching for species in haloarchaea. Proc. Natl. Acad. Sci. U.S.A. *104*, 14092–14097.

Perez-Losada, M., Browne, E.B., Madsen, A., Wirth, T., Viscidi, R.P., and Crandall, K.A. (2006). Population genetics of microbial pathogens estimated from multilocus sequence typing (MLST) data. Infect. Genet. Evol. *6*, 97–112.

Posada, D. (2008). jModelTest: phylogenetic model averaging. Mol. Biol. Evol. *25*, 1253–1256.

Posada, D., and Buckley, T.R. (2004). Model selection and model averaging in phylogenetics: advantages of Akaike information criterion and bayesian approaches over likelihood ratio tests. Syst. Biol. *53*, 793–808.

Posada, D., and Crandall, K.A. (1998). MODELTEST: testing the model of DNA substitution. Bioinformatics *14*, 817–818.

Posada, D., and Crandall, K.A. (2001). Selecting the best-fit model of nucleotide substitution. Syst. Biol. *50*, 580–601.

Posada, D., and Crandall, K.A. (2002). The effect of recombination on the accuracy of phylogeny estimation. J. Mol. Evol. *54*, 396–402.

Pruesse, E., Quast, C., Knittel, K., Fuchs, B.M., Ludwig, W., Peplies, J., and Glockner, F.O. (2007). SILVA: a comprehensive online resource for quality checked and aligned ribosomal RNA sequence data compatible with ARB. Nucleic Acids Res. *35*, 7188–7196.

R Development Core Team (2009). R: A Language and Environment for Statistical Computing. http://www.R-project.org.

Rambaut, A. (2009). FigTree v1.2.3. Available from http://tree.bio.ed.ac.uk/software/figtree/.

Rambaut, A., and Drummond, A.J. (2007). Tracer v1.4. Available from http://beast.bio.ed.ac.uk/Tracer

Ramette, A., and Tiedje, J.M. (2007). Biogeography: an emerging cornerstone for understanding prokaryotic diversity, ecology, and evolution. Microb. Ecol. *53*, 197–207.

Rasko, D.A., Rosovitz, M.J., Myers, G.S., Mongodin, E.F., Fricke, W.F., Gajer, P., Crabtree, J., Sebaihia, M., Thomson, N.R., Chaudhuri, R., Henderson, I.R., Sperandio, V., and Ravel, J. (2008). The pangenome structure of *Escherichia coli*: comparative genomic analysis of *E. coli* commensal and pathogenic isolates. J. Bacteriol. *190*, 6881–6893.

Reid, S.D., Herbelin, C.J., Bumbaugh, A.C., Selander, R.K., and Whittam, T.S. (2000). Parallel evolution of virulence in pathogenic *Escherichia coli*. Nature *406*, 64–67.

Rokas, A., Williams, B.L., King, N., and Carroll, S.B. (2003). Genome-scale approaches to resolving incongruence in molecular phylogenies. Nature *425*, 798–804.

Ronquist, F., and Huelsenbeck, J.P. (2003). MrBAYES 3:Bayesian phylogenetic inference under mixed models. Bioinformatics *19*, 1572–1574.

Rosenberg, N.A. (2002). The probability of topological concordance of gene trees and species trees. Theor. Popul. Biol. *61*, 225–247.

Rosenberg, N.A., and Nordborg, M. (2002). Genealogical trees, coalescent theory and the analysis of genetic polymorphisms. Nature Rev. Genet. *3*, 380–390.

Saitou, N., and Nei, M. (1987). The neighbor-joining method: A new method for reconstructing phylogenetic trees. Mol. Biol. Evol. *4*, 406–425.

Schierup, M.H., and Hein, J. (2000). Consequences of recombination on traditional phylogenetic analysis. Genetics *156*, 879–891.

Schmidt, H.A., Strimmer, K., Vingron, M., and von Haeseler, A. (2002). TREE-PUZZLE: maximum likelihood phylogenetic analysis using quartets and parallel computing. Bioinformatics *18*, 502–504.

Schubert, S., Darlu, P., Clermont, O., Wieser, A., Magistro, G., Hoffmann, C., Weinert, K., Tenaillon, O., Matic, I., and Denamur, E. (2009). Role of intraspecies recombination in the spread of pathogenicity islands within the *Escherichia coli* species. PLoS Pathog. *5*, e1000257.

Shimodaira, H. (2002). An approximately unbiased test of phylogenetic tree selection. Syst. Biol. *51*, 492–508.

Shimodaira, H., and Hasegawa, M. (1999). Multiple comparisons of log-likelihoods with applications to phylogenetic inference. Mol. Biol. Evol. *16*, 1114–1116.

Shimodaira, H., and Hasegawa, M. (2001). CONSEL: for assessing the confidence of phylogenetic tree selection. Bioinformatics *17*, 1246–1247.

Silva, C., Vinuesa, P., Eguiarte, L.E., Souza, V., and Martínez-Romero, E. (2005). Evolutionary genetics and biogeographic structure of *Rhizobium gallicum sensu lato*, a widely distributed bacterial symbiont of diverse legumes. Mol. Ecol. *14*, 4033–4050.

So, R.B., Ladha, J.K., and Young, J.P. (1994). Photosynthetic symbionts of *Aeschynomene* spp. form a cluster with bradyrhizobia on the basis of fatty acid and rRNA analyses. Int. J. Syst. Bacteriol. *44*, 392–403.

Soltis, P.S., Soltis, D.E., and Chase, M.W. (1999). Angiosperm phylogeny inferred from multiple genes as a tool for comparative biology. Nature *402*, 402–404.

Stackebrandt, E., Frederiksen, W., Garrity, G.M., Grimont, P.A., Kämpfer, P., Maiden, M.C., Nesme, X., Rosselló-Mora, R., Swings, J., Trüper, H.G., Vauterin, L., Ward, A.C., and Whitman, W.B. (2002). Report of the *ad hoc* committee for the re-evaluation of the species definition in bacteriology. Int. J. Syst. Evol. Microbiol. *52*, 1043–1047.

Stackebrandt, E., and Goebel, B.M. (1994). Taxonomic note: A place for DNA–DNA reassociation and 16S rRNA sequence analysis in the present species definition in bacteriology. Int. J. Syst. Bacteriol. *44*, 846–849.

Stajich, J.E., Block, D., Boulez, K., Brenner, S.E., Chervitz, S.A., Dagdigian, C., Fuellen, G., Gilbert, J.G., Korf, I., Lapp, H., Lehväslaiho, H., Matsalla, C., Mungall, C.J., Osborne, B.I., Pocock, M.R., Schattner, P., Senger, M., Stein, L.D., Stupka, E., Wilkinson, M.D., and Birney, E. (2002). The Bioperl toolkit: Perl modules for the life sciences. Genome Res. *12*, 1611–1618.

Stamatakis, A. (2006). RAxML-VI-HPC: maximum likelihood-based phylogenetic analyses with thousands of taxa and mixed models. Bioinformatics *22*, 2688–2690.

Stamatakis, A., Hoover, P., and Rougemont, J. (2008). A rapid bootstrap algorithm for the RAxML Web servers. Syst. Biol. *57*, 758–771.

Strimmer, K., and Rambaut, A. (2002). Inferring confidence sets of possibly misspecified gene trees. Proc. Biol. Sci. *269*, 137–142.

Sullivan, J.T., and Ronson, C.W. (1998). Evolution of rhizobia by acquisition of a 500-kb symbiosis island that integrates into a phe-tRNA gene. Proc. Natl. Acad. Sci. U.S.A. *95*, 5145–5149.

Suyama, M., Torrents, D., and Bork, P. (2006). PAL2NAL: robust conversion of protein sequence alignments into the corresponding codon alignments. Nucleic Acids Res. *34*, W609–612.

Swofford, D.L. (2002). PAUP*: Phylogenetic Analysis Using Parsimony and Other Methods (software). (Sunderland, MA: Sinuauer Associates).

Swofford, D.L., Olsen, G.J., Waddel, P.J., and Hillis, D.M. (1996). Phylogenetic inference. In Molecular Systematics, D.M. Hillis, C. Moritz, and B.K. Mable, eds. (Sunderland, MA: Sinauer Associates), pp. 407–514.

Tamura, K., Dudley, J., Nei, M., and Kumar, S. (2007). MEGA4: Molecular Evolutionary Genetics Analysis (MEGA) software version 4.0. Mol. Biol. Evol. *24*, 1596–1599.

Tettelin, H., Masignani, V., Cieslewicz, M.J., Donati, C., Medini, D., Ward, N.L., Angiuoli, S.V., Crabtree, J., Jones, A.L., Durkin, A.S., Deboy, R.T., Davidsen, T.M., Mora, M., Scarselli, M., Margarit y Ros, I., Peterson, J.D., Hauser, C.R., Sundaram, J.P., Nelson, W.C., Madupu, R., Brinkac, L.M., Dodson, R.J., Rosovitz, M.J., Sullivan, S.A., Daugherty, S.C., Haft, D.H., Selengut, J., Gwinn, M.L., Zhou, L., Zafar, N., Khouri, H., Radune, D., Dimitrov, G., Watkins, K., O'Connor, K.J., Smith, S., Utterback, T.R., White, O., Rubens, C.E., Grandi, G., Madoff, L.C., Kasper, D.L., Telford, J.L., Wessels, M.R., Rappuoli, R., and Fraser, C.M. (2005). Genome analysis of multiple pathogenic isolates of *Streptococcus agalactiae*: implications for the microbial "pan-genome". Proc. Natl. Acad Sci. USA *102*, 13950–13955.

Tettelin, H., Riley, D., Cattuto, C., and Medini, D. (2008). Comparative genomics: the bacterial pan-genome. Curr. Opin. Microbiol. *11*, 472–477.

Touchon, M., Hoede, C., Tenaillon, O., Barbe, V., Baeriswyl, S., Bidet, P., Bingen, E., Bonacorsi, S., Bouchier, C., Bouvet, O., Calteau, A., Chiapello, H., Clermont, O., Cruveiller, S., Danchin, A., Diard, M., Dossat, C., Karoui, M.E., Frapy, E., Garry, L., Ghigo, J.M., Gilles, A.M., Johnson, J., Le Bouguénec, C., Lescat, M., Mangenot, S., Martinez-Jéhanne, V., Matic, I., Nassif, X., Oztas, S., Petit, M.A., Pichon, C., Rouy, Z., Ruf, C.S., Schneider, D., Tourret, J., Vacherie, B., Vallenet, D., Médigue, C., Rocha, E.P., and Denamur, E. (2009). Organised genome dynamics in the *Escherichia coli* species results in highly diverse adaptive paths. PLoS Genet. *5*, e1000344.

van Rossum, D., Schuurmans, F.P., Gillis, M., Muyotcha, A., Van Verseveld, H.W., Stouthamer, A.H., and Boogerd, F.C. (1995). Genetic and phenetic analyses of *Bradyrhizobium* strains nodulating peanut (*Arachis hypogaea* L.) roots. Appl. Environ. Microbiol. *61*, 1599–1609.

Vandamme, P., Pot, B., Gillis, M., de Vos, P., Kersters, K., and Swings, J. (1996). Polyphasic taxonomy, a consensus approach to bacterial systematics. Microbiol. Rev. *60*, 407–438.

Villesen, P. (2007). FaBox: an online toolbox for fasta sequences. Mol. Ecol. Notes *7*, 965–968.

Vinuesa, P., León-Barrios, M., Silva, C., Willems, A., Jarabo-Lorenzo, A., Pérez-Galdona, R., Werner, D., and Martínez-Romero, E. (2005a). *Bradyrhizobium canariense* sp. nov., an acid-tolerant endosymbiont that nodulates endemic genistoid legumes (Papilionoideae: Genisteae) growing in the Canary Islands, along with *B. japonicum* bv. *genistearum*, *Bradyrhizobium* genospecies a and *Bradyrhizobium* genospecies b. Int. J. Syst. Evol. Microbiol. *55*, 569–575.

Vinuesa, P., Rojas-Jimenez, K., Contreras-Moreira, B., Mahna, S.K., Prasad, B.N., Moe, H., Selvaraju, S.B., Thierfelder, H., and Werner, D. (2008). Multilocus sequence analysis for assessment of the biogeography and evolutionary genetics of four *Bradyrhizobium* species that nodulate soybeans on the Asiatic continent. Appl. Environ. Microbiol. *74*, 6987–6996.

Vinuesa, P., Silva, C., Lorite, M.J., Izaguirre-Mayoral, M.L., Bedmar, E.J., and Martínez-Romero, E. (2005b). Molecular systematics of rhizobia based on maximum likelihood and Bayesian phylogenies inferred from *rrs*, *atpD*, *recA* and *nifH* sequences, and their use in the classification of *Sesbania* microsymbionts from Venezuelan wetlands. Syst. Appl. Microbiol. *28*, 702–716.

Vinuesa, P., Silva, C., Werner, D., and Martínez-Romero, E. (2005c). Population genetics and phylogenetic inference in bacterial molecular systematics: the roles of migration and recombination in *Bradyrhizobium* species cohesion and delineation. Mol. Phylogenet. Evol. *34*, 29–54.

Waterhouse, A.M., Procter, J.B., Martin, D.M., Clamp, M., and Barton, G.J. (2009). Jalview Version 2 – a multiple sequence alignment editor and analysis workbench. Bioinformatics *25*, 1189–1191.

Welch, R.A., Burland, V., Plunkett, G., 3rd, Redford, P., Roesch, P., Rasko, D., Buckles, E.L., Liou, S.R., Boutin, A., Hackett, J., Stroud, D., Mayhew, G.F., Rose, D.J., Zhou, S., Schwartz, D.C., Perna, N.T., Mobley, H.L., Donnenberg, M.S., and Blattner, F.R. (2002). Extensive mosaic structure revealed by the complete genome sequence of uropathogenic *Escherichia coli*. Proc. Natl. Acad. Sci. U.S.A. *99*, 17020–17024.

Wernersson, R., and Pedersen, A.G. (2003). RevTrans: Multiple alignment of coding DNA from aligned amino acid sequences. Nucleic Acids Res. *31*, 3537–3539.

Wertz, J.E., Goldstone, C., Gordon, D.M., and Riley, M.A. (2003). A molecular phylogeny of enteric bacteria and implications for a bacterial species concept. J. Evol. Biol. *16*, 1236–1248.

Whitaker, R.J., Grogan, D.W., and Taylor, J.W. (2003). Geographic barriers isolate endemic populations of hyperthermophilic archaea. Science *301*, 976–978.

Wirth, T., Falush, D., Lan, R., Colles, F., Mensa, P., Wieler, L.H., Karch, H., Reeves, P.R., Maiden, M.C., Ochman, H., and Achtman, M. (2006). Sex and virulence in *Escherichia coli*: an evolutionary perspective. Mol. Microbiol. *60*, 1136–1151.

Xia, X., and Xie, Z. (2001). DAMBE: Software package for data analysis in molecular biology and evolution. J. Hered. *92*, 371–373.

Young, J.P., Crossman, L.C., Johnston, A.W., Thomson, N.R., Ghazoui, Z.F., Hull, K.H., Wexler, M., Curson, A.R., Todd, J.D., Poole, P.S., Mauchline, T.H., East, A.K., Quail, M.A., Churcher, C., Arrowsmith, C., Cherevach, I., Chillingworth, T., Clarke, K., Cronin, A., Davis, P., Fraser, A., Hance, Z., Hauser, H., Jagels, K., Moule, S., Mungall, K., Norbertczak, H., Rabbinowitsch, E., Sanders, M., Simmonds, M., Whitehead, S., and Parkhill, J. (2006). The genome of *Rhizobium leguminosarum* has recognizable core and accessory components. Genome Biol. *7*, R34.

# Molecular Phylogeny of Microorganisms: Is rRNA Still a Useful Marker?

4

Wolfgang Ludwig

Abstract

The introduction of comparative rRNA sequence analysis certainly represents a major milestone in the history of microbiology. The current taxonomy of prokaryotes as well as modern probe and chip based identification methods are mainly based upon rRNA derived phylogenetic conclusions. The significance of single gene based phylogenetic inference was evaluated by including alternative global markers such as elongation and initiation factors, RNA polymerase subunits, DNA gyrases, heat shock and recA proteins. Although the comparative analyses are hampered by the generally low phylogenetic information content, and different resolution power, and multiple copies of the individual markers, the domain and prokaryotic phyla concept is globally supported.

## Introduction

The introduction of the rRNA approach by Carl Woese (Fox *et al.*, 1977) allowed for the first time in the history of microbiology comprehensive phylogenetic studies of the living world. Furthermore, rRNA based studies revolutionized the way that prokaryotic taxonomy has been practiced during the last three decades (Konstantinidis and Tiedje, 2007). After decades of artificial systematics, rRNA based clustering provided the basis for establishing a hierarchical framework of taxa – at least roughly – reflecting a natural relationship of the organisms. This has been nicely documented by the changes in the 'Approved lists of bacterial names' (Sneath and Brenner, 1992) and the editions of *Bergey's Manual* (Whitman, 2009) and other publications on prokaryotic taxonomy. Numerous taxa have been newly described, transferred, or emended, and these processes are still going on. Thus, the current taxonomy of prokaryotes, as documented in the most recent edition of *Bergey's Manual of Systematic Bacteriology* (Whitman, 2009) is based upon the phylogenetic framework deduced from small subunit rRNA data. Furthermore, comparative rRNA sequence analysis and rRNA targeted probing became standard techniques in microbial identification. Cultivation independent formats of the rRNA technology opened the door for studying the thus far hidden microbial world (Amann *et al.*, 1995). Rapid high throughput sequencing technologies contributed to a small subunit rRNA sequence data set which is the largest data set currently available for a gene or gene product. Although the advantages of this phylogenetic marker are well documented with respect to information content and comprehensiveness of the available data set (Ludwig *et al.*, 1998; Ludwig and Klenk, 2001), it is known and meanwhile widely accepted that single marker-based studies may resolve gene phylogenies but can only roughly reflect the evolutionary history of the respective organisms. Including alternative phylogenetic markers may help to come closer to a more realistic model of the phylogeny of the organisms. In the pre-genomics era a limited number of studies focused on other markers such as elongation factors, ATPase subunits and RNA polymerases (Ludwig *et al.*, 1993; Klenk *et al.*, 1994). These studies already showed the limitations concerning the phylogenetic information content of the selected markers. Although similar overall tree topologies were indicated by the investigations, marker specific discrepancies were commonly found. It was discussed earlier that such discrepancies had to be expected. Given the generally limited information content, the individual markers can only document small parts of the evolutionary time. Consequently, different markers may have preserved information on different time spans of evolution (Ludwig and Klenk, 2001). Nevertheless, the rapid progress in full genome sequencing gave rise to great hopes that further markers would be detected which could help resolving the evolutionary history of the organisms more correctly. However, already the comparative analysis of the first fully sequenced genomes indicated a rather limited number of potential phylogenetic markers which fulfil the criteria of universal occurrence, functional constancy and sufficient sequence conservation as well as complexity (Huynen and Bork, 1998). Although new methodologies, algorithms, and software for full genome comparison were and continuously are developed (Bapteste *et al.*, 2005: Delsuc *et al.*, 2005; Simonson *et al.*, 2005), it remains fact that in comparison with the richness in genetic information of even the smallest prokaryotic genomes, the part that can be used for comprehensive phylogenetic studies of the three domains of life is minimal.

Given the comprehensiveness of the available data, the current taxonomy as well as the history of prokaryotic phylogeny analysis, the small subunit rRNA based picture is used as the basis for comparison with data obtained from other potential markers. Interestingly, this procedure is followed by many other authors irrespectively whether they support the current rRNA-based view of phylogeny or propose other models of evolution.

## Information content of marker molecules

Before discussing the similarities and discrepancies of phylogenetic conclusions based upon individual or concatenated rRNA and non-rRNA markers, some general remarks concerning the potential significance of such comparisons have to be made. Against the background of three to four billion years of evolution of cellular life, the phylogenetic information content of any of the commonly used marker molecules is generally limited (Ludwig *et al.*, 1998; Ludwig and Klenk, 2001; Gupta and Griffiths, 2002). Sequence conservation is one of the primary characteristics of a universal phylogenetic marker. Consequently, only part of the primary structure positions is variable and hence informative at all. Owing to functional constraints and selective pressure the potential information content is further reduced, given that the variable positions differ with respect to the number of allowed character states (the four nucleotides or the 20 amino acids plus the insertion/deletion events). Thus, in comparison to the theoretically huge space of primary structure variation, only a minute number of changes can be found and measured in real sequence data. Consequently, the real information content is considerably reduced. The remaining information may be obscured by a burden of noise especially at highly variable sites. Typical examples are 'false' identities resulting from multiple changes during the course of evolution. Conservation profiles in combination with correction and optimization functions – implemented in commonly used phylogenetic methods – allow the estimation of evolutionary events which are not directly documented by character changes in addition to those which can directly be measured. However, in comparison with three to four billion years of evolution only an insignificant number of events can be measured or estimated based upon sequence comparison. Thus, only a spot check of evolutionary changes is possible.

## Significance of tree topology

The commonly followed way visualizing phylogenetic relationships is tree reconstruction based upon primary structure alignments. Although 'tree thinking' has been criticized recently and alternative methods such as principal component analysis and/or heat maps are available (Bapteste *et al.*, 2005, 2008), tree or dendrogram visualization is certainly justified as the more human readable version. However, the robustness and resolution power of phylogenetic trees based on marker sequence data are often overestimated. The data-inherent problems discussed above, in combination with the shortcomings of the data analysis and tree reconstruction tools, remarkably reduce the significance of local tree topologies (Ludwig *et al.*, 1998; Ludwig and Klenk, 2001). Any tree reconstruction method relies on evolutionary models, which may be correct or appropriate for only part of the data or positions included. Given the huge tree space and the limited power of even the currently available supercomputers, exhaustive testing of all potential topologies is not possible. Therefore, heuristics and assumptions have to be included in commonly used tree reconstruction software. Furthermore, tree topologies are heavily influenced by data and parameter selection. Even when keeping methods and parameters constant, adding new data may change the topology not only around the insertion points of the new branches but also at distant regions of the trees as a result of branch attraction effects. Changing parameters and position selection filters usually allows recognizing branch attraction effects as such. Using comprehensive and balanced data sets often helps to reduce them. However, using comprehensive data sets often results in bush-like topologies, making resampling techniques for tree evaluation obsolete. Modern rapid maximum likelihood and maximum parsimony approaches visualize significance by branch lengths, facilitating a sound topology interpretation (Stamatakis, 2006). Generally, tree topologies should be critically evaluated and interpreted. The quintessence is that a range of uncertainty (Ludwig *et al.*, 1998) has to be assumed for local branching orders in trees. Recommendations for data selection and quality check as well as the generation and application of filters and parameters for tree reconstruction and topology evaluation are given by Peplies *et al.* (2008).

## Tree comparison

As mentioned above, basing phylogenetic conclusions on multiple informative markers is nowadays well accepted and widely in use. However, a further consequence of the limited information content of each individual marker is that different markers may carry or lack the information on different eras of evolution. Thus, locally different tree topologies have to be expected when comparing phylogenetic conclusions based upon different alternative markers. Assessing such differences, it has to be taken into account that the significance of local tree topologies, especially when characterized by relatively short internodes (branches), is often low (Ludwig and Klenk, 2001). Thus the range of uncertainty mentioned above (Ludwig *et al.*, 1998) has to be taken into consideration when assessing similarities or discrepancies of trees based upon alternative phylogenetic markers.

## Marker selection

The most important prerequisites of phylogenetic markers are ubiquitous presence and functional constancy, besides size and structure variation. Even now in the age of transcriptomics and proteomics functional constancy often is not known or tested. Another important problem of phylogenetic studies based upon alternative markers concerns orthology of the respective molecules. The use of orthologous (direct common ancestor) genes or gene products is essential for the delineation of the monophyletic status of groups of molecules or organisms, whereas paralogous markers (indirect common ancestor, derived from duplicated genes or acquired by horizontal gene transfer) can be used for the relative rooting of or within monophyletic (sub)trees. Multiple variants (within one organism) have been documented for almost all commonly used markers (Ludwig and Schleifer, 2005). In many cases primary structure similarities are high, but more and more highly diverged polymorphisms are discovered. Thus, discrepancies of tree topologies could result from 'illegitimate' inclusion of paralogous markers, reconstructing trees that should reflect 'vertical' phylogeny. Paralogous data can only be recognized as such if the respective data set contains examples which assign both copies to a given organism. The available fully annotated genomes provide extremely valuable information in this context. However, defining the orthologous variants often remains difficult or even impossible. Any phylogenetic conclusions may be complicated or misleading if two or more copies are maintained but one of the products changed its function, or if one of the copies was lost during the course of evolution. The classical example is the case of proton translocating ATPase subunits (Ludwig *et al.*, 1993; Zhaxybayeva *et al.*, 2005).

## Sequence quality and alignment

The quality of the sequence data and their alignment are also heavily influencing tree reconstruction and phylogenetic conclusions. Despite the enormous progress in sequencing technology and data analysis tools, the primary data quality did not adequately improve. This might be partially imputed to the – against the background of the data flood – inevitable automation of data analyses. In the case of protein markers the *in silico* translation of the DNA sequences provides some system-inherited quality check. Concerning the rRNA data the situation is worse. Among the more than 345,000 sequence entries for *Bacteria* in the SILVA small subunit rRNA Ref database (Prüsse *et al.,* 2007) only 23,000 are 95% complete. Only 10,000 sequences successfully passed a restrictive quality analysis creating the basis for a *Bacteria* tree shown in Fig. 4.1. After careful data analysis applying various filters and methods using the ARB package (Ludwig *et al.,* 2004), sequences of lower quality – especially those of missing type taxa – were added. A special ARB treeing tool preventing topology influences by the lower quality sequences was applied during this procedure. Databases and trees for the validly described type taxa of the Prokaryotes are provided by the 'All-Species Living Tree Project' (Yarza *et al.,* 2008).

Besides sequence quality an optimal alignment is another crucial requirement for sound phylogenetic analyses. Given that the higher order structure of marker molecules is connected to its function and hence a target of the evolutionary process, higher order structure information has to be taken into consideration putting (most likely) homologous sequence positions in common columns. Three decades of individual marker based phylogenetic analyses have shown that despite the improvements of aligning software tools the higher order structure aspects usually are not adequately recognized by such tools, and 'manual' intervention by specialists still is needed. In the case of rRNA integrated databases maintained by specialists are available providing processed primary structure and additional descriptive information (DeSantis *et al.,* 2006; http://greengenes.lbl.gov; Cole *et al.,* 2008; http://rdp.cme.msu.edu/; Prüsse *et al.,* 2008; http://www.arb-silva.de/). For generally conserved protein markers some authors provide databases of aligned primary structures.

## The rRNA based domain and phylum concept

### Small subunit rRNA

The current view of small subunit rRNA derived phylogeny clearly supports the three-domain concept of Bacteria, Archaea and Eucarya. Given that paralogous markers are not available for the small subunit rRNA, a significant positioning of a root is not possible. The second edition of *Bergey's Manual of Systematic Bacteriology* (Garrity, 2001) distinguishes 23 bacterial phyla containing at least a few culturable bacteria. However, sequence data obtained by applying cultivation independent techniques indicate the presence of a number of additional groups to which the phylum status could be assigned. It is well known that phylogenetic trees are only models of the evolutionary affiliations of the organisms and may be heavily influenced and changed whenever the underlying data base is extended. Most phyla were defined during the early phase of comparative rRNA sequencing based upon limited data sets. Trees showing monophyletic clusters well separated by 'long' internal branches facilitated phylum distinction at that time. However, meanwhile the situation changed towards bush-like topologies. In many cases the definition of a significant borderline separating individual phyla or even the monophyletic status of the subgroups of some phyla became more difficult or questionable. Thus, in volume 3 of the current edition of *Bergey's Manual of Systematic Bacteriology* a potential independent phylum status for some taxa traditionally assigned to the *Firmicutes* phylum is discussed (Ludwig *et al.,* 2009). The reverse was found for the previously separated *Fusobacteria* phylum, which based upon the updated database and applying modern powerful maximum likelihood methods tends to be found among the *Firmicutes* (Yarza *et al.,* 2008). For most of the phyla a significant relative branching order cannot be defined. A slightly deeper branching of the phyla *Aquificae, Thermotogae,* and *Dictyoglomi* can be seen using data sets modified according to the 50% positional conservation convention (Ludwig

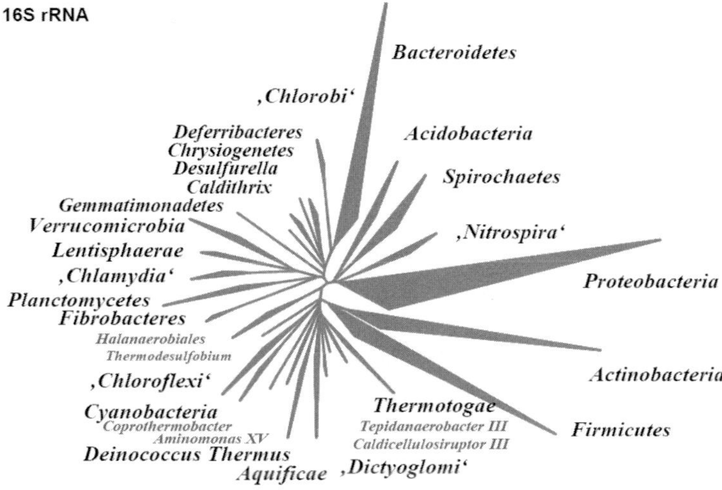


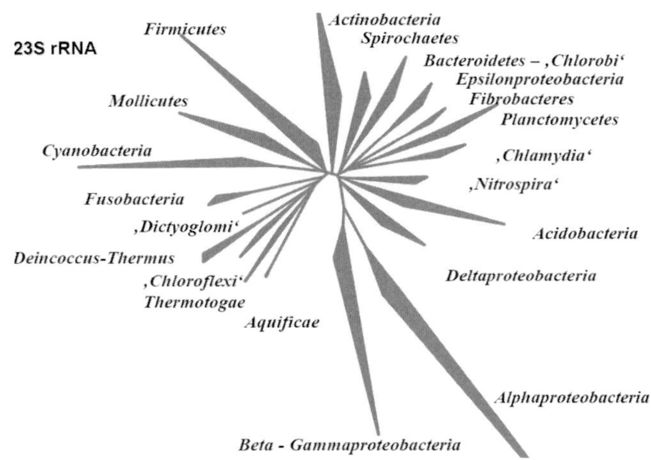


**Figure 4.1** Top: 16S rRNA based tree of the phyla of *Bacteria*. The underlying data set of all available full length and quality checked sequences was extended by including all type strain sequences. Alignment positions invariant in at least 50% of all bacterial sequences were used for tree reconstruction. The tree topology is based on maximum parsimony and maximum likelihood analyses using the ARB software package. Bottom: 23S rRNA based tree of the phyla of *Bacteria*. Procedures and parameters for tree reconstruction were as used for the 16S rRNA based tree.

*et al.*, 1998) if rooted by *Archaea* and *Eucarya* sequences. Similarly, there is a tendency of a common group comprising '*Chlamydia*', *Fibrobacteres*, *Lentisphaerea*, *Planctomycetes* and *Verrucomicrobia*. The *Bacteroidetes* and '*Chlorobi*' represent another phylum cluster. All other phylum clusters are rather unstable when slightly changing the parameters or the underlying data set for tree reconstruction. The *Firmicutes* (including the *Fusobacteria* and *Mollicutes*) represent an independent cluster like the phyla *Actinobacteria, Nitrospirae, Spirochaetes, Fibrobacteres, Bacteroidetes, Cyanobacteria, Deferribacteres, Fusobacteria,* and *Proteobacteria.*

## Large subunit rRNA

The large subunit rRNA is probably the most informative phylogenetic marker. Its primary structure is at least as conserved as that of the small subunit rRNA, but contains more and longer stretches of informative positions. Although the LSU rRNA gene sequences provide more information, the major drawback is currently the limited data base which comprises only about 14000 (at least 60% complete) sequences. Minor local differences can be seen when SSU and LSU rRNA derived trees are compared (Ludwig and Klenk, 2001). The phyla are still clearly defined and separated (Fig. 4.1). A deeper branching is supported for *Aquificae*, and *Thermotogae* when rooted including *Archaea* and *Eucarya*. The '*Chloroflexi*' and the *Thermus–Deinococcus* phylum share a common root. *Chlamydiae* and *Planctomycetes* as well as *Bacteroidetes* and *Chlorobi*, respectively, represent two phylum clusters.

In principle, the SSU and LSU rRNA genes fulfil all the requirements of a phylogenetic marker molecule. However, the resolution power is limited especially at levels of closer relationships. Due to functional constraints sequence changes occur gradually rather than in a continuous process. A direct correlation to a time scale cannot be postulated. Branching patterns in the periphery of a tree do not reliably reflect the phylogenetic relatedness. There is a low phylogenetic resolving power of closely related organisms (>97% similarity), i.e. rRNA sequence data analyses are usually not sufficient to distinguish prokaryotes at the species level.

## Alternative marker molecules that support the phyla concept

A small number of protein markers could be used to evaluate the rRNA based global phylogenetic conclusions (Ludwig and Schleifer, 2005). Among those which according to the currently available full genome sequence data belong to the – at least domain-wide – universal core of the genome are genes encoding translation elongation and initiation factors, RNA polymerase subunits, (proton translocating) ATPase subunits, DNA gyrases, recA, ribosomal and heat shock proteins. The respective data sets were recently updated especially with respect to the genomes currently available from genome reviews at the EMBL database (Sterk *et al.*, 2006; http://www.ebi.ac.uk/GenomeReviews/). Phylogenetic trees were reconstructed for the individual markers based on aligned amino acid sequences applying various position filters and maximum parsimony as well as maximum likelihood treeing approaches. Tree comparisons showed general support of the rRNA based phylum concept but also local topology differences. In many cases the phylum status is supported by common clusters in trees based upon the alternative markers used for this comparative study. However, a common relative branching order of the phyla is rarely supported. Furthermore, for some phyla a common origin of its members cannot be seen in part of the individual marker trees. Nevertheless no clear disproof of the phylum concept by e.g. intermingling of representatives of different phyla at levels of high sequence similarity could be shown. The local topology differences probably result from different or lower resolution power of the individual markers. As mentioned above, given the generally limited information content, different markers may have preserved information on different evolutionary spaces of time and long periods may not be documented at all. As already seen with the rRNA trees, the significance for a relative branching order of the phyla or other major cluster is low. Comparative analysis is further complicated by the many cases of gene duplications causing orthology problems. The results of tree comparison will be briefly discussed in the following for a selection of markers and the bacterial phyla.

### RNA polymerase

The largest subunits of the RNA polymerase of Bacteria (β, β′), Archaea (A′, A″, B′, B″) and Eucarya (RPB1, RPB2) are highly conserved and ubiquitous. These subunits represent the largest of the 'universal' markers. The respective molecules of *Escherichia coli* comprise 1342 and 1407 amino acids, respectively. This may indicate a higher information content. However, there are a number of highly variable stretches in these markers not containing useful information for the phylum and lower intra-phylum levels. The comparative sequence data analysis of the bacterial subunits of RNA polymerase and their archaeal and eucaryal counterparts show a clear separation of the three domains. Trees based upon the β and β′ subunits nicely support the rRNA based phylum concept (Fig. 4.2). Only the *Epsilonproteobacteria* are separated from the other subclasses of the *Proteobacteria*. Both the '*Chlamydia*'–*Planctomycetes*–*Verrucomicrobia* and the *Bacteroidetes*–*Chlorobi* phylum groups are

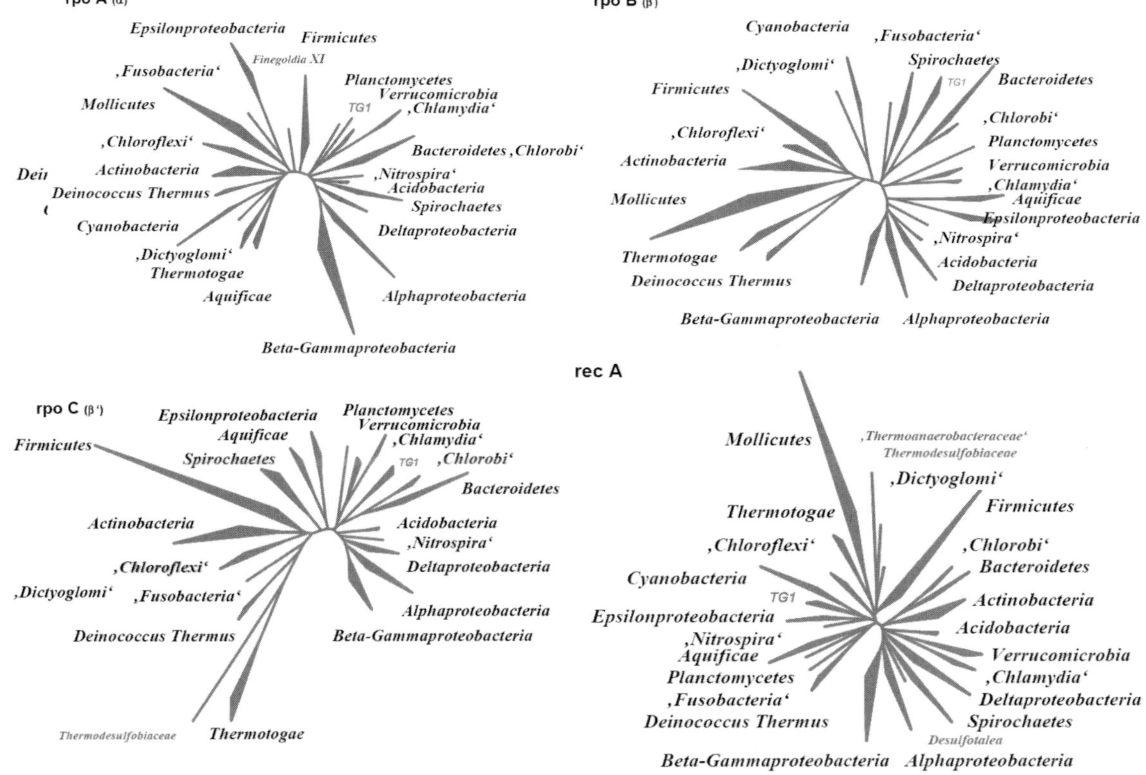


**Figure 4.2** Top left: RNA polymerase alpha-subunit-based tree of the phyla of *Bacteria*. The underlying data set comprises all respective predicted amino acid sequences from published full genome data as well as additional sequences available in public databases. Only taxa for which full genome data are available are shown. The topology was reconstructed applying maximum likelihood methods using the ARB software package. Alignment positions invariant in at least 30% of all bacterial sequences were used for tree reconstruction. Top right: RNA polymerase subunit beta based tree of the phyla of *Bacteria*. Procedures and parameters for tree reconstruction were as used for the RNA polymerase alpha-subunit-based tree. Bottom left: RNA polymerase subunit beta prime based tree of the phyla of *Bacteria*. Procedures and parameters for tree reconstruction were as used for the RNA polymerase alpha-subunit-based tree. Bottom right: recA protein based tree of the phyla of *Bacteria*. Procedures and parameters for tree reconstruction were as used for the RNA polymerase alpha-subunit-based tree.

supported as well. Even the small bacterial alpha-subunit-based analyses (Fig. 4.2) provide good support by a similar tree clustering situation as discussed for the larger subunits.

## Initiation and elongation factors

Initiation factors IF2 and elongation factors EF-G and EF-Tu as well as their archaeal or eukaryotic counterparts share a common origin and presumably represent descendants of early gene duplications (Fig. 4.3). The three domain concept is clearly supported by those markers and the phylum concept supported roughly.

As for many other protein markers the Initiation factor IF2 tree (Fig. 4.3) does not support a monophyletic structure of the *Proteobacteria*. The *Delta-* and *Epsilonproteobacteria* are separated from an *Alpha-, Beta-, Gammaproteobacteria* cluster along with a further split of part of the *Deltaproteobacteria.* Similar but not significant tendencies have already been observed in rRNA – especially large subunit rRNA – studies. A separate status for some new phylum candidates traditionally assigned to the *Firmicutes* as mentioned in the rRNA section is supported as are *Bacteroides–Chlorobi* and *Chlamydia–Verrucomicrobium* phylum groups.

Further splitting of *Proteobacteria* was indicated by EF-G analyses (Fig. 4.3). In addition the '*Bacilli*' and *Clostridia* classes of the *Firmicutes* are separated, as well as some representatives of new candidate phyla currently assigned to the *Firmicutes*. A *Bacteroides – Chlorobi* phylum cluster is supported. The situation is further complicated by clusters of EF-G like markers within the intraphylum lineages radiation. These clusters contain sequences from organisms also represented by 'true' EF-G sequences. This may indicate further intraphylum gene duplications.

The smallest marker of this group is elongation factor Tu, comprising only 394 amino acids in the case of *E. coli*. The smaller size and hence lower information content is reflected by reduced resolution power at the phylum levels (Fig. 4.3). Besides the *Proteobacteria* also *Actinobacteria, Bacteroidetes* and *Spirochaetes* do not appear as monophyletic groups, and *Alpha-* and *Deltaproteobacteria* are further disrupted. However, the *Firmicutes* phylum is maintained except the new phylum candidates. Examples of gene duplications indicate potential paralogy problems. Duplicated diverged genes are documented for *Streptomyces coelicolor* and *S. ramocissimus* (van Wezel *et al.*, 1995). One of the variants appears as the deepest branch among the Bacteria. *Streptomyces ramocissimus* contains a third copy which is less diverged and clusters among the *Actinobacteria*. This gene triplication demonstrates that multiple diverged genes or gene products may hamper phylogenetic analyses at various levels of relationships. Obviously, it is difficult to decide whether differences in rRNA and EF-Tu-based trees indicate different resolution capacity, different history, or reflect the relationships of paralogous markers.

### DNA gyrases

DNA gyrases are often used in multi locus sequence analysis (Maiden *et al.*, 1998), assuming a higher resolution power (than rRNA) at levels of closer relationship. This is compensated by an apparently reduced resolution at the interphylum levels.

In the gyrase A tree (Fig. 4.4) the *Proteobacteria* phylum is somewhat disrupted by intermingling of *Proteobacteria* subclasses and other phyla. The *Firmicutes* are also split in several lineages. However, the '*Chlamydia*'–*Planctomycetes–Verrucomicrobia* as well as the *Bacteroidetes–Chlorobi* phylum groups are nicely supported. Most notably, representatives of the *Archaea* are not clearly separated from the *Bacteria* and root at the inter-phylum level. The picture is further complicated by topoisomerase IV clusters which are positioned at the inter- and – in the case of *Cyanobacteria*–intraphylum levels, again indicating gene duplications at different levels of relationships or evolutionary time. Interestingly, within these clusters the rRNA based concept is often supported again. The situation within the *Cyanobacteria* phylum is shown as an example (Fig. 4.4). A similar situation can be seen in the gyrase B tree (Fig. 4.4).

### Heat shock proteins

The domains of the *Archaea* and *Bacteria* are well defined and separated when Hsp60 protein sequences (*Bacteria*: GroEL; *Archaea*: Tf-55) are phylogenetically analysed. Most of the bacterial phyla represented in the data set appear as monophyletic clusters. A relative branching order of the phyla is only partially resolved and with low significance (Fig. 4.3). A monophyletic status of the *Proteobacteria* is not supported. Although the *Betaproteobacteria*, the *Gammaproteobacteria* and the *Alphaproteobacteria* are unified in a common group, the *Deltaproteobacteria* and the *Epsilonproteobacteria* represent their own lineages. In the case of the *Firmicutes* again some new phylum candidates are separated from the cluster. The '*Chlamydia*' – *Planctomycetes* – *Verrucomicrobia* as well as the *Bacteroidetes* – *Chlorobi* phylum groups are supported. Also gene duplication and/or lateral transfer effects may be of importance as indicated by the multiple variants of *Chlamydophila* genes. One of the copies clusters among the '*Chlamydia*' – *Planctomycetes* – *Verrucomicrobia* phylum group, whereas the additional versions of Hsp60 like molecules represent deeper branches in the tree (not shown).

The Hsp70 (DnaK) heat shock protein data do not allow a distinct separation of the domains *Archaea* and *Bacteria*. Moreover, no closer relationships of individual representatives of the different domains could be observed. The

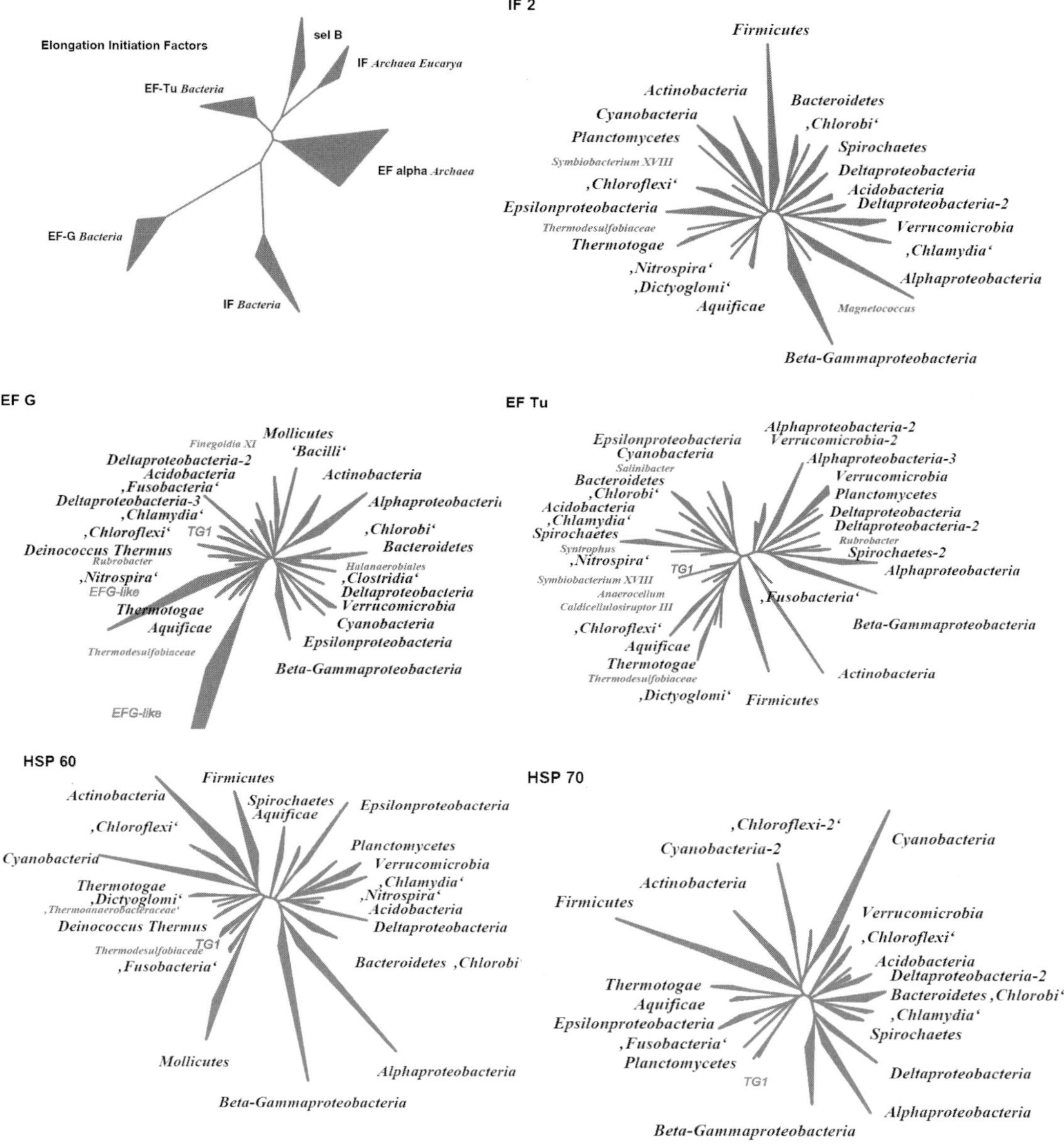


**Figure 4.3** Top left: The common ancestry of translation initiation and elongation factors. Procedures and parameters for tree reconstruction were as used for the RNA polymerase alpha-subunit-based tree. Top right: Initiation factor 2 based tree of the phyla of *Bacteria*. Procedures and parameters for tree reconstruction were as used for the RNA polymerase alpha-subunit-based tree. Middle left: Elongation factor G based tree of the phyla of *Bacteria*. Procedures and parameters for tree reconstruction were as used for the RNA polymerase alpha-subunit-based tree. Middle right: Elongation factor Tu based tree of the phyla of *Bacteria*. Procedures and parameters for tree reconstruction were as used for the RNA polymerase alpha-subunit-based tree. Bottom left: Heat shock protein HSP60 based tree of the phyla of *Bacteria*. Procedures and parameters for tree reconstruction were as used for the RNA polymerase alpha-subunit-based tree. Bottom right: Heat shock protein HSP70 based tree of the phyla of *Bacteria*. Procedures and parameters for tree reconstruction were as used for the RNA polymerase alpha-subunit-based tree.

archaeal and bacterial lineages appear intermixed at a rather low level of relationship (not shown). Many of the bacterial phyla are supported by the Hsp70 tree (Fig. 4.3). The *Epsilonproteobacteria* as well as part of the *Deltaproteobacteria* subclasses are remote from the other subclasses. Gene duplications are indicated for representatives of *Cyanobacteria* and *Chloroflexi*.

### RecA

The RecA protein sequence data also indicate the existence of the bacterial phyla (Fig. 4.2). However, a monophyletic status is not documented for the *Proteobacteria,* and in the case of the *Firmicutes* some new phylum candidates are separated. As can be expected for a smaller marker (the *E. coli* protein comprises 353 amino acids), the short edges forming the branching pattern of the major lineages indicate low resolving power, and the pattern may also be hampered by undetected paralogy effects. Multiple protein sequences sharing only moderate sequence identity were reported for several organisms such as *Myxococcus*.

## Markers that do not clearly support the phyla concept

The history of (proton translocating) ATPase based analyses provides the classical example of phylogenetic confusion resulting from early gene duplication. The catalytic beta subunit of the $F_1F_0$-type ATPase was tested first as an alternative phylogenetic marker and the results supported the contemporary small subunit rRNA based view of bacterial phylogeny (Amann *et al.*, 1988). Later, the catalytical subunit of vacuolar type of ATPase was regarded as the eucaryal and archaeal counterpart. A closer relationship of the latter two is documented by overall sequence identity of more than 60%, whereas only moderate relationship to the bacterial version is indicated by identities of less than 24%. Based upon sequence comparisons it was postulated that the catalytical and noncatalytical subunits of both types of ATPase ($F_1F_0$ and vacuolar) originate from a common ancestor as descendants of the products of an early gene duplication (Hilario and Gogarten, 1998; Ludwig *et al.*, 1998). It was generally assumed that the vacuolar type of ATPase is a characteristic component of the members of the latter two domains, whereas the $F_1F_0$-type was regarded as unique for the *Bacteria* domain. Later, it was shown that some representatives of the *Bacteria* domain contained the genes and proteins for both, the $F_1F_0$ as well as the vacuolar type of ATPase (Neumaier, 1996). A similar situation was shown for an archaeon (Sumi *et al.*, 1997). Thus not only the alpha and beta subunits originate from early gene duplications but also their $F_1F_0$ and vacuolar type versions resulted from independent duplication events. For *Enterococcus hirae* the $F_1F_0$ type ATPase was shown to be responsible for proton translocation, whereas the vacuolar type works as a sodium pump (Kakinuma *et al.*, 1999). These data and the finding that for example the $F_1F_0$ type ATPase was always present in a selection of closely related *Enterococcus* species whereas the vacuolar type could only be found in a few of them (Neumaier, 1996) illustrate that changing the function of one of the ATPases also changes the selection pressure. Apparently, the additional vacuolar type in certain bacteria is no longer essential for the organisms and may be changed at a higher rate or completely lost during the course of evolution. Consequently, the phylogenetic positioning of such copies among the essential variants may be problematic or misleading. The presence of additional vacuolar ATPase genes besides the $F_1F_0$ ATPase could be shown for representatives of different bacterial phyla. However, there are bacterial taxa – members of the phyla *Deinococcus-Thermus, Spirochetes* and *Chlamydia* – most of which carry only the vacuolar version. This may indicate that the V-ATPase became the essential version in the time of their diversification. Both types were found in part of the representatives of the *Spirochaetes.*

Within the $F_1F_0$ type ATPase alpha as well as beta subunit based trees (Fig. 4.5) the *Actinobacteria, Firmicutes* and *Proteobacteria* appear not monophyletic, whereas the *Bacteroidetes–Chlorobi* phylum group is supported. Most notably, full genome analyses documented the presence of a third type besides the $F_1F_0$ and vacuolar types for some organisms such as Planctomycetes or *Proteobacteria* (Glöckner *et al.*, 2003). Including the respective sequences for phylogenetic treeing resulted in deep rooting of the third type on the *Bacteria* branch (Fig. 4.3), thus indicating further early gene duplication

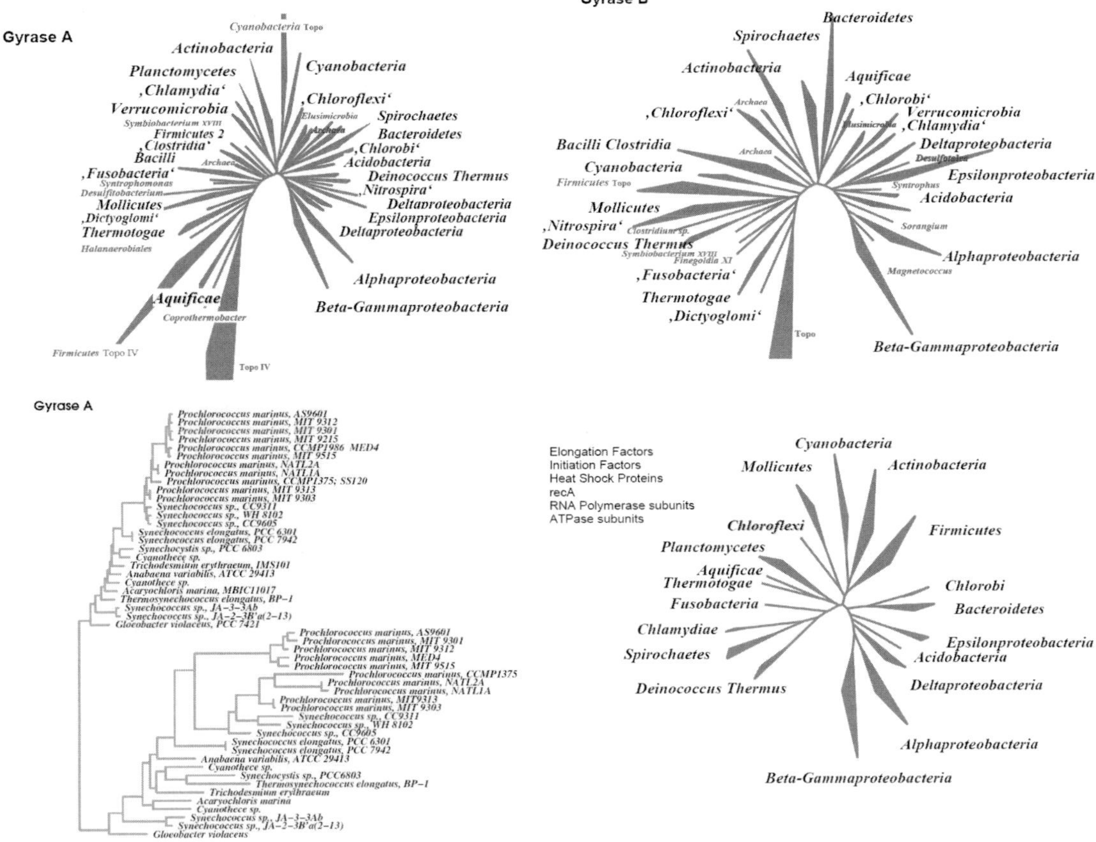


**Figure 4.4** Top left: DNA gyrase subunit A based tree of the phyla of *Bacteria*. Procedures and parameters for tree reconstruction were as used for the RNA polymerase alpha-subunit-based tree. Top right: DNA gyrase subunit A based tree of the phyla of *Bacteria*. Procedures and parameters for tree reconstruction were as used for the RNA polymerase alpha-subunit-based tree. Bottom left: Detailed dendrogram for DNA gyrase subunit A documenting intra-*Cyanobacteria* phylum gene duplication. Procedures and parameters for tree reconstruction were as used for the RNA polymerase alpha-subunit-based tree. top right: Tree of bacterial phyla based on concatenated alignments of the protein markers. Bottom right: recA protein based tree of the phyla of *Bacteria*. Procedures and parameters for tree reconstruction were as used for the RNA polymerase alpha-subunit-based tree.

events. In all cases of multiple genes described for the alpha subunit the respective pendant was also found for the beta subunit.

Using the vacuolar data and excluding those 'bacterial' sequences which are duplicates, a picture emerges of moderately related *Archaea* and *Bacteria* separated from the *Eucarya* (Fig. 4.5). Although the *Chlamydiae* and *Spirochetes* are separated from the *Archaea*, the *Thermus-Deinococcus* phylum deeply branches among the archaeal lines. However, this intermingling is caused by a small number of positions with a low positional variability (with respect to the number of different character states).

## Concatenated data sets

When performing multiple marker-based phylogenetic analyses, the processed primary data are nowadays preferentially concatenated to base the analyses on a single alignment (Brown *et al.*, 2001). The intention is to visualize the consensus of the individual markers while 'smoothing' the difference, assuming this might come closer reflecting the organism's relationships.

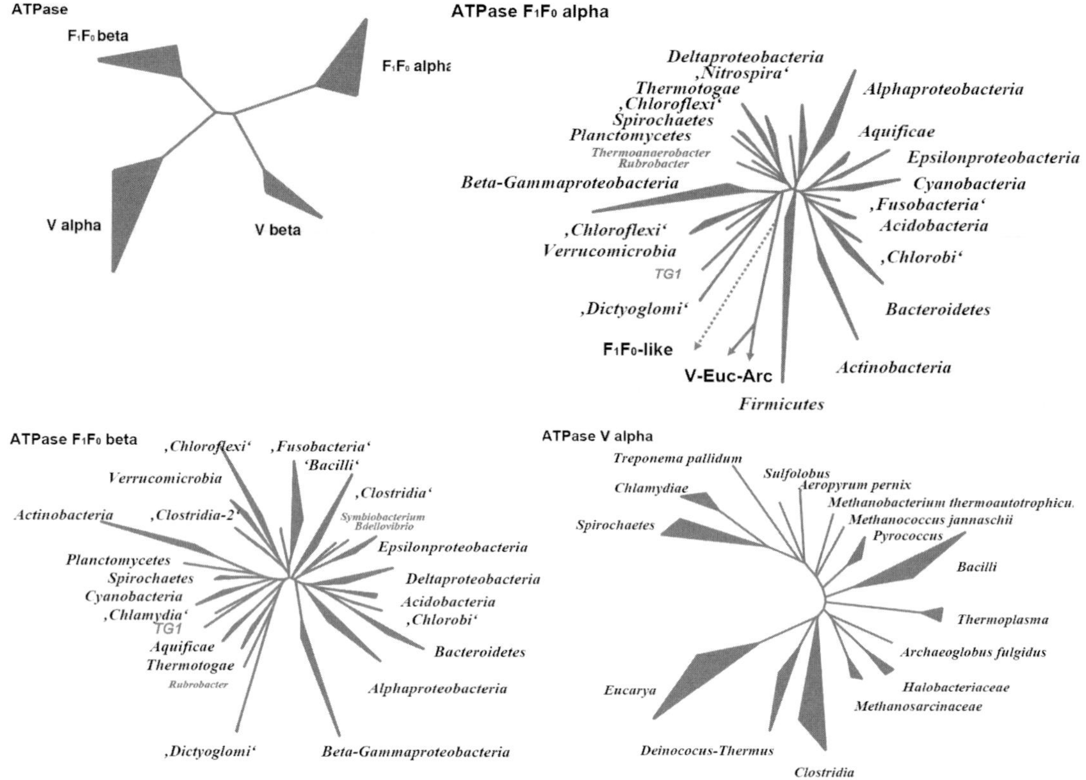


**Figure 4.5** Top left: The common ancestry of $F_1F_0$ and vacuolar type ATPase subunits alpha and beta. Procedures and parameters for tree reconstruction were as used for the RNA polymerase alpha-subunit-based tree. Top right: $F_1F_0$ ATPase alpha-subunit-based tree of the phyla of *Bacteria*. The positions of the homologues from *Archaea* and *Eucarya* are indicated by arrows. The dotted arrow line indicates the branch for $F_1F_0$-like markers found in some *Bacteria*. Procedures and parameters for tree reconstruction were as used for the RNA polymerase alpha-subunit-based tree. Bottom left: $F_1F_0$ ATPase subunit beta based tree of the phyla of *Bacteria*. Procedures and parameters for tree reconstruction were as used for the RNA polymerase alpha-subunit-based tree. Bottom right: Vacuolar type ATPase alpha-subunit-based tree of the phyla of *Archaea* and *Bacteria*. Procedures and parameters for tree reconstruction were as used for the RNA polymerase alpha-subunit-based tree.

This procedure, however, violates the principle of basing phylogenetic analyses upon orthologues only. Furthermore, smoothing by concatenating also means hiding information which is carried by only single or a minor part of individual markers. As mentioned above individual markers may have preserved information on different eras of evolution. Nevertheless, the concatenation approach facilitates multiple marker based comparative analyses. However, one should always be aware of losing valuable information or obtaining misleading results (Bapteste *et al.*, 2007). An example is given for the bacterial phyla concatenating the markers described above including the ATPase subunits. An 'artifical' grouping of *Chlamydiae, Deinococus-Thermus,* and *Spirochaetes* can be seen in the tree (Fig. 4.4), presumably resulting from branch attraction caused by including paralogous ATPase subunits.

## Whole genome approaches

Along with the rapidly increasing number of full genome sequences new methods were developed and applied to put phylogenetics on a broader base. These approaches are also often called phylogenomics. However, where is the threshold? Isn't the procedure of directed marker selection and comparative interpretation of the individual results described above also phylogenomics? The great advantage of full or partial genome versus single gene comparisons concerns the sampling size. The sampling error and potential saturation problems are certainly reduced by increasing the number of characters used for phylogeny inference (Rokas, 2006). At the beginning of the age of genomics data interpretation (annotation) was a specialist task, but it was soon replaced by automated bioinformatics tool pipelines. A similar development is seen in phylogenomics. Automation is certainly essential to encompass the current data flood. However, in terms of phylogeny, both the big sampling size as well as automation carry a burden of system-inherited shortcomings or biases. As mentioned above, the orthologous character of markers used for comparison is one of the basic requisites in phylogenetics. Since there are many cases of gene duplications and horizontal gene transfers as indicated by the currently available genome data and as postulated in many publications, there is a high risk of including paralogues especially by applying automated tools. Three decades of phylogenetic analyses based on single or a few distinct genes taught us that sequence similarity not necessarily means orthology. Despite the enormous information content of genomes, only a minimal part of it can be used to reconstruct global phylogeny at the domain and interphylum levels.

The following paragraphs present a number of phylogenomics methods that are currently in use (Delsuc *et al.*, 2005).

### Character strings

An approach derived from whole genome features is based on the distribution of oligonucleotide (Campbell *et al.*, 1999; Lin and Gerstein, 2000; Pride *et al.*, 2003; Edwards *et al.*, 2002) or amino acid strings (Qi *et al.*, 2004). Determination of genome-wide oligonucleotide frequencies was shown to provide characteristic signatures. Although some phylogenetic signal can be extracted from this oligonucleotide 'word usage' (Delsuc *et al.*, 2005), its relevance for comprehensive inference remains questionable, given that the approach is independent of homology or orthology assessments. Differences from – (Campbell *et al.* 1999; Lin and Gerstein, 2000; Pride *et al.*, 2003) and support of (Pride *et al.*, 2003; Qi *et al.*, 2004) the rRNA based phylum concept were reported, depending on the parameters applied.

### Gene content

Measuring the presence or absence of genes (annotated as such) or pathways provides valuable information on gene sets shared or not by different genomes (Fitz-Gibbon and House, 1999; Snel *et al.*, 1999; Tekaia *et al.*, 1999; Lin and Gerstein., 2000; Wolf *et al.*, 2001; Korbel *et al.*, 2002; Heymans and Singh, 2003; Simonson *et al.*, 2005; Zhang *et al.*, 2006). Smaller data sets – with respect to species diversity – supported the rRNA based phyla concept (Snel *et al.*, 1999). Partial support of the phylum concept was shown by some authors (Wolf *et al.*, 2001; Korbel *et al.*, 2002; Zhang *et al.*, 2006), whereas in other publications the phyla concept was not supported at all (Tekaia *et al.*, 1999; Lin and Gerstein, 2000).

### Gene order

Besides determining the fraction of shared genes alternative approaches based on relative order or neighbourhood of genes on the genome are in use (Blanchette *et al.*, 1999; Wolf *et al.*, 2001; Korbel *et al.*, 2002). Similarly, context networks were established and used for phylogenetic genome comparison (Ding *et al.*, 2008). In this phylogenic profiling method, pairs of proteins with similar patterns of presence and absence across genomes are identified. In a gene neighbours and fusion method, pairs of genes that fused or clustered together during evolution are detected. According to Ding *et al.* (2008) the phyla concept was basically supported.

### Rare genomic changes

In the rare genomic changes approach, genomes are studied and clustered according to shared complex characters that have very low probability of being the result of convergence (Delsuc *et al.*, 2005). Insertions and deletions (indels), intron positions, retroposon integrations, and gene fusion and fission events are used as characteristics to infer phylogenetic relationships (Philippe and Laurent, 1998; Rokas and Holland, 2000; Gupta and Griffith, 2006; Lake *et al.*, 2009; Valas and Bourne, 2009). The orthology aspect was not taken into account in earlier studies that simply counted presence or absence of indels. Later, the primary and even higher order structure environments were included to ensure comparison of orthologous characteristics. Some support of the phyla concept was shown in these studies (Valas and Bourne, 2009).

### Average character identity

Among the methods relying on genome derived parameters average nucleotide (ANI) or amino acid (AAI) identity of all conserved genes or proteins between two genomes (Konstantinidis and Tiedje, 2005, 2007; Goris *et al.*, 2007) represent tools bridging between gene content and direct sequence analysis approaches. These identity values allow clustering of genomes. The methodology is intended being appropriate to replace traditional genomic DNA–DNA hybridization (Konstantinidis and Tiedje, 2007; Richter and Rossello-Mora, 2009).

### Multigene-based phylogenetic analyses

Various selections of genes or proteins taken from full genome data according to varying selection criteria (e.g. orthologues or core, character, or accessory genes) were used to reconstruct trees of life. Either concatenated sequence data provided the base for treeing or individual marker based trees were combined to super trees applying respective bioinformatic tools (Brown *et al.*, 2001; Wolf *et al.*, 2001; Brochier *et al.*, 2002; Daubin *et al.*, 2002; Ciccarelli *et al.*, 2006; Gophna *et al.*, 2005; Bapteste *et al.*, 2008). Depending on data selection and parameters applied the phyla concept was more or less supported in all cited publications.

## Conclusions

A selection of conserved proteins was used as alternative phylogenetic markers to evaluate the current view of rRNA-based relationships. The comparative analyses were generally hampered by the imbalance of the available sequence data sets. Despite the reasonable number of full genome data, none of them is nearly comparable to the comprehensive small-subunit rRNA sequence database. Comprehensiveness not only with respect to numbers of different sequences but also to representation of diverse levels of relationship is needed to recognize variable and conserved sequence positions as such, and to estimate the significance of separation or unification of the molecules or organisms in the respective trees. The tree comparisons shown above as examples were confined to the verification of the rRNA defined bacterial phyla. For balance of the data sets of non-rRNA markers, only those organisms are represented in the trees for which full genome data are available. Consequently, many supports as well as paralogy problems from incomplete marker data sets are not shown.

The rRNA based bacterial phylum concept is at least partly supported by all alternative markers. However, in most cases subclasses of *Proteobacteria* do not appear as monophyletic phyla. Whereas the *Beta-* and *Gammaproteobacteria* are almost always unified in a common group that in many cases includes the *Alphaproteobacteria*, the *Delta-* and especially the *Epsilonproteobacteria* more often represent their own lineages. A tendency of splitting off the *Epsilonproteobacteria* from the other subgroups is also seen with 23S rRNA sequence databases. The potential separate phylum status of some taxa traditionally assigned to the *Firmicutes* phylum as indicated by rRNA data (Ludwig *et al.*, 2009) is also supported by part of the protein markers.

It is well known that phylogenetic trees are not static but dynamic structures that may change their local topologies when new data become available. Thus, many of the inconsistencies and discrepancies which currently are supported by a few sequences only may change or probably be resolved in the near future.

Another problem apparently often underestimated in the past are multiple evolutionary diverged versions of homologous markers. Many examples of duplications with different degrees of sequence divergence (phylum to genus level) were found in the databases of phylogenetic markers included in the present study. Examples of multiple genes or proteins are so far known for all alternative markers except for the RNA polymerases. Even in the case of rRNA a few cases of diverged multiple copies have been reported (Miller *et al.*, 2005). It is generally difficult to decide whether these paralogous genes resulted from duplications or from horizontal gene transfer. The resulting problems hampering sound phylogenetic analysis are diverse, e.g. to recognize paralogous markers as such, to define the orthologous versions, to recognize the copies sharing the same function and selection pressure, to differentiate clonal duplication and lateral gene transfer. Although new methods for detecting lateral gene transfer have more recently been developed (Gogarten and Townsend, 2005; Poptsova and Gogarten, 2007), this differentiation remains a difficult task. Given that 'ancient' events of duplication or potential gene transfer have to be recognized as such, it is highly likely that potential peculiarities have been 'normalized' during the course of time. Furthermore, the assumption that a relaxed selection pressure acting upon non-essential versions should be expressed by a higher rate of change in comparison with that of the essential component is not necessarily correct. The function of the respective product may still require sequence conservation, while the loss of the gene may happen at a higher frequency. Anyhow, currently it is in many cases impossible to unambiguously decide whether the discrepancies observed are substantial or rather result from 'illegitimate' comparison of paralogous markers for reconstructing 'vertical' phylogeny. In this context it has to be considered that multiple rRNA genes are common among bacteria, however, a survey of multiple small and large subunit rRNA sequences from representatives of the different phyla showed that the range of divergence usually does not exceed 1.5–2% (unpublished). This is within the normal 'noise' of rRNA based phylogenetic trees (Ludwig and Klenk, 2001). A few exceptions are known so far such as *Thermobispora bispora* and *Haloarcula marismortui* for which, respectively, 6.4% or 5% sequence divergence of small subunit rRNA genes have been reported (Wang *et al.*, 1997; Mylvaganam and Dennis, 1992). Also cases of hybrid small subunit rRNA have been postulated (Parker, 2001; Miller *et al.*, 2005). Up to now, these are rare exceptions.

The difficulties and shortcomings of phylogenetic markers and analyses concerning potential paralogy, limited information content, noise in the data, models of evolution, treeing methods, compositional bias, general and long-branch attraction, heterotachy as well as significance of results and conclusions which are evident at the single marker level, persist even if the analyses are based on a broader fundament by multiple marker and full-genome comparisons. Moreover, the paralogy problem might be raised to a higher power applying full-genome approaches. Addressing the mandatory requirements of orthology and functional constancy in genome-based phylogenetic analyses is often difficult, inadequate, or even impossible. Given that many gene duplications and events of lateral gene transfer have been shown or postulated from genome comparisons, defining sets of orthologues according to sequence identity not necessarily unifies markers of identical function. Even now, in the age of high throughput methods, routine analysis of expression and function of all potential markers of a genome might not be realistic.

Thus, the phylogenetic relevance of genome approaches heavily depends on the accuracy of (orthologous) target selection. Character string-based methods provide differentiating genome characteristics but do not care about orthology. The relevance of approaches analysing gene content and order as well as average character identity (ANI, AAI) correlates with marker selection under the criteria of orthology and conserved function. Even if based on an appropriate marker selection, only part of the available information is used, given that only presence/absence or average identity is measured. The rare genomic changes methods such as determining the presence or absence of indels are only justified if the respective target molecules share direct common ancestry (orthology). However, their phylogenetic significance is often overestimated. Performing comparative sequence analysis the significance of the information provided by an indel is that of a meaningful single base or amino acid change. The concatenation or super-treeing procedures, in combination with careful sequence alignment according to higher order structure, positional variability analysis and filtering as well as significance tests, certainly extract more of the information preserved by the individual markers. However, one has to be aware that orthologous data sets are combined or compared which in many cases are not homologous at all. Thus compositional bias, general and long-branch attraction effects, heterotachy as well as hiding information by smoothing may affect the quality and significance of phylogenetic conclusions.

Certainly, genome evolution is only partially clonal but driven largely by gene acquisition or loss. There still seems to be agreement that character, accessory or life style genes are preferably subject of the latter evolutionary processes, whereas informational and other core genes are less frequently concerned. Thus, probably all genomes might be chimeric with respect to foreign DNA received and established during the course of evolution. Therefore, major discrepancies have to be expected for potentially clonal and acquired gene phylogenies. However, in terms of the organism's phylogeny, does this mean that it is a progeny of the xenologous DNAs source organism? Nevertheless, stably established foreign DNA may provide useful – now clonal – markers for phylogeny reconstruction above this level of relationship. Besides or along with horizontal gene transfer recombination may have obscured phylogenetic signals by disrupting vertical heredity. For closely related organisms it was postulated that genomic similarity results from frequent exchange of genes rather than from common ancestry (Feil *et al.*, 2001). With respect to transfer and recombination ironically the highly conserved genes commonly used for phylogenetic investigations should be preferred targets. At least in microbial communities of organisms living in close proximity, the primary structures of these genes should be rapidly equalized. Obviously, that is not the case as indicated by microbial diversity commonly found applying rRNA sequence or probe based techniques in environmental studies. Clonality could still be abandoned if these genes are frequently transferred replacing the residing homologue without effective recombination. If this would frequently occur, more incongruities would have to be expected comparing the marker genes. Even if recombination events concern short fragments of conserved genes (Ueda *et al.*, 1999; Wang and Thang, 2000; Parker, 2001), this shouldn't strongly affect the overall picture as shown for the bacterial phyla and major subgroups. As discussed above, a significant detailed resolution cannot be expected anyway. Consequently, finding sets of orthologous markers for the respective levels of relationship certainly is among the most important steps of genome based phylogeny inference. Although, sophisticated bioinformatics tools are available for evaluating the plausibility of genome markers, models, trees, and networks, the directed approach searching universal core gene markers might still be appropriate. Anyhow, the set of core genes or proteins which seem to fulfil the criteria for global phylogenetic markers carrying information for the domain, phyla and lower levels is rather small.

Against this background, the ongoing controversial debate concerning the existence of any clonality, trees, rings, networks and root of life (Gogarten *et al.*, 2002; Bapteste *et al.*, 2005) loses some of its explosive nature. For almost any perception of genome phylogeny there are supporting regions in the genome. Nevertheless, there exists a small set of conserved markers preserving a genealogical trace – certainly obscured by a burden of noise and missing information – that goes back in time to the universal ancestor state. The rRNAs are certainly among them. Despite all the progress in high throughput sequencing full genome analyses will unlikely become the routine standard in prokaryotic taxonomy and identification in the next future. At the current stage of knowledge and database situation the rRNAs still seem to be among the most informative molecules for phylogenetic analyses, and in combination with the big data set it is certainly justified to base the bacterial taxonomy at the higher ranks upon the results of comparative rRNA sequence analysis.

## Acknowledgements

The author dedicates this chapter to Professor Karl Heinz Schleifer in gratitude.

## References

Amann, R., Ludwig, W., and Schleifer, K.H. (1988). β-subunit of ATP-synthase: a useful marker for studying the phylogenetic relationship of bacteria. J. Gen. Microbiol. *134*, 2815–2821.

Amann, R., Ludwig, W., and Schleifer, K.H. (1995). Phylogenetic identification and *in situ* detection of individual microbial cells without cultivation. Microbiol. Rev. *59*, 143–169.

Bapteste, E., Susko, E., Leigh, J., MacLeod, D., Charlebois, R.L., and Doolittle, W.F. (2005). Do orthologous gene phylogenies really support tree-thinking? BMC Evol. Biol. *5*, 33–43.

Bapteste, E., Susko, E., Leigh, J., Ruiz-Trillo, I., Bucknam, J., and Doolittle, W.F. (2008). Alternative methods for concatenation of core genes indicate a lack of resolution in deep nodes of the prokaryotic phylogeny. Mol. Biol. Evol. *25*, 83–91.

Blanchette, M., Kunisawa, T., and Sankoff, D. (1999). Gene order breakpoint evidence in animal mitochondrial phylogeny. J. Mol. Evol. *49*, 193–203.

Brochier, C., Bapteste, E., Moreira, D., and Philippe, H. (2002). Eubacterial phylogeny based on translational apparatus proteins. Trends Genet. *18*, 1–5.

Brown, J.R., Douady, C.J., Italia, M.J., Marshall, W.E., and Stanhope, M.J. (2001). Universal trees based on large combined protein sequence data sets. Nature Genet. *28*, 281–285.

Campbell, A., Mrázek, J., and Karlin, S. (1999). Genome signature comparisons among prokaryote, plasmid, and mitochondrial DNA. Proc. Natl. Acad. Sci. U.S.A. *96*, 9184–9189.

Ciccarelli, F.D., Doerks, T., von Mering, C., Creevey, C.J., Snel, B., and Bork, P. (2006). Toward automatic reconstruction of a highly resolved tree of life. Science *311*, 1283–1286.

Cole, J.R., Wang, Q., Cardenas, E., Fish, E., Chai, B., Farris, R.J., Kulam-Syed-Mohideen, A.S., McGarrell, D.M., Marsh T., Garrity, G.M., and Tiedje. J.M. (2008). The Ribosomal Database Project: improved alignments and new tools for rRNA analysis. Nucl. Acids Res. *37*, D141-D145.

Daubin, V., Gouy, M., and Perrière, G. (2002). A phylogenomic approach to bacterial phylogeny: evidence of a core of genes sharing a common history. Genome Res. *12*, 1080–1090.

Delsuc, F., Brinkmann, H., and Philippe, H. (2005). Phylogenomics and the reconstruction of the tree of life. Nature Rev. Gen. *6*, 361–375.

DeSantis, T.Z., Hugenholtz, P., Larsen, N., Rojas, M., Brodie, E.L., Keller, K., Huber, T., Dalevi, D., Hu, P., and Anderse, G.L. (2006). Greengenes, a chimera-checked 16S rRNA gene database and workbench compatible with ARB. Appl. Environ. Microbiol. *72*, 5069–5072.

Ding, G., Yu, Z., Zhao, J., Wang, Z., Li, Y., Xing, X., Wang, C., Liu, L., and Li, Y. (2008). Tree of Life based on genome context networks. PLoS ONE *3(10)*: e3357.

Edwards, S.V., Fertil, B., Giron, A., and Deschavanne, P.J. (2002). A genomic schism in birds revealed by phylogenetic analysis of DNA strings. Syst. Biol. *51*, 599–613.

Feil, E.J., Holmes, E.C., Bessen, D.E., Chan, M-S., Day, N.P.J., Enright, N.C., Goldstein, R., Hood, D.E., Kalia, A., Moore, C.E., Zhou, J., and Spratt, B.G. (2001). Recombination within natural populations of pathogenic bacteria: Short-term empirical estimates and long-term phylogenetic consequences. Proc. Natl. Acad. Sci. U.S.A. *98*, 182–187.

Fitz-Gibbon, S.T., and House, C.H. (1999). Whole genome-based phylogenetic analysis of free-living microorganisms. Nucl. Acids Res. *27*, 4218–4222.

Fox, G.E., Pechman, K.R., and Woese, C.R. (1977). Comparative cataloging of 16S ribosomal ribonucleic acid: molecular approach to procaryotic systematics. Int. J. Syst. Bacteriol. *27*, 44–57.

Garrity, G.M. (ed.) (2001). *Bergey's Manual* of Systematic Bacteriology, 2nd ed. Vol. 1. (New York: Springer).

Glöckner, F.O., Kube, M., Bauer, M., Teeling, H., Lombardot, T., Ludwig, W., Gade, D., Beck, A., Borzym, K., Heitmann, K., Rabus, R., Schlesner, H., Amann, R., and Reinhardt, R. (2003). Complete genome sequence of the marine planctomycete *Pirellula* sp. strain 1. Proc. Natl. Acad. Sci. U.S.A. *100*, 8298–8303.

Gogarten, J.P., and Townsend, J.P. (2005). Horizontal gene transfer, genome innovation and evolution. Nature Rev. Microbiol. *3*, 679–687.

Gogarten, J.P., Doolittle, W.F., and Lawrence, J.G. (2002). Prokaryotic evolution in light of gene transfer" molecular biology and evolution. Mol. Biol. Evol. *19*, 2226–2238.

Gophna, U., Doolittle, W.F., and Charlebois, R.L. (2005). Weighted genome trees: refinements and applications. J. Bacteriol. *187*, 1305–1316.

Goris, J., Konstantinidis, K.T., Klappenbach, J.A., Coenye, T., Vandamme, P., and Tiedje, J.M. (2007). DNA–DNA hybridization values and their relation to whole genome sequence similarities. Int. J. Syst. Evol. Microbiol. *57*, 81–91.

Gupta, R.S., and Griffiths, E. (2002). Critical issues in bacterial phylogeny. Theor. Popul. Biol. *61*, 423–434.

Gupta, R.S., and Griffiths, E. (2006). Chlamydiae-specific proteins and indels: novel tools for studies. Trends Microbiol. *14*: 527–535.

Heymans, M., and Singh, A.K. (2003). Deriving phylogenetic trees from the similarity analysis of metabolic pathways. Bioinformatics *19*, i138-i146.

Hilario, E., and Gogarten, J.P. (1998). The prokaryote-to-eukaryote transition reflected in the evolution of the V/F/A-ATPase catalytic and proteolipid subunits. J. Mol. Evol. *46*, 703–715.

Huynen, M.A., and Bork, P. (1998). Measuring genome evolution. Proc. Natl. Acad. Sci. U.S.A. *95*, 5849–5856.

Kakinuma, Y., Yamato, I., and Murata, T. (1999). Structure and function of vacuolar $Na^+$-translocating ATPase in *Enterococcus hirae*. J. Bioenerg. Biomembr. *31*, 7–14.

Klenk, H.P., Palm, P., and Zillig, W. (1994). DNA-dependent RNA polymerases as phylogenetic marker molecules. Syst. Appl. Microbiol. *16*, 638–647.

Konstantinidis, K.T., and Tiedje, J.M. (2005). Genomic insights into the species definition for prokaryotes. Proc. Natl. Acad. Sci. U.S.A. *102*, 2567–2572.

Konstantinidis, K.T., and Tiedje, J.M. (2007). Prokaryotic taxonomy and phylogeny in the genomic era: advancements and challenges ahead. Curr. Opin. Microbiol. *10*, 504–509.

Korbel, J.O., Snel, B., Huynen, M.A., and Bork, P. (2002). SHOT: a web server for the construction of genome phylogenies. Trends Genet. *18*, 158–162.

Lake, J.A., Skophammer, R.G., Herbold, C.W., and Servin, J.A. (2009). Genome beginnings: rooting the tree of life. Phil. Trans. R. Soc. Lond. B. *364*, 2177–2185.

Lin, J., and Gerstein, M. (2000). Whole-genome trees based on the occurrence of folds and orthologs: implications for comparing genomes on different levels. Genome Res. *10*, 808–818.

Ludwig, W., and Klenk, H.P. (2001). Overview: a phylogenetic backbone and taxonomic framework for procaryotic systematics. In *Bergey's Manual* of Systematic Bacteriology, 2nd edition, Vol. 1, G.M. Garrity, ed. (New York: Springer), pp. 49–65.

Ludwig, W., and Schleifer, K.H. (1999). Phylogeny of *Bacteria* beyond the 16S rRNA standard. ASM News *65*, 752–757.

Ludwig, W., and Schleifer, K.H. (2005). Molecular phylogeny of bacteria based on comparative sequence analysis of conserved genes. In Microbial Phylogeny and Evolution, Concepts and Controversies. Sapp, J. ed. (Oxford University Press), pp. 70–98.

Ludwig, W., Neumaier, J., Klugbauer, N., Brockmann, E., Roller, C., Jilg, S., Reetz, K., Schachtner, I., Ludvigsen, A., Bachleitner, M., Fischer, U., and Schleifer, K.H. (1993). Phylogenetic relationships of bacteria based on comparative sequence analysis of elongation factor Tu and ATP-synthase β-subunit genes. Antonie van Leeuwenhoek *64*, 285–305.

Ludwig, W., Strunk, O., Klugbauer, S., Klugbauer, N., Weizenegger, M., Neumaier, J., Bachleitner, M., and Schleifer, K.H. (1998). Bacterial phylogeny based on comparative sequence analysis. Electrophoresis *19*, 554–568.

Ludwig, W., Strunk, O., Westram, R., Richter, L., Meier, H., Yadhukumar, Buchner, A., Lai, T., Steppi, S., Jobb, G., Förster, W., Brettske, I., Gerber, S., Ginhart, A.W., Gross, O., Grumann, S., Hermann, S., Jost, R., König, A., Liss, T., Lüßmann, R., May, M., Nonhoff, B., Reichel, B., Strehlow, R., Stamatakis, A., Stuckmann, N., Vilbig, A., Lenke, M., Ludwig, T., Bode, A., and Schleifer, K.H. (2004). ARB: a software environment for sequence data. Nucl. Acids Res. *32*, 1363–1371.

Ludwig, W., Schleifer, K.H., and Whitman, W.B. (2009). Revised road map to the phylum Firmicutes. In *Bergey's Manual* of Systematic Bacteriology, 2nd edition, Vol. 3, Whitman, W.B., ed. (New York, Springer), pp. 1–13.

Maiden, M.C., Bygraves, J.A., Feil, E., Morelli, G., Russell, J.E., Urwin, R., Zhang, Q., Zhou, J., Zurth, K., Caugant, D.A., Feavers, I.M., Achtman, M., and Spratt, B.G. (1998). Multilocus sequence typing: a portable approach to the identification of clones within populations of pathogenic microorganisms. Proc. Natl. Acad. Sci. U.S.A. *95*, 3140–3145.

Miller, S.R., Augustine, S., Olson, T.L., Blankenship, R.E., Selker, J., and Wood, A.M. (2005). Discovery of a free-living chlorophyll *d*-producing cyanobacterium with a hybrid proteobacterial/cyanobacterial small-subunit rRNAgene. Proc. Natl. Acad. Sci. U.S.A. *102*, 850–855.

Mylvaganam, S., and Dennis, P.P. (1992). Sequence heterogeneity between the two genes encoding 16S rRNA from the halophilic archaebacterium *Haloarcula marismortui*. Genetics *130*, 399–410.

Neumaier, J. (1996). Gene der katalytischen Untereinheit der V-Typ- und der $F_1F_0$-ATPase bei Bakterien: Vorkommen und vergleichende Sequenzanalyse. Thesis, Technische Universität München, Germany.

Parker, M.W. (2001). Case of localized recombination in 23S rRNA genes from divergent *Bradyrhizobium* lineages associated with neotropical legumes. Appl. Environ. Microbiol. *67*, 2076–2082.

Peplies, J., Kottmann, R., Ludwig, W., and Glöckner, F.O. (2008). A standard operating procedure for phylogenetic inference (SOPPI) using (rRNA) marker genes. Syst. Appl. Microbiol. *31*, 251–257.

Philippe, H., and Laurent, J. (1998). How good are deep phylogenetic trees? Curr. Opin. Genet. Dev. *8*, 616–623.

Poptsova, M.S., and Gogarten, J.P. (2007). The power of phylogenetic approaches to detect horizontally transferred genes. BMC Evol. Biol. *7*, 45.

Pride, D.T., Meinersmann, R.J., Wassenaar, T.M., and Blaser, M.J. (2003). Evolutionary implications of microbial genome tetranucleotide frequency biases. Genome Res. *13*, 145–158.

Pruesse, E., Quast, C., Knittel, K., Fuchs, B., Ludwig, W., Peplies, J., and Glöckner, F.O. (2007). SILVA: a comprehensive online resource for quality checked and aligned ribosomal RNA sequence data compatible with ARB. Nucl. Acids Res. *35*, 7188–7196.

Qi, J., Wang, B., and Hao, B.I. (2004). Whole proteome prokaryote phylogeny without sequence alignment: a K-string composition approach. J. Mol. Evol. *58*, 1–11.

Richter, M. and Rosselló-Mora, R (2009) Shifting the genomic gold standard for the prokaryotic species definition. Proc. Natl. Akad. Sci. USA. *106*, 19126–19131.

Rokas, A. (2006) Genomics and the Tree of Life. Science. *313*, 1897–1898.

Rokas, A., and Holland, P.W. (2000). Rare genomic changes as a tool for phylogenetics. Trends Ecol. Evol. *15*, 454–459.

Simonson, A.B., Servin, J.A., Skophammer, R.G., Herbold, C.A., Rivera, M.C., and Lake, J.A. (2005). Decoding the genomic tree of life. Proc. Natl. Acad. Sci. U.S.A. *102*, 6608–6613.

Sneath, P.H.A., and Brenner, D.J. (1992) "Official" nomenclature lists. ASM News *58*, 175.

Snel, B., Bork, P., and Huynen, M.A. (1999). Genome phylogeny based on gene content. Nature Genet. *21*, 108–110.

Sterk P., Kersey P.J., and Apweiler R. (2006). Genome reviews: standardizing content and representation of information about complete genomes. OMICS *10*, 114–118.

Sumi, M., Yohda, M., Koga, Y., and Yoshida, M. (1997). $F_0F_1$-ATPase genes from an archaebacterium, *Methanosarcina barkeri*. Biochem. Biophys. Res. Commun. *241*, 427–433.

Stamatakis, A. (2006) RAxML-VI-HPC: maximum likelihood-based phylogenetic analyses with thousands of of taxa and mixed models. Bioinformatics *22*, 2688–2690.

Tekaia, F., Lazcano, A., and Dujon, B. (1999). The genomic tree as revealed from whole proteome comparisons. Genome Res. *9*, 550–557.

Ueda, K., Seki, T., Kudo, T., Yshida, T., and Katoka, M. (1999). Two distinct mechanisms cause heterogeneity of 16S rRNA. J. Bacteriol. *181*, 78–82.

Valas, R.E., and Bourne, P.E. (2009). Structural analysis of polarizing indels: an emerging consensus on the root of the tree of life. Biology Direct *4*, 30–46.

Van Wezel, G.P., Takano, E., Vijgenboom, E., Bosch, L., and Bibb, M.J. (1995). The tuf3 Gene of *Streptomyces coelicolor* A3(2) encodes an inessential elongation factor Tu that is apparently subject to positive stringent control. Microbiology *141*, 2519–2528.

Wang, Y.Z., and Thang, Z. (2000). Comparative sequence analyses reveal frequent occurrence of short segments containing an abnormally high number of non-random base variations in bacterial rRNA genes. Microbiology *146*, 2845–2854.

Wang, Y.Z., Thang, Z., and Ramanan, N. (1997). The actinomycete *Thermobispora bispora* contains two distinct types of transcriptionally active 16S rRNA genes. J. Bacteriol. *179*, 3270–3276.

Whitman, W.B. (ed.) (2009). *Bergey's Manual* of Systematic Bacteriology, 2nd edition (New York: Springer).

Wolf, Y.I., Rogozin, I.B., Grishin, N.V., Tatusov, R.L., and Koonin, E.V. (2001) Genome trees constructed using five different approaches suggest new major bacterial clades. BMC Evol. Biol. *1*, 8.

Yarza, P., Richter, M., Peplies, J., Euzéby, J., Amann, R., Schleifer, K.H., Ludwig, W., Glöckner, F.O., and Rosselló-Mora, R. (2008). The All-Species Living Tree project: A 16S rRNA-based phylogenetic tree of all sequenced type strains. Syst. Appl. Microbiol. *31*, 241–250.

Zhang, Y., Li, S., Skogerbo, G., Zhang, Z., Zhu, X., Zhang, Z., Sun, S., Lu, H., Shi, B., and Chen, R. (2006). Phylophenetic properties of metabolic pathway topologies as revealed by global analysis. BMC Bioinformatics 7, 252.

Zhaxybayeva, O., Lapierre, P., and Gogarten, J.P. (2005). Ancient gene duplications and the root(s) of the tree of life. Protoplasma *227*, 53–64.

# The Phyla of Prokaryotes – Cultured and Uncultured

5

Aharon Oren

Abstract

There is no official classification of prokaryotes. For the higher taxa there even is no official nomenclature: the rules of the International Code of Nomenclature of Prokaryotes do not cover taxa above the rank of class. The most commonly accepted division of the prokaryotes in two 'subkingdoms' or 'domains' (Bacteria and Archaea) and the classification of their species with validly published names in respectively 27 and 2 'phyla' or 'divisions' (as of November 2009) is primarily based on 16S rRNA sequence comparisons. This type of classification was adopted in the latest edition of *Bergey's Manual of Systematic Bacteriology.* Alternative classifications have been proposed as well, based e.g. on the structure of the cell wall. Some 16S rRNA sequence-based phyla unite prokaryotes of similar physiological properties (for example Cyanobacteria, Chlorobi, Thermotogae); others (Euryarchaeota, Proteobacteria, Flavobacteria) contain organisms with highly disparate lifestyles. Some phyla based on deep 16S rRNA lineages are currently represented by one or a few species only. Environmental genomics/metagenomics approaches suggest existence of many more phyla based on the deep lineages of 16S rRNA gene sequences recovered. To obtain the organisms harbouring these sequences and to study their properties is a major challenge of microbiology today.

> *I do not deny but nature, in the constant production of particular beings, makes them not always new and various, but very much alike and of kin one to another: but I think it nevertheless true, that the boundaries of the species, whereby men sort them, are made by men; since the essences of the species, distinguished by different names, are, as has been proved, of man's making, and seldom adequate to the internal nature of the things they are taken from. So that we may truly say, such a manner of sorting of things is the workmanship of men.*
>
> (John Locke, *An Essay Concerning Humane Nature*, 1690)

## Introduction

At the time this book was written, over 8200 different species of prokaryotes (Bacteria and Archaea combined) had been described and named. Table 5.1, based on the 'List of Prokaryotic Names with Standing in Nomenclature' (Euzéby, 1997; http://www.bacterio.cict.fr), summarizes the numbers of names of species, genera, families, orders, and classes with standing in the prokaryote nomenclature.

The question how all these prokaryote species, with their tremendous morphological and physiological diversity (Madigan *et al.*, 2009; Zinder and Dworkin, 2006), should be classified in higher taxa has occupied microbiologists since the days of Christian Ehrenberg and Ferdianand Cohn (see Chapter 1). As clearly pointed out by the philosopher John Locke in the quotation given above, the 'manner of sorting of things is the workmanship of men', and not something inherent to the nature of the organisms. There is an official nomenclature of prokaryotes, regulated by internationally agreed-upon rules fixed in the

**Table 5.1** The number of different species of prokaryotes described (Bacteria and Archaea combined) with names with standing in prokaryote nomenclature, and the number of higher taxa in which these are classified, as of 3 November 2009. Derived from www.bacterio.cict.fr

| | |
|---|---|
| Rank of kingdom[1] | 1 |
| Rank of subkingdom[1] | 2 |
| Rank of division or phylum[1,2] | 26 |
| Rank of class | 70 |
| Rank of order | 116[3] |
| Rank of family | 253[4] |
| Rank of genus | 1,732[5] |
| Rank of species | 8,226[6] |

[1]Categories not covered by the Rules of the Bacteriological Code (Lapage *et al.*, 1992)

[2]Phyla proposed in the Approved Lists of Bacterial Names (Skerman *et al.*, 1980) and in the International Journal of Systematic Bacteriology/International Journal of Systematic and Evolutionary Microbiology.

[3]117, of which one is illegitimate.

[4]259, of which six are illegitimate.

[5]1852, of which about 99 are considered as synonyms, and 20 are illegitimate.

[6]9787, of which 31 are later homotypic synonyms cited in the Approved Lists of Bacterial Names, 1187 are new combinations, 11 are *nomina nova*, about 265 are considered as later heterotypic synonyms, and 67 are illegitimate.

International Code of Nomenclature of Bacteria (The Bacteriological Code) (Lapage *et al.*, 1992), now renamed as the International Code of Nomenclature of Prokaryotes, but there is no official classification of prokaryotes. Concepts about how to sort the organisms in groups are constantly changing according to the development of new ideas and more advanced scientific methods. This was clearly stated by Vandamme *et al.* (1996): 'There will never be a definitive classification of bacteria'.

Still, classification of the ever increasing number of known prokaryotes is essential for identification purposes and other applications. Thus, Table 5.1 shows that (as of November 2009) the species were classified within over 1700 genera, over 250 families, more than 110 orders, and 70 classes. The ranks of species (*species*), genus (*genus*), family (*familia*), order (*ordo*) and class (*classis*) are covered by the nomenclature rules of the Bacteriological Code, as well as the ranks of subgenus, subtribe, tribe, subfamily, suborder and subclass.

Names of higher taxa above the rank of class are widely used in the classification of prokaryotes. Table 5.1 thus lists 26 divisions or phyla and two subkingdoms (Bacteria and Archaea). However, their names are not covered by the rules of the Bacteriological Code, and they therefore have no official standing in the nomenclature. The International Code of Botanical Nomenclature knows the rank of 'Phylum' or 'Division' (*Divisio*). No such category appears in the Bacteriological Code, but the rank of 'Kingdom' is given in an example to Rule 8 ('Kingdom – *Procaryotae*') (Lapage *et al.*, 1992). Although the issue of the nomenclature of higher taxa above that of Class has been raised from time to time in the meetings of the International Committee on Systematics of Bacteria/International Committee on Systematics of Prokaryotes, it is not to be expected that the nomenclature of these higher taxa will be included in the provisions of the Code in the near future. Ranks such as 'domain', 'kingdom', 'phylum', etc. have therefore no official status.

This chapter aims at providing an overview of the higher taxa of prokaryotes, not only as proposed by *Bergey's Manual*, but in alternative classification schemes as well. Rather than attempting a balanced approach in which the different groups are discussed according to their importance in nature or according to the number of species, genera, etc. currently assigned to each of these higher taxa (the pages of *Bergey's Manual* are an excellent source of information here, as are the chapters in *The Prokaryotes* (Dworkin *et al.*, 2006)), special emphasis will be on recently discovered or proposed higher taxa and groups that have aroused special interest in recent years for various reasons.

## The definition of phyla and other higher taxa of prokaryotes

Even today, in spite of our advanced understanding of the diversity, the physiology, and also the genomics of prokaryotic microorganisms, there still is no scientifically based species concept for the prokaryotes. The operational definition that is widely accepted describes a prokaryote species as 'a monophyletic and genomically coherent cluster of individual organisms that show a high degree of overall similarity in many independent characteristics, and is diagnosable by a discriminative phenotypic property' (Rosselló-Mora and Amann, 2001). This is the so-called phylophenetic species concept. The delineation of species is ideally based on a 'polyphasic' approach, in which as many characteristics, phenotypical as well as genotypical, are determined (Vandamme *et al.*, 1996). However, there is no consensus on how many distinguishing features are sufficient to create a species.

Probably the most widely accepted pragmatic definition of a prokaryotic species defines a species as a group of strains, including the type strain, that share at least 70% total genome DNA–DNA hybridization and having less than 5°C $\Delta T_m$ (= the difference in the melting temperature between the homologous and the heterologous hybrids formed under standard conditions). Thus, total genome DNA–DNA hybridization values are the key parameters in this species delineation, and DNA reassociation values are an indirect expression of the real genome sequence identity. Values from 30–70% DNA relatedness reflect a moderate degree of relationship. This operational definition was proposed more than 20 years ago based on the conclusions of an *ad hoc* committee (Wayne *et al.*, 1987), and was ratified based on renewed discussions on the subject (Stackebrandt *et al.*, 2002; see also Oren, 2004a). It should be realized that the delineation value of 70% is artificial, but it has proven satisfactory in most cases.

No ways of delineating higher taxa were proposed by the different committees that have tried to deal with the issue how to delineate prokaryotic species. One can define a genus as a collection of species with many characters in common, a family as a collection of genera that share many characters, a class as a collection of orders with many common properties, etc., but how many and which characters should be in common is largely a matter of personal judgment, and clear guidelines are lacking. Therefore it not surprising that in the course of time the schemes for the classification of the higher taxa have been subject to changes, based on the advancement of scientific insights and often on the personal preferences of the authors as well. Few attempts have been made to better define the higher ranks. The introductory assay on 'Defining taxonomic ranks' written by Erko Stackebrandt for the latest edition of the comprehensive handbook *The Prokaryotes* (Stackebrandt, 2006) mostly deals with delineation of species, hardly mentions the higher taxa, and does not provide any clear guidelines on the latter.

When Murray (1984) wrote his chapter entitled 'The higher taxa, or, A place for everything ...?' in the first edition of *Bergey's Manual of Systematic Bacteriology*, he rightfully considered that 'At the higher taxonomic levels some sort of serviceable scheme that can recognize and accommodate to a degree the possibilities will be of service ...'. The scheme he proposed divided the Kingdom Procaryotae into four divisions:

Division I: Gracilicutes (basically Gram-negatives, with anoxygenic and oxygenic phototrophs)
Division II: Firmicutes
Division III: Tenericutes (Mollicutes)
Division IV: Mendosicutes (a group that contains the Archaea)

This classification scheme was based on the earlier division of the prokaryotes into Gracilicutes, Firmacutes, Mollicutes, and Mendocutes, as based primarily on the properties and structure of the cell wall (Gibbons and Murray, 1978). It replaced the division of the Kingdom Procaryotae into Division I, 'Photobacteria' (phototrophic prokaryotes), and Division II, 'Scotobacteria' (prokaryotes indifferent to light), as proposed in the 1974 edition of *Bergey's Manual of Determinative Bacteriology*, the predecessor of *Bergey's Manual of Systematic Bacteriology* (Murray, 1974).

A survey of the taxonomic literature of the past decade shows that the rank of phylum is most widely used to designate the major groups of the Bacteria and the Archaea. Its popularity can in part be attributed to its use in the 2nd edition of *Bergey's Manual of Systematic Bacteriology*, the first volume of which was published in 2001. The new edition of the manual uses a phylogenetic approach for its classification, primarily based on 16S rRNA sequence comparisons (Garrity and Holt, 2001a; Ludwig and Klenk, 2001; Krieg and Garrity, 2001). A number of new phyla were described in the chapters of *Bergey's Manual*, and names for other new phyla were proposed in the taxonomic literature based on the concepts outlined by the manual, such as the Thaumarchaeota (Brochier-Armanet *et al.*, 2008), the Nanoarchaeota (Huber *et al.*, 2002), the Synergistetes (Jumas-Bilak *et al.*, 2009), the Caldiserica (Mori *et al.*, 2009), the Gemmatimonadetes (Zhang *et al.*, 2003), and others (Cavalier-Smith, 2002). Earlier the term 'Division' was sometimes used, such as in the description of the Verrucomicrobia (Hedlund *et al.*, 1997). Table 5.2 presents the major divisions of the prokaryote world, based on the domains, phyla, and classes as given in the classification proposed in the second edition of *Bergey's Manual* (Garrity and Holt, 2001a) and including later additions. Only those groups are listed that contain at least one species whose name was validly published according to the rules of the Bacteriological Code. The number of phyla listed in Table 5.2 is 28 and not 26 as given in Table 5.1 which is based on Euzéby's 'List of Bacterial Names with Standing in Nomenclature'. The two entries not included in the list in Table 5.1 are 'Deinococcus–Thermus' and Crenarchaeota; the latter name has standing in the nomenclature only as the name of a class, as an earlier homotypic synonym of the name Thermoprotei.

## Proposals of higher taxa of prokaryotes based on criteria other than 16S rRNA sequence information

The phylogenetic classification of prokaryotes as based on 16S rRNA sequence comparison and advanced by the latest edition of *Bergey's Manual* has become so widely accepted that it is by many considered as some kind of an 'official' classification. It cannot be sufficiently stressed that such an official classification does not exist (Brenner *et al.*, 2001). There sure is room for other opinions, and alternative classification schemes can be designed and used if desired. One such alternative way to classify the prokaryotes into higher taxa was proposed by Cavalier-Smith (2002) (Table 5.3).

The Cavalier-Smith scheme is based on a single kingdom, the Bacteria, with two subkingdoms – the Negibacteria (prokaryotes that have a cell envelope with two distinct membranes) and the Unibacteria (organisms with a single membrane). The latter group includes the Posibacteria (organisms with a Gram-positive cell wall) and the Archaebacteria (more commonly referred to as Archaea) (see also Gupta, 1998). Altogether the classification includes eight divisions and 29 classes (12 of which have names that were not validly published under the rules of the Bacteriological Code). In Cavalier-Smith's classification scheme, the Archaebacteria together with the Eukaryotes form the clade Neomura, as they share many common characters.

## To what extent are the higher taxa phenotypically coherent?

The phyla of prokaryotes, as listed in Table 5.2, were established primarily on the basis of 16S rRNA gene comparisons (Woese and Fox, 1977; Woese, 1987; Woese *et al.*, 1990; Ludwig and Schleifer, 1997; Ludwig and Klenk, 2001). If indeed the 16S rRNA based phylogenetic trees reflect prokaryote evolution, the higher taxa should be expected to contain organisms that share many common properties, even to the extent of a similar life style. In some cases this is indeed the case. The Archaea share many features

**Table 5.2** The major divisions of the prokaryote world – the domains, phyla, and classes as given in the classification proposed in the second edition of *Bergey's Manual of Systematic Bacteriology* (Garrity and Holt, 2001a) and including later additions. Names with standing in the prokaryote nomenclature as of 3 November 2009 are printed in italic type

| Subkingdom ('domain')[1] | Division (phylum)[1] | Class | Orders |
|---|---|---|---|
| Archaea | | | |
| | Crenarchaeota[2] | | |
| | | Thermoprotei | Thermoproteales<br>Desulfurococcales<br>Sulfolobales<br>Acidilobales |
| | Euryarchaeota | | |
| | | Methanobacteria | Methanobacteriales |
| | | Methanothermea (Methanococci)[3] | Methanococcales<br>Methanomicrobiales<br>Methanosarcinales<br>Methanocellales |
| | | Halomebacteria (Halobacteria)[4] | Halobacteriales |
| | | Protoarchaea (Thermococci)[5] | Thermococcales |
| | | Thermoplasmata | Thermoplasmatales |
| | | Picrophilea | Picrophilales |
| | | Archaeoglobi | Archaeoglobales |
| | | Methanopyri | Methanopyrales |
| Bacteria | | | |
| | Aquificae | | |
| | | Aquificae | Aquificales |
| | Thermotogae | | |
| | | Thermotogae | Thermotogales |
| | Thermodesulfobacteria | | |
| | | Thermodesulfobacteria | Thermodesulfobacteriales |
| | 'Deinococcus–Thermus' | | |
| | | Deinococci | Deinococcales<br>Thermales |
| | Chrysiogenetes | | |
| | | Chrysiogenetes | Chrysiogenales |
| | Chloroflexi | | |
| | | 'Chloroflexi'<br>Anaerolineae<br>Caldilineae | 'Chloroflexales'<br>'Herpetosiphonales'<br>Anaerolineales<br>Caldilineales |
| | (yet to be assigned) | | |
| | | Ktedonobacteria | Ktedonobacterales |

**Table 5.2** continued

| Subkingdom ('domain')[1] | Division (phylum)[1] | Class | Orders |
|---|---|---|---|
| | Thermomicrobia | | |
| | | Thermomicrobia | Thermomicrobiales |
| | Nitrospirae | | |
| | | 'Nitrospira' | 'Nitrospirales' |
| | Deferribacteres | | |
| | | Deferribacteres | Deferribacterales |
| | | Ferrobacteria | Geovibriales |
| | Synergistetes | | |
| | | Synergistia | Synergistales |
| | Cyanobacteria | | |
| | | 'Cyanobacteria' | [orders with names with standing under the International Code of Botanical Nomenclature] |
| | | | Prochlorales |
| | Chlorobi | | |
| | | Chlorobea | Chlorobiales |
| | | Ignavibacteria[6] | Ignavibacteriales[6] |
| | Proteobacteria | | |
| | | Alphaproteobacteria | Rhodospirillales<br>Rickettsiales<br>Rhodobacterales<br>Sphingomonadales<br>Caulobacterales<br>'Rhizobiales'[7]<br>Hyphomicrobiales<br>Kiloniellales<br>Kordiimonadales<br>Sneathiellales |
| | | Betaproteobacteria | Burkholderiales<br>Hydrogenophilales<br>Methylophilales<br>Neisseriales<br>Nitrosomonadales<br>Rhodocyclales<br>Spirillales |

**Table 5.2** continued

| Subkingdom ('domain')[1] | Division (phylum)[1] | Class | Orders |
|---|---|---|---|
| | | Gammaproteobacteria | Chromatiales<br>'Xanthomonadales'[8]<br>Lysobacterales<br>Cardiobacteriales<br>'Thiotrichales'[9]<br>Legionellales<br>Beggiatoales<br>Methylococcales<br>Oceanospirillales<br>Pseudomonadales<br>Alteromonadales<br>'Vibrionales'<br>Aeromonadales<br>'Enterobacteriales'<br>Pasteurellales<br>Acidithiobacillales |
| | | Deltabacteria (Deltaproteobacteria)[10] | Desulfurellales<br>Desulfovibrionales<br>Desulfobacterales<br>Desulfuromonadales<br>Syntrophobacterales<br>Bdellovibrionales<br>Myxococcales<br>Desulfarculales |
| | | Epsilonproteobacteria | Campylobacterales<br>Nautiliales |
| | Firmicutes | | |
| | | 'Clostridia' | Clostridiales<br>'Thermoanaerobacteriales'<br>Halanaerobiales<br>Natranaerobiales |
| | | 'Mollicutes'[11] | Mycoplasmatales<br>Entomoplasmatales<br>Acholeplasmatales<br>Anaeroplasmatales |
| | | Firmibacteria (Teichobacteria; 'Bacilli')[12] | Bacillales<br>'Lactobacillales' |
| | | Thermolithobacteria | Thermolithobacterales |
| | (yet to be assigned) | | |
| | | (yet to be assigned) | Haloplasmatales |

**Table 5.2** continued

| Subkingdom ('domain')[1] | Division (phylum)[1] | Class | Orders |
|---|---|---|---|
| | Actinobacteria | | |
| | | 'Actinobacteria'[13] | Acidimicrobiales<br>Rubrobacterales<br>Solirubrobacterales<br>Thermoleophilales<br>Coriobacterales<br>Sphaerobacterales<br>Actinomycetales<br>Actinoplanales<br>Streptomycetales<br>Bifidobacteriales<br>Nitriliruptorales |
| | Planctomycetes | | |
| | | Planctomycea | Planctomycetales |
| | Chlamydiae | | |
| | | Chlamydiae | Chlamydiales |
| | Spirochaetes | | |
| | | Spirochaetes | Spirochaetales |
| | Fibrobacteres | | |
| | | 'Fibrobacteres' | 'Fibrobacterales' |
| | Gemmatimonadetes | | |
| | | Gemmatimonadetes | Gemmatimonadales |
| | Acidobacteria | | |
| | | Acidobacteria | Acidobacteriales |
| | | Holophagae | Holophagales<br>Acanthopleuribacterales |
| | Bacteroidetes | | |
| | | 'Bacteroides' | 'Bacteroidales' |
| | | Flavobacteria | Cytophagales<br>('Flavobacteriales') |
| | | 'Sphingobacteria' | 'Sphingobacteriales' |
| | Fusobacteria | | |
| | | 'Fusobacteria' | 'Fusobacteriales' |
| | Verrucomicrobia | | |
| | | Verrucomicrobiae<br>Opitutae | Verrucomicrobiales<br>Opitutales<br>Puniceicoccales |

**Table 5.2** continued

| Subkingdom ('domain')[1] | Division (phylum)[1] | Class | Orders |
|---|---|---|---|
| | Lentisphaerae | | |
| | | (yet to be assigned) | Lentisphaerales<br>Victivallales |
| | Dictyoglomi | | |
| | | 'Dictyoglomi' | 'Dictyoglomales' |
| | Caldiserica | | |
| | | Caldisericia | Caldisericales |

[1]The ranks of subkingdom, domain, division, and phylum are not covered by the rules of the Bacteriological Code.

[2]The name Crenarchaeota has standing in the nomenclature only as the name of a class, as an earlier homotypic synonym of the name *Thermoprotei*.

[3]Methanococci is a later homotypic synonym of *Methanothermea*.

[4]*Halobacteria* is a later homotypic synonym of *Halomebacteria*.

[5]Thermococci is a later homotypic synonym of *Protoarchaea*.

[6]In press at the time of writing (Iino *et al.*, 2010).

[7]'Rhizobiales' Kuykendall 2006 is illegitimate because it contains the genus *Hyphomicrobium* which is the type of the order *Hyphomicrobiales*.

[8]'Xanthomonadales' is illegitimate because it contains the genus *Lysobacter* which is the type of the order *Lysobacterales*.

[9]'Thiotrichales' is illegitimate because it contains the genus *Beggiatoa* which is the type of the order *Beggiatoales*.

[10]*Deltabacteria* is an earlier homotypic synonym of *Deltaproteobacteria*.

[11]The name 'Mollicutes' is illegitimate as it was proposed without the designation of a nomenclatural type.

[12]*Firmibacteria* is an earlier homotypic synonym of *Teichobacteria*.

[13]'Actinobacteria' is illegitimate because it was proposed without the designation of a nomenclatural type.

**Table 5.3** The higher taxa of prokaroytes and their classification as proposed by Cavalier-Smith (2002)

| Kingdom | Subkingdom | Infrakingdom | Division | Classes | Examples of genera |
|---|---|---|---|---|---|
| Bacteria | | | | | |
| | Negibacteria | | | | |
| | | Eubacteria | | | |
| | | | Eobacteria | Chlorobacteria[1]<br>Hadobacteria[1] | *Chloroflexus*<br>*Thermus* |
| | | Glycobacteria | | | |
| | | | Cyanobacteria | Gloeobacteria[1]<br>Chroobacteria[1]<br>Hormogoneae[1] | *Chroococcus*[2]<br>*Oscillatoria*[2]<br>*Nostoc*[2] |
| | | | Spirochaetae | Spirochaetes | Spirochaeta |
| | | | Sphingobacteria | Flavobacteria<br>*Chlorobea* | *Cytophaga*<br>*Chlorobium* |
| | | | Planctobacteria | Planctomycea<br>Verrucomicrobiae<br>Chlamydiae | *Planktomyces*<br>*Verrucomicrobium*<br>*Chlamydia* |
| | | | Proteobacteria | Chromatibacteria[1]<br>Alphabacteria[1]<br>Deltabacteria<br>Epsilobacteria[1]<br>Ferrobacteria<br>Acidobacteria | *Escherichia*<br>*Rickettsia*<br>*Myxococcus*<br>*Helicobacter*<br>*Geovibrio*<br>*Acidobacterium* |
| | Unibacteria | | | | |
| | | | Posibacteria | Togobacteria[1]<br>Teichobacteria[1]<br>Mollicutes[1]<br>Arthrobacteria<br>Arabobacteria<br>Streptomycetes | *Thermotoga*<br>*Bacillus*<br>*Mycoplasma*<br>*Arthrobacter*<br>*Corynebacterium*<br>*Streptomyces* |
| | | | Archaebacteria | Methanotherma<br>Archaeoglobea[1]<br>Halomebacteria<br>Protoarchaea<br>Picrophilea<br>Crenarchaeota | *Methanococcus*<br>*Archaeoglobus*<br>*Halobacterium*<br>*Thermococcus*<br>*Picrophilus*<br>*Thermoproteus* |

[1]Names that have no standing in the prokaryote nomenclature according to the Rules of the Bacteriological Code as they are illegitimate. Note that the nomenclature of the categories 'Kingdom', 'Subkingdom', 'Infrakingdom' and 'Division' is not regulated by the rules of the Code.

[2]Names validly published under the rules of the International Code of Botanical Nomenclature.

such as isoprenoid ether lipids, lack of a peptidoglycan cell wall, etc.; the cyanobacteria are all oxygenic phototrophs; the Spirochaetes all have spirally shaped cells with a characteristic mode of motility due to the unusual way the flagella are inserted, and the cultured members of the Thermotogae and the Aquificae phyla consist entirely of thermophiles. However, there is some evidence for the existence of non-thermophilic members of the Thermotogae (Nesbø *et al.*, 2006).

Other phyla (as well as classes and orders classified within those phyla), however, are very heterogeneous from the aspect of the physiology of their members. Most of the five classes of the phylum Proteobacteria contain obligatory and facultative aerobic, obligatory anaerobic, photoautotrophic, photoheterotrophic, and chemolithotrophic organisms of highly diverse morphology, and these use a wide range of substrates as carbon and energy sources. The only reason those organisms were grouped together is the similarity in their 16S rRNA gene sequences, as this is the sole property on which this particular classification scheme is based. Also the phyla Firmicutes and Bacteroidetes contain species that differ greatly in their physiological and other phenotypic properties. As far as modes of energy generation are concerned, all known types of metabolism are realized in the Bacteria domain with one major exception: methanogenesis is known only in the Euryarchaeota branch of the domain Archaea. One niche in which the Archaea still appear to have a monopoly is the extremely high temperature environment. No known members of the Bacteria or the Eucarya grow at temperatures above 95°C (probably due to the limited thermostability of their lipid membranes); all life above that temperature up to 115–120°C and possibly even higher is archaeal.

Horizontal gene transfer may to a large extent be responsible for the apparent lack in physiological consistency of many phyla and other higher taxa. Genetic exchange among phylogenetically unrelated prokaryotes is extensive (Doolittle, 1999). As more and more complete genome sequences became available it became increasingly clear that each genome is essentially a mosaic of genetic elements derived from different sources. The more the genes are moving around by horizontal gene transfer, the less it can be expected that organisms that are related on the basis of ribosomal RNA sequence comparison will also lead a similar life style. It is therefore possible that higher taxa that were earlier considered homogeneous with respect to the physiological properties of their member species will prove to be much more divergent when more representatives will be isolated and characterized in the future. This already has happened in some cases. The Planctomycea were until recently a small group of mostly aerobic heterotrophic budding bacteria of unusual cellular structure, having membrane-enclosed intracellular compartments (species such *Planctomyces* and *Pirellula*). The discovery of the anaerobic bacteria responsible for autotrophic ammonia oxidation with nitrite as electron acceptor (the anammox process), mediated by representatives of the phylum (Strous *et al.*, 2006) that share the multiple-compartment prokaryotic cell structure with that of the earlier described heterotrophic aerobes but have an altogether different physiology, shows that the group is far more diverse than assumed earlier. Another obvious example is the discovery of *Salinibacter*, an extremely halophilic representative of the Bacteroidetes (Antón *et al.*, 2002): this organism shares many properties (also on the genetic level) with the order Halobacteriales (Euryarchaeota) (Mongodin *et al.*, 2005).

The discovery that at least some non-thermophilic Crenarchaeota make a living by chemoautotrophic oxidation of ammonia (Könneke *et al.*, 2005) shows that types of metabolism that were earlier known only from the Bacteria may be functional in some Archaea as well. The occurrence of processes such as autotrophic ammonia oxidation ('*Candidatus* Nitrosopumilus') and dissimilatory sulfate reduction (*Archaeoglobus*) among the Archaea as well the Bacteria (*Nitrosomonas, Desulfovibrio*, etc.) shows that there is little correlation between phylogenetic position in the rRNA-based tree and the type of metabolism used for energy generation. The assembly of the different metabolic pathways may well have been mediated by lateral gene transfer (Boucher *et al.*, 2003).

One property that is notably absent from the archaeal domain is that of chlorophyll-based photosynthesis, a feature found in many branches of the domain Bacteria. No fewer than five of its phyla contain phototrophs. Chlorophyll *a*-based oxygenic photosynthesis is found in the Cyanobacteria, and bacteriochlorophyll-based anoxygenic photosynthesis is encountered in all known representatives of the Chlorobi, in most species of the Chloroflexi, in many representatives of the Proteobacteria (within the classes Alphaproteobacteria, Betaproteobacteria and Gammaproteobacteria – classes that all contain very heterogeneous assemblages of species as far as their physiology is concerned), as well as in the family Heliobacteraceae within the phylum Firmicutes. Recently, bacteriochlorophyll-based anoxygenic photosynthesis was also found in '*Candidatus* Chloroacidobacterium thermophilum', a representative of the Acidobacteria phylum (Bryant *et al.*, 2007). Whether bacteriochlorophyll-based use of light energy is indeed absent in the Archaea remains to be ascertained, as it was recently reported that a yet uncultured representative of the Crenarchaeota contains a functional bacterochlorophyll *a* synthase (Meng *et al.*, 2009).

Some extremely halophilic representatives of the family Halobacteriaceae (Euryarchaeota) are able to use light energy based on the absorption of photons by bacteriorhodopsin with the formation of a transmembrane proton gradient. Light can thus provide energy for cell maintenance and even drive photoheterotrophic growth, especially in situations where oxygen for respiration is in short supply. In the past, photoheterotrophic life driven by the retinal protein bacteriorhodopsin was considered unique to a few extremely halophilic Euryarchaeota (*Halobacterium* and relatives), but the discovery of similar retinal proteins in different branches of the Bacteria (proteorhodopsin in the some cultured and uncultured representatives of the Alphaproteobacteria and the Gammaproteobacteria, xanthorhodopsin in *Salinibacter* (Bacteroidetes), actinorhodopsin in some Actinobacteria) shows that such a mode of life may be realized in phylogenetically unrelated branches of the tree.

## The phyla of the Archaea – an overview

In the classification scheme of *Bergey's Manual* (Garrity and Holt, 2001a) the domain Archaea is subdivided into two phyla: the Crenarchaeota with one class, and the Euryarchaeota with seven classes (Table 5.2). Cavalier-Smith (2002) classified the division Archaebacteria within the subkingdom Unibacteria. With the exception of most methanogens (Euryarchaeota), all cultured representatives of the domain Archaea are extremophiles that inhabit environments with extremely high temperatures (many methanogens are thermophilic as well), often combined with growth at low pH, or environments characterized by high salt concentrations (often up to saturation), sometimes combined with extremely alkaline conditions. Physiologically both phyla are very heterogeneous.

Three more phyla have been proposed within the Archaea: the Nanoarchaeota, the Thaumarchaeota and the Korarchaeota. At the time of writing (November 2009), these phyla were not represented by cultured representatives with names validly published according to the rules of the Bacteriological Code.

When the 16S rRNA of 'Nanoarchaeum equitans', a parasite/symbiont of the extremely thermophilic *Ignicoccus hospitalis* (Crenarchaeota, Thermoprotei) was analysed, the establishment of a new phylum – the Nanoarchaeota – was proposed, based on the low similarity of the sequence with any of the known organisms (Hohn *et al.*, 2002; Huber *et al.*, 2002; Branciamore *et al.*, 2008). Related 16S rRNA sequences have also been recovered from low-temperature hypersaline environments in Inner Mongolia and South Africa (Casanueva *et al.*, 2008). It was even suggested that the tree of life may be rooted in the domain Archaea in the branch leading to the phylum Nanoarchaeota (Di Giulio, 2007). However, an alternative view is that these 'Nanoarchaea' may represent a fast-evolving euryarchaeal lineage related to the Thermococcales (Brochier *et al.*, 2005).

The proposal for the establishment of the Thaumarchaeota as the third phylum within the archaeal domain was made by Brochier-Armanet *et al.* (2008) to accommodate a number of lineages of non-extremophilic marine organisms earlier classified with the Crenarchaeota, and represented mainly by environmental 16S rRNA gene sequences (DeLong, 1998; Karner *et al.*, 2001). Two representatives of the group have been studied in some depth. One is 'Cenarchaeum symbiosum', a symbiont of a marine sponge (Preston *et al.*, 1996). Based on the properties of this unusual organism the establishment of a new order, the Cenarchaeales, has been proposed (Cavalier-Smith, 2002); however, this name is illegitimate as the name of the proposed type genus Cenarchaeum was never validly published. Another representative of the Thaumarchaeota is '*Candidatus* Nitrosopumilus maritimus', the recently isolated marine autotrophic ammonia oxidizing prokaryote that has a 16S rRNA sequence similar to one of the earlier recognized groups of marine mesophilic Crenarchaeota based on environmental genomics studies (Könneke *et al.*, 2005).

The proposed phylum Korarchaeota consists of thermophilic Archaea. 16S rRNA gene sequences assigned to the group were first recovered from Obsidian Pool in Yellowstone National Park (Barns *et al.*, 1994), and related sequences were found in other thermal environments (Auchtung *et al.*, 2006). No representatives of the group are yet available in pure culture, but establishment of an enrichment culture has enabled the sequencing of the genome of '*Candidatus* Korachaeum cryptofilum'. Based on the genomic information it was inferred that the organism probably makes a living by fermentation of peptides. Many of its genes show an affiliation with the Crenarchaeota, but several conserved cellular systems such as cell division and DNA replication resemble those of Euryarchaeota counterparts (Elkins *et al.*, 2008).

### The Crenarchaeota

With the exception of the mesophilic marine Archaea ('Cenarchaeum', 'Nitrosopumilus') previously classified within the Crenarchaeota but recently moved to the newly proposed third phylum with the Archaea, the Thaumarchaeota (Brochier-Armanet *et al.*, 2008), all cultured members of the Crenarchaeota are extremely thermophilic. They have been isolated from hydrothermal vents, hot springs, and other thermal environments. The phylum currently has one class (Thermoprotei), divided into four orders: Thermoproteales, Sulfolobales, Desulfurococcales, and Acidilobales, a recently added order of anaerobic thermoacidophiles (Prokofeva *et al.*, 2009). The Crenarchaeota phylum contains aerobes as well as anaerobes. Many representatives obtain their energy by oxidizing hydrogen or organic compounds while using elemental sulfur as terminal electron acceptor in respiration; others reduce sulfurur compounds with hydrogen as energy source. Some are chemoautotrophs; others require organic carbon sources.

### The Euryarchaeota

In the phylum Euryarchaeota we find a far greater phenotypical diversity than in the Crenarchaeota.

Three out of the eight classes presently recognized (the Methanobacteria, the Methanothermea, and the Methanopyri) consist of methanogenic anaerobes. Thermophiles can be found in all three classes.

The Halomebacteria (a name that has priority over the more widely used name Halobacteria) generally have an aerobic lifestyle, and their habitat is restricted to hypersaline environments, typically from 150 g/l salt up to saturation. Many are alkaliphilic as well as halophilic. The recently proposed new name Haloarchaeales for the class (DasSarma and DasSarma, 2008) is illegitimate (Oren, 2008).

The four remaining classes [Thermoplasmata, Picrophilaea, Protoarchaea (a name validly published before the more widely used Thermococci) and Archaeoglobi] are all thermophilic, and most live as anaerobic or facultative aerobic heterotrophs. Many representatives require low pH as well as high temperatures for growth. The Archaeoglobi are heterotrophs or chemoautotrophs, and perform anaerobic respiration with sulfate or nitrate as terminal electron acceptors. Environmental genomics studies of the oceans have indicated that large and diverse communities of Euryarchaeota are present in the sea as well. No representatives of mesophilic marine aerobic members of the phylum have yet been brought into culture, and their mode of metabolism is as yet unknown.

## The phyla of the Bacteria – an overview

Currently the Bacteria are divided into 27 phyla. Some of these, such as the Proteobacteria, the Firmicutes and the Actinobacteria, are very diverse with respect to cell morphology, physiological properties and life style of their members. Other phyla such as the Chlamydiae, the Spirochaetes, and the Chlorobi have less species and are physiologically more coherent. Extreme cases are the phyla Caldiserica, represented thus far by a single cultured and named species, *Caldisericum exile* (Mori *et al.*, 2009), and the Gemmatimonadetes with *Gemmatimonas aurantiaca* as the sole species (Zhang *et al.*, 2003). Such phyla have obtained their special status on the basis of their highly divergent 16S rRNA gene sequences.

The following sections provide a short overview of the phyla of the Bacteria and their properties.

## The Aquificae

The phylum Aquificae currently contains a single class (Aquificae) with a single order (Aquificales). Among its genera are *Aquifex* and *Hydrogenobacter*. All members of phylum are thermophilic to hyperthermophilic chemoorganotrophs that oxidize hydrogen or reduced sulfurur compounds with oxygen or nitrate as an electron acceptor.

## The Thermotogae

Also the Thermotogae phylum is small and coherent, both morphologically and physiologically. All members of the single single class (Thermotogae) and order (Thermotogales) are extremely thermophilic anaerobes with an outer sheath-like envelope.

## The Thermodesulfobacteria

The deep-branching bacterial phylum Thermodesulfobacteria is presently represented by a single genus *Thermodesulfobacterium*, consisting of sulfate-reducing anaerobes that use compounds such as lactate, pyruvate and ethanol as electron donor. The genus is unusual as the hydrophobic chains in its lipids are bound to glycerol by ether bonds, a property characteristic of the Archaea are but not common in the domain Bacteria.

## The 'Deinococcus–Thermus' group

The 'Deinococcus–Thermus' group, with the single class Deinococci, harbours chemoorganotrophic aerobes of highly disparate properties. The order Deinococcales includes the genus *Deinococcus*, a genus of cocci widespread in soil (including desert soil) and aquatic habitats. Some representatives were isolated from air samples. Most species are highly resistant to radiation (light, UV, radioactivity), thanks to an unusually effective mechanism for repair of radiation-induced DNA damage. The second order, the Thermales, consists of thermophilic bacteria that abound in hot springs and other high-temperature environments.

## The Chrysiogenetes

The phylum Chrysiogenetes forms a deep lineage within the Bacteria, and is currently represented by a single genus *Chrysiogenes* with a single species, *Chrysiogenes arsenatis* that grows anaerobically by oxidizing acetate with arsenate as electron acceptor (Many *et al.*, 1996; Garrity and Holt, 2001b). How diverse this lineage may further be is yet unknown.

## The Chloroflexi

The best known representatives of the Choroflexi phylum belong to the class 'Chloroflexi', a group of Gram-negative bacteria showing gliding motility. Their physiology varies greatly: some members (*Chloroflexus*) are anoxygenic phototrophs, others (*Herpetosiphon*) are aerobic heterotrophs.

Recently two new classes have been proposed within this phylum, the Anaerolineae and the Caldilineae, to accommodate the moderately thermophilic filamentous anaerobes of the genus *Anaerolinea* (two species) and of the genus *Levilinea* (one species) (Yamada *et al.*, 2006).

The newly described class Ktedonobacteria, with as yet a single species, *Ktedonobacter racemifer*, is loosely affiliated with the Chloroflexi. This interesting filamentous, branched mycelia forming mesophilic aerobe-microaerophile from soil produces spores similar to the Actinobacteria (Cavaletti *et al.*, 2006).

## The Thermomicrobia

The phylum Thermomicrobia is currently represented by a single genus *Thermomicrobium* with a single species of aerobic thermophiles. *Thermomicrobium* does not have peptidoglycan in its cell wall, and its lipids are highly unusual. These are 1,2-dialcohols that have neither ester nor ether linkages.

## The Nitrospirae

The Nitrospirae phylum is physiologically very heterogeneous: it includes nitrite-oxidizing autotrophs (*Nitrospira*), iron-oxidizers (*Leptospirillum*), as well as thermophilic sulfate reducers (*Thermodesulfovibrio*). No names of classes and orders belonging to this group have as yet been validly published.

## The Deferribacteres

The phylum Deferribacteres (classes Deferribacteres and Ferrobacteria) includes a small number of heterotrophic anaerobes (*Deferribacter, Geovibrio*) which use oxidized iron, manganese, or nitrate as electron acceptor.

## The Synergistetes

The most recent addition to the list of phyla of prokaryotes is the Synergistetes with the class Synergistia and the order Synergistales (Jumas-Bilak *et al.*, 2009). This phylum was created to accommodate different genera of anaerobic amino acid-degrading bacteria isolated from humans, animals, and terrestrial and oceanic habitats (*Synergistes, Jonquetella, Aminiphilus, Aminobacterium,* and others) that form a deep lineage phylogenetically unrelated with other groups.

## The Cyanobacteria

The Cyanobacteria form a physiologically coherent group of aerobic oxygenic phototrophs that possess chlorophyll *a* in their photosynthetic reaction centres. The group contains unicellular as well as filamentous types, and includes many species that fix molecular nitrogen. In some filamentous nitrogen-fixing types, special differentiated cells (heterocysts) may be formed to accommodate the nitrogen fixation machinery. Other kinds of differentiated cells can occur as well, such as akinetes and baeocytes for survival and dispersion.

Only few genera and species of cyanobacteria have been named under the rules of the Bacteriological Code. The only order whose name has standing in the prokaryote nomenclature is the Prochlorales. Different classification schemes exist, and their nomenclature is primarily based on the provisions of the International Code of Botanical Nomenclature. The names of classes (Hormogoneae, Gloeobacteria, Chroobacteria) and orders (Chroococcales, Gloeobacterales, Oscillatoriales, Stigonematales, Nostocales, Pleurocapsales) proposed by Cavalier-Smith (2002) are all illegitimate under the Bacteriological Code. More information on nomenclature issues is given by Oren (2004b) and Oren and Tindall (2005).

Well-known species include the unicellular *Synechococcus* and *Microcystis* and filamentous types such as *Oscillatoria, Phormidium, Anabaena,* and *Nostoc.* Some are moderately thermophilic and can perform oxygenic photosynthesis up to a temperature of 73–74°C.

## The Chlorobi

The phylum Chlorobi mostly consists of obligatory anaerobic anoxygenic phototrophs that use reduced sulfur compounds as electron donor for the autotrophic fixation of carbon dioxide (class Chlorobea, order Chlorobiales; representative species *Chlorobium, Prosthecochloris*).

A recently proposed new class, the Ignavibacteria with as yet a single species (*Ignavibacterium album*, a moderately thermophilic anaerobic bacterium isolated from microbial mats at a terrestrial hot spring), represents a novel lineage at the periphery of the Chlorobi phylum (Iino *et al.*, 2010).

## The Proteobacteria

The phylum Proteobacteria is not only the phylum with the most described species, it is also the most heterogeneous of all phyla. Nearly all major types of metabolism (with the exception of methanogenesis found only in the Euryarchaeota and oxygenic photosynthesis restricted to the Cyanobacteria) are encountered: aerobic, anaerobic; chemoorganotrophic, chemolithotrophic and anoxygenic phototrophic. Also morphologically the phylum is very diverse. The majority of bacteria of medical, industrial, and agricultural importance belong to the Proteobacteria.

Five classes are recognized within the Proteobacteria: the Alphaproteobacteria with 10 orders, the Betaproteobacteria with eight orders, the Gammaproteobacteria with 14 orders, the Deltabacteria (Deltaproteobacteria) with eight orders, and the Epsilonproteobacteria with two orders (Table 5.2). A sixth class was proposed as '*Candidatus* Zetaproteobacteria', thus far represented only by 'Mariprofundus ferrioxidans', a neutrophilic marine Fe(II)-oxidizing bacterium associated with hydrothermal venting of iron-rich fluids associated with seamounts in the world's oceans (Emerson *et al.*, 2008). At the time of writing (November 2009) the name of this species was not yet validly published.

The Alphaproteobacteria form a large and very heterogeneous group of Gram-negative bacteria, including anoxygenic photoheterotrophs (*Rhodospirillum, Rhodobacter*), aerobic heterotrophs (*Paracoccus, Caulobacter, Acetobacter*), obligate intracellular parasites/pathogens (*Rickettsia*), plant symbionts (*Rhizobium*), chemoautotrophs (*Nitrobacter*), and many others. The Betaproteobacteria class is another large and very heterogeneous group which includes heterotrophs (*Spirillum, Zoogloea*), chemoautotrophs (*Thiobacillus, Nitrosomonas*), and anoxygenic photoheterotrophs (*Rhodoferax*). The third class, the Gammaproteobacteria, is no less heterogeneous than the first two. Among its members are anoyxgenic photoautotrophic bacteria (e.g. *Chromatium*), chemolithotrophic ammonium oxidizers (e.g. *Nitrosococcus*), chemolithotrophic sulfur oxidizers (*Beggiatoa*), heterotrophic aerobes (*Pseudomonas, Azotobacter, Vibrio*), fermentative anaerobes (*Ruminobacter*), the enteric bacteria (*Escherichia, Salmonella, Proteus*), and many others. The Deltabacteria (Deltaproteobacteria) includes most dissimilatory sulfate reducing bacteria (*Desulfovibrio, Desulfobacter*), certain nitrifying chemoautotrophs (*Nitrospina*), iron-reducing bacteria (*Geobacter*), anaerobic obligatory syntrophic bacteria (*Syntrophobacter*), predatory bacteria (*Bdellovibrio*), the gliding myxobacteria with their complex life cycles, and others. Finally, the Epsilonproteobacteria class contains relatively few cultured representatives yet, but among those are some interesting organisms such as *Helicobacter* – the causative agent of gastric ulcers, the sulfur-oxidizing bacterium *Thiovulum*, *Campylobacter* – a genus of intestinal bacteria, and *Wolinella* – a rumen bacterium that performs anaerobic respiration with a number of electron acceptors.

Anoxygenic photosynthesis based on bacteriochlorophyll *a* or *b* is widespread among the Proteobacteria, and for that reason the Proteobacteria lineage was originally designated 'Purple Bacteria' in Woese's phylogenetic trees. Purple photoheterotrophs are found in the Alphaproteobacteria (e.g. *Rhodospirillum, Rhodopseudomonas*) and in the Betaproteobacteria (*Rhodocyclus*); all purple sulfur bacteria (*Chromatium* and relatives) belong to the Gammaproteobacteria. Bacteriochlorophyll is also found in some genera that do not primarily lead a phototrophic life style.

Most autotrophic nitrifying bacteria are found in the Proteobacteria; we find organisms that oxidize ammonia to nitrite in the Betaproteobacteria and the Gammaproteobacteria, and nitrite oxidizers in the Alphaproteobacteria, Gammaproteobacteria and Deltabacteria. Autotrophic nitrification is, however, not restricted to the Proteobacteria as shown by the recent discovery of the nitrifying mesophilic Archaea classified in the candidate phylum Thaumarchaeota and the existence of nitrite oxidizers in the Nitrospirae phylum, unrelated to the Proteobacteria.

Other types of chemolithotrophy are also widespread among the Proteobacteria: use of reduced sulfur compounds is encountered in the Alphaproteobacteria (*Starkeya*), Betaproteobacteria (*Thiobacillus*), Gammaproteobacteria (*Acidithiobacillus, Beggiatoa*), and Epsilonproteobacteria (*Thiovulum*); aerobic hydrogen oxidizing autotrophs are also found in several classes.

A type of metabolism that is apparently limited to the Proteobacteria phylum is methanotrophy. All 'type I' methanotrophs (organisms that possess intracellular membranes in bundles distributed throughout the cell, have an incomplete citric acid cycle and use the ribulose monophosphate pathway for carbon fixation) belong to the Gammaproteobacteria; all 'type II' methanotrophs (organisms that have intracellular paired membranes running along the periphery of the cell, have a complete citric acid cycle, and use the serine pathway for carbon assimilation) are all found in the Alphaproteobacteria class.

Most sulfate-reducing anaerobes described belong to the Deltabacteria. Exceptions are the endospore-forming *Desulfotomaculum* (Firmicutes), the bacterial thermophilic sulfate reducers *Thermodesulfobacterium* (phylum Thermodesulfobacteria) and *Thermodesulfovibrio* (phylum Nitrospirae), and the archaeal thermophilic sulfate reducing genus *Archaeoglobus* (Euryarchaeota).

The ability to fix molecular nitrogen is widespread among the Proteobacteria, examples being *Azotobacter* and *Rhizobium* (Alphaproteobacteria), *Azoarcus* (Betaproteobacteria), and many sulfate-reducing anaerobes within the Deltabacteria.

## The Firmicutes

Most species of the Firmicutes are characterized by a Gram-positive type cell wall. However, the wall-less pathogens of the class Mollicutes (*Mycoplasma* and relatives) are classified within the Firmicutes as well on the basis of their phylogenetic position as apparent from their 16S rRNA gene sequences. Currently the phylum is divided into four classes.

The class Clostridia consists of anaerobic fermentative Gram-positive bacteria, many of which form endospores. The group includes genera such as *Clostridium, Ruminococcus,* and *Peptostreptococcus,* endospore-forming dissimilatory sulfate reducers of the genus *Desulfotomaculum,* and anoxygenic phototrophs (*Heliobacterium*). Some are thermophilic (*Thermoanaerobacterium*) or halophilic (*Halanaerobium* and relatives) and there even are anaerobic haloalkaliphilic thermophiles (*Natranaerobius*). The Mollicutes (orders Mycoplasmatales, Entomoplasmatales, Acholeplasmatales and Anaeroplasmatales) is a group of mostly wall-less pathogens. The class Firmibacteria (Teichobacteria) contains the genus *Bacillus* and related genera of aerobic Gram-positive endospore-forming organisms, as well as the lactic acid bacteria (*Lactobacillus, Leuconostoc, Streptococcus*). The fourth class currently recognized, the Thermolithobacteria, is as yet represented by a single genus *Thermolithobacter* with two described species of chemolithoautotrophic thermophilic anaerobic Fe(III) reducers (Sokolova *et al.*, 2007).

## The Actinobacteria

The phylum Actinobacteria contains a large group of Gram-positive, mostly aerobic heterotrophic microorganisms, many of them growing in the form of mycelia (*Streptomyces, Actinomyces*), pathogens such as *Corynebacterium* and *Mycobacterium,* fermentative anaerobes (*Propionibacterium*), symbiotic nitrogen fixers (*Frankia*), and others.

## The Planctomycetes

The Planctomycetes are a small group of mostly aerobic heterotrophic bacteria of unusual cellular structure. *Planctomyces* and *Pirellula* are the best-known genera. Their cells are characterized by the presence of a membrane-enclosed intracellular compartment, and they multiply not by binary fission like most prokaryotes but by budding. The group includes the anaerobic chemoautotrophic organisms that oxidize ammonium ions with nitrite as electron acceptor – the 'anammox' bacteria. These have not yet been grown in pure culture and the provisional names given to some cultured species still have no standing in the prokaryote nomenclature.

## The Chlamydiae

The phylum Chlamydiae consists of a group of intracellular parasites and pathogens, including the causative agents of trachoma and psittacosis.

## The Spirochaetes

The Spirochaetes are aerobic or anaerobic prokaryotes with an unusual mode of motility. They have multiple polar flagella ('endoflagella') that fold back from each pole and remain located in the periplasm of the cell. The group includes free-living species (*Spirochaeta*), symbionts (*Cristispira*), as well as pathogens (*Treponema, Borrelia*).

## The Fibrobacteres

The Fibrobacteres form a deep lineage within the Bacteria. Presently the phylum consists of a single genus, *Fibrobacter,* of anaerobic fermentative organisms found in the rumen of ruminant animals.

## The Gemmatimonadetes

Thus far the only cultured representative of the deep lineage for which the phylum name Gemmatimonadetes was proposed is *Gemmatimonas aurantiaca,* a Gram-negative, aerobic, polyphosphate-accumulating bacterium isolated from a biological wastewater treatment system (Zhang *et al.*, 2003).

### The Acidobacteria

The phylum Acidobacteria, class Acidobacteria, encompasses a small group of acidophilic organotrophs (*Acidobacterium*) and iron reducers (*Geothrix*). A new class, the Holophagae, was recently added to the phylum to accommodate the homoacetogenic *Holophaga foetida* and the newly isolated *Acanthopleuribacter pedis*, isolated from a chiton (Fukunaga *et al.*, 2008). The finding of the anaerobic thermophilic '*Candidatus* Chloroacidobacterium thermophilum' in a hot spring in Yellowstone National Park, an anoxygenic photosynthetic bacterium with bacteriochlorophyll and chlorosomes (Bryant *et al.*, 2007), shows that the metabolic diversity within the Acidobacteria phylum may be much greater than previously assumed.

### The Bacteroidetes

The Bacteroidetes phylum contains a disparate group of prokaryotes, highly diverse in their physiology. The 'Bacteroides' class harbours anaerobic fermentative bacteria. Most species within the class Flavobacteria are aerobic, often pigmented heterotrophs. The 'Sphingobacteria' include heterotrophic bacteria involved in aerobic degradation, often of polymeric substrates (*Cytophaga, Flexibacter*), thermophiles (*Rhodothermus*), as well as red extremely halophilic bacteria (*Salinibacter*) that physiologically resemble the Halobacteriales (Euryarchaeota).

### The Fusobacteria

The phylum Fusobacteria consists of a small group of fermentative anaerobes of different metabolic types that form a phylogenetically deep lineage within the Bacteria. The group includes the butyrate-forming *Fusobacterium* found in the oral cavity and digestive systems of humans and animals, and *Propionigenium* that ferments succinate to propionate and carbon dioxide.

### The Verrucomicrobia

The phylum Verrucomicrobia was created to classify species of *Verrucomicrobium* and *Prosthecobacter,* a small group of prosthecate budding bacteria that form a deep linage within the phylogenetic tree of the Bacteria (Hedlund *et al.*, 1997). A new class, the *Opitutae,* was recently added to this phylum to accommodate the genus *Opitutus* (obligatory anaerobic fermentative bacteria isolated from rice paddy soil) and the marine *Puniceicoccus* (Choo *et al.*, 2007).

### The Lentisphaerae

The phylum *Lentisphaerae* thus far contains two genera only, both with a single species: *Lentisphaera araneosa*, an exopolymer-producing marine bacterium, and *Victivallis vadensis*, a sugar-fermenting anaerobe isolated from human faeces.

### The Dictyoglomi

The phylum *Dictyoglomi* is thus far represented only by *Dictyoglomus,* a genus with two species of anaerobic thermophilic organotrophs.

### The Caldiserica

The phylum *Caldiserica* is a recent addition to the list of bacterial phyla. Earlier it was a candidate phylum ('phylum OP5'), recognized on the basis of environmental 16S rRNA sequences (Hugenholtz *et al.*, 1998b). It currently contains a single species: *Caldisericum exile,* an anaerobic, thermophilic, filamentous bacterium isolated from a hot spring in Japan (Mori *et al.*, 2009).

## The yet uncultured phyla

Culture-independent studies of 16S rRNA genes amplified from DNA extracted from different ecosystems (marine, freshwater, soil, thermal environments, etc.) have shown that the true microbial diversity in nature is much larger than that represented by the organisms thus far brought into culture. Only seldom did the sequences recovered from the environment fully match those of cultured microorganisms, and in most cases it could be inferred that they belonged to yet undescribed species, often belonging to new genera and families (DeLong and Pace, 2001; Handelsman, 2004; Hugenholtz *et al.*, 1998a,b; Oren, 2004a). In many cases did the 'environmental' 16S rRNA genes recovered differ so much from those of the known Bacteria and Archaea that the organisms harbouring these sequences should be classified in new orders, classes, and even phyla, on the basis of their deep 16S rRNA lineages (Barns *et al.*, 1994; Hugenholtz *et al.*, 1998a,b). The more sequences were collected from environmental DNA, the more such candidate phyla and other higher taxa were proposed. There are major groups of microorganisms in nature that are present in large numbers in the most common ecosystems, but of which we do not have a single representative in culture yet.

Based on the rRNA sequence information it is possible to design probes for the detection of the novel group of organisms by techniques such as fluorescence in situ hybridization, and this approach provides information on the distribution and morphology of the organisms of yet unknown properties (Ludwig *et al.*, 1977). The world's oceans contain large amounts of Archaea: both Crenarchaeota (or the new phylum of the Thaumarchaeota created to accommodate those marine mesophilic Archaea affiliated with the Crenarchaeota), as well as Euryarchaeota (DeLong, 1998). Fluorescence *in situ* hybridization using Archaea- and Bacteria-specific probes has shown that about 30% of all prokaryotes in the oceans belong to the archaeal domain (Karner *et al.*, 2001). However, we have very little information on the physiology of these extremely abundant marine Archaea, and the first cultures of mesophilic marine Crenarchaeota have only recently been obtained.

It is especially rewarding when, as happens from time to time, a representative of such a group previously known from environmental sequences is brought into culture, so that the properties of the organism itself can be studied as well. Recent examples of such breakthroughs are:

- The isolation of the marine autotrophic ammonia-oxidizing archaeon 'Nitrosopumilus maritimus' (Könneke *et al.*, 2005). The name of the species was not yet validly published; in the future it may be classified within the newly proposed phylum Thaumarchaeota (Brochier-Armanet *et al.*, 2008).
- The characterization of the anaerobic, thermophilic, filamentous bacterium *Caldisericum exile* with a 16S rRNA sequence affiliated with the earlier defined candidate phylum OP5 (Mori *et al.*, 2009).
- The isolation of *Gemmatimonas aurantiaca* as the first cultured representative of the new phylum Gemmatimonadetes (Zhang *et al.*, 2003).
- The characterization of *Methanocella paludicola*, the first isolate of the archaeal lineage 'Rice Cluster I', previously known from environmental sequences only (Sakai *et al.*, 2008).
- The isolation of 'Elusimicrobium minutum', the first cultivated representative of the phylum 'Elusimicrobia' (formerly Termite Group1) (Herlemann *et al.*, 2009).

Increased efforts toward the cultivation and characterization of representatives of other yet uncultured groups will be necessary to provide a better insight into the true microbial diversity present in nature.

## Final comments

The 16S rRNA sequence based division into higher taxa (subkingdoms or domains, phyla, classes and orders), as adopted by the 2nd edition of *Bergey's Manual of Systematic Bacteriology* and as summarized in Table 5.2, is currently the most widely used classification scheme for the prokaryotes. The system is based on the known 16S rRNA gene sequences of nearly all type strains of the prokaryotic species whose names have been validly published (see also Yarza *et al.*, 2008). The system is convenient to use: as soon as the 16S rRNA sequence of a new isolate (or environmental clone) has been determined, a simple BLAST search immediately shows its place within the system.

Some of the phyla defined on the basis of 16S rRNA sequence similarity are physiologically homogeneous, while others consist of organisms of highly disparate lifestyles. In some cases it may therefore be possible to predict from the 16S rRNA sequence of a yet uncultured bacterium or archaeon what its properties might be (and, as a consequence, methods may be designed that may lead to its isolation and characterization). In other cases, however, the 16S rRNA sequence does not give any clue about the possible way of life of an unknown organism. Two recent examples suffice to demonstrate this. When the elusive 'anammox' bacteria were first isolated ('Brocadia anammoxidans', 'Scalindua kuenenii', etc.) and it had become clear that phylogenetically these intriguing bacteria belong to the Planctomycetes phylum, nothing within our earlier information on the nature of the Planctomycetes group, by then consisting of aerobic heterotrophs (*Planctomyces*, *Pirellula*, etc.), could have led to an educated guess that this phylum may contain such a group of unique anaerobic autotrophs as well. A second example relates to the isolation of the marine autotrophic ammonia-oxidizing '*Candidatus* Nitrosopumilus maritimus' (Könneke *et al.*, 2005). Although the abundance of crenarchaeal 16S rRNA sequences in the marine environment was known from the early 1990s onwards (DeLong, 1998; Karner *et al.*, 2001), the information on the physiology of the first cultured representative of the group (now proposed to be classified in a new phylum of Archaea, the Thaumarchaeota; Brochier-Armanet *et al.*, 2008) came as a tremendous surprise, as thus far autotrophic aerobic ammonia oxidation was only known from the bacterial Proteobacteria phylum.

We now have over 8000 species of prokaryotes whose names have been validly published under the rules of the Bacteriological Code (Lapage *et al.*, 1992). In the past 5 years, 600 new species on the average have annually been added to the list. Thanks to the well-ordered framework established by rules of the Bacteriological Code, the centralized registration and indexing of the new names in a single journal, and the establishment and maintenance of Jean Euzéby's website 'List of Prokaryotic Names with Standing in Nomenclature' (Euzéby, 1997; www.bacterio.cict.fr), one can at any moment obtain a complete overview of the number of prokaryotes described and their correct names.

As already stressed earlier in this chapter, there is no official classification of Bacteria and Archaea, and 16S rRNA sequence-based schemes such as exemplified in Table 5.2 represent just one way of sorting the prokaryotes. Alternative classification schemes such as the scheme proposed by Cavalier-Smith (2002) (Table 5.3), although not very popular among microbial taxonomists, are well argued.

## References


Antón, J., Oren, A., Benlloch, S., Rodríguez-Valera, F., Amann, R., and Rosselló-Mora, R. (2002). *Salinibacter ruber* gen. nov., sp. nov., a novel extreme halophilic member of the *Bacteria* from saltern crystallizer ponds. Int. J. Syst. Evol. Microbiol. *52*, 485–491.

Auchtung, T.A., Takacs-Vesbach, C.D., and Cavanaugh, C.M. (2006). 16S rRNA phylogenetic investigation of the candidate division "*Korachaeota*". Appl. Environ. Microbiol. *72*, 5077–5082.

Branciamore, S., Gallori, E., and Di Giulio, M. (2008). The basal phylogenetic position of *Nanoarchaeum equitans* (Nanoarchaeota). Front. Biosci. *13*, 6886–6892.

Barns, S.M., Fundyga, R.E., Jeffries, M.W., and Pace, N.R. (1994). Remarkable diversity detected in a Yellowstone National Park hot spring environment. Proc. Natl. Acad. Sci. U.S.A. *91*, 1609–1613.

Boucher, Y., Douady, C.J., Papke, R.T., Walsh, D.A., Boudreau, M.E.R., Nesbø, C.L., Case, R.J., and Doolittle, W.F. (2003). Lateral gene transfer and the origins of prokaryotic groups. Ann. Rev. Genet. *37*, 283–328.

Brenner, D.J., Staley, J.T., and Krieg, N.R. (2001). Classification of procaryotic organisms and the concept of bacterial speciation, In *Bergey's Manual* of Systematic Bacteriology, 2nd ed., Vol. 1. The Archaea and the Deeply Branching and Phototrophic Bacteria, D.R. Boone, R.W. Castenholz, and G.M. Garrity, eds. (New York: Springer), pp. 27–31.

Brochier, C., Gribaldo, S., Zivanovic, Y., Confalonieri, F., and Forterre, P. (2005). Nanoarchaea: representatives of a novel archaeal phylum or a fast-evolving euryarchaeal lineage related to Thermococcales? Genome Biol. *6*, R42.

Brochier-Armanet, C., Boussau, B., Gribaldo, S., and Forterre, P. (2008). Mesophilic crenarchaeota: proposal for a third archaeal phylum, the Thaumarchaeota. Nature Rev. Microbiol. *6*, 245–252.

Bryant, D.A., Garcia Costas, A.M., Maresca, J.A., Gomez Maqueo Chew, A., Klatt, C.G., Bateson, M.M., Tallon, L.J., Hostetler, J., Nelson, W.C., Heidelberg, J.F., and Ward, D.M. (2007). *Candidatus* Chloroacidobacterium thermophilum: an anaerobic phototrophic acidobacterium. Science *317*, 523–526.

Casanueva, A., Galada, N., Baker, C.G., Grant, W.D., Heaphy, S., Jones, B., Ma, Y., Ventosa, A., Blamey, J., and Cowan, D.A. 2008. Nanoarchaeal 16S rRNA sequences are widely dispersed in hyperthermophilic and mesophilic halophilic environments. Extremophiles *12*, 651–656.

Cavaletti, L., Monciardini, P., Bamonte, R., Schumann, P., Rohde, M., Sosio, M., and Donadio, S. (2006). New lineage of filamentous, spore-forming, Gram-positive bacteria from soil. Appl. Environ. Microbiol. *72*, 4360–4369.

Cavalier-Smith, T. (2002). The neomuran origin of Archaebacteria, the negibacterial root of the universal tree and bacterial megaclassification. Int. J. Syst. Evol. Microbiol. *52*, 7–76.

Choo, Y.J., Lee, K., Song, J., and Cho, J.-C. (2007). *Puniceicoccus vermicola* gen. nov., sp. nov., a novel marine bacterium, and description of *Puniceicoccaceae* fam. nov., *Puniceicoccales* ord. nov., *Opitutaceae* fam. nov., *Opitutales* ord. nov. and *Opitutae* classis nov. in the phylum '*Verrucomicrobia*'. Int. J. Syst. Evol. Microbiol. *57*, 532–537.

DasSarma, P., and DasSarma, S. (2008). On the origin of prokaryote "species": the taxonomy of halophilic Archaea. Saline Syst. *4*, 5.

DeLong, E.F. (1998). Everything in moderation: Archaea as "nonextremophiles". Curr. Opin. Genet. Devel. *8*, 649–654.

DeLong, E.F., and Pace, N.R. (2001). Environmental diversity of bacteria and archaea. Syst. Biol. *50*, 470–478.

Di Giulio, M. (2007). The tree of life might be rooted in the branch leading to Nanoarchaeota. Gene *401*, 108–113.

Doolittle, W.F. (1999). Phylogenetic classification and the universal tree. Science *284*, 2124–2128.

Dworkin, M., Falkow, S., Rosenberg, E., Schleifer, K.-H., and Stackebrandt, E., eds. (2006). The Prokaryotes. A Handbook on the Biology of Bacteria: Ecophysiology and Biochemistry (New York: Springer).

Elkins, J.G., Podar, M., Graham, D.E., Makarova, K.S., Wolf, Y., Randau, L., Hedlund, B.P., Brochier-Armanet, C., Kunin, V., Anderson, I., Lapidus, A., Goltsman, E., Barry, K., Koonin, E.V., Hugenholtz, P., Kyrpides, N., Wanner, G., Richardson, P., Keller, M., and Stetter, K.O. (2008). A korarchaeal genome reveals insights into the evolution of the Archaea. Proc. Natl. Acad. Sci. U.S.A. *105*, 8102–8107.

Emerson, D., Rentz, J.A., Lilburn, T.G., Davis, R.E., Aldrich, H., Chan, C., and Moyer, C.L. (2008). A novel lineage of Proteobacteria involved in formation of marine Fe-oxidizing microbial mat communities. PLoS One *2(8)*, e667.

Euzéby, J.P. (1997). List of bacterial names with standing in nomenclature: a folder available on the internet. Int. J. Syst. Bacteriol. *47*, 590–592. (Online: http://www.bacterio.cict.fr).

Fukunaga, Y., Kurahashi, M., Yanagi, K., Yokota, A., and Harayama, S. (2008). *Acanthopleuribacter pedis* gen. nov., sp. nov., a marine bacterium isolated from a chiton, and description of *Acanthopleuribacteraceae* fam. nov., *Acanthopleuribacterales* ord. nov., *Holophagaceae* fam. nov., *Holophagales* ord. nov. and *Holophagae* classis nov. in the phylum *'Acidobacteria'*. Int. J. Syst. Evol. Microbiol. *58*, 2597–2601.

Garrity, G.M., and Holt, J.G. (2001a). The road map to the Manual. In Bergey's Manual of Systematic Bacteriology, 2nd edition, Vol. 1. The *Archaea* and the deeply branching and phototrophic *Bacteria*, D.R. Boone, R.W. Castenholz, and G.M. Garrity, eds. (New York: Springer), pp. 119–166.

Garrity, G.M., and Holt, J.G. (2001b). Class I. *Chrysiogenetes* class. nov. In Bergey's Manual of Systematic Bacteriology, 2nd edition, Vol. 1. The *Archaea* and the deeply branching and phototrophic *Bacteria*, D.R. Boone, R.W. Castenholz, and G.M. Garrity, eds. (New York: Springer), p. 421.

Gibbons, N.E., and Murray, R.G.E. (1978). Proposals concerning the higher taxa of bacteria. Int. J. Syst. Bacteriol. *28*, 1–6.

Gupta, R.S. (1998). What are archaebacteria: life's third domain or monoderm prokaryotes related to Gram-positive bacteria? A new proposal for the classification of prokaryotic organisms. Mol. Microbiol. *29*, 695–707.

Handelsman, J. (2004). Metagenomics: Application of genomics to uncultured microorganisms. Microbiol. Mol. Biol. Rev. *68*, 669–685.

Hedlund, B.P., Gosink, J.J., and Staley, J.T. (1997). *Verrucomicrobia* div. nov., a new division of the *Bacteria* containing three new species of *Prosthecobacter*. Antonie van Leeuwenhoek *72*, 29–38.

Herlemann, D.P.R., Geissinger, O. Ikeda-Ohtsubo, W. Kunin, V. Sun, H. Lapidus, A., Hugenholtz, P. and Brune, A. (2009). Genomic analysis of "Elusimicrobium minutum," the first cultivated representative of the phylum "Elusimicrobia" (formerly Termite Group1). Appl. Environ. Microbiol. *75*, 2841–2849.

Hohn, M.J., Hedlund, B.P., and Huber, H. (2002). Detection of 16S rDNA sequences representing the novel phylum 'Nanoarchaeota': indication for a wide distribution in high-temperature biotopes. Syst. Appl. Microbiol. *25*, 551–554.

Huber, H., Hohn, M.J., Rachel, R., Fuchs, T., Wimmer, V.C., and Stetter, K.O. (2002). A new phylum of Archaea represented by a nanosized hyperthermophilic symbiont. Nature *417*: 63–67.

Hugenholtz, P., Goebel, B.M., and Pace, N.R. (1998a). Impact of culture-independent studies on the emerging phylogenetic view of bacterial diversity. J. Bacteriol. *180*: 4765–4774.

Hugenholtz, P., Pituelle, C., Hershberger, K.L., and Pace, N.R. (1998b). Novel division-level bacterial diversity in a Yellowstone hot spring. J. Bacteriol. *180*, 366–376.

Iino, T., Mori, K., Uchino, Y., Nakagawa, T., Harayama, S., and Suzuki, K.-i. (2010). *Ignavibacterium album* gen. nov., sp. nov., a moderately thermophilic anaerobic bacterium isolated from microbial mats at a terrestrial hot spring, and proposal of *Ignavibacteria* classis nov. for a novel lineage at the periphery of the green sulfur bacteria. Int. J. Syst. Evol. Microbiol., in press (doi: ijs.0.012484–0).

Jumas-Bilak, E., Roudière, L., and Marchandin, H. (2009). Description of *'Synergistetes'* phyl. nov. and emended description of the phylum *'Deferribacteres'* and of the family *Syntrophomonadaceae*, phylum *'Firmicutes'*. Int. J. Syst. Evol. Microbiol. *59*, 1028–1035.

Karner, M.B., DeLong, E.F., and Karl, D.M. (2001). Archaeal dominance in the mesopelagic zone of the Pacific Ocean. Nature *409*, 507–510.

Könneke, M., Bernhard, A.E., de la Torre, J.R., Walker, C.B., Waterbury, J.B., and Stahl, D.A. (2005). Isolation of an autotrophic nitrifying marine archaeon. Nature *437*, 543–546.

Krieg, N.R., and Garrity, G.M. (2001). On using the Manual. In *Bergey's Manual* of Systematic Bacteriology, 2nd. ed, vol. 1, D.R. Boone, R.W. Castenholz, and G.M. Garrity, eds. (New York: Springer), pp. 15–19.

Lapage, S.P., Sneath, P.H.A., Lessel, E.F., Skerman, V.B.D., Seeliger, H.P.R., and Clark, W.A., eds. (1992). International Code of Nomenclature of Bacteria (1990 Revision). Bacteriological Code (Washington DC: American Society for Microbiology).

Locke, J. (1690). An Essay Concerning Humane Understanding (4th. ed., 1700) (London: Awnsham, John Churchill, and Samuel Manship).

Ludwig, W., and Klenk, H.-P. (2001). Overview: a phylogenetic backbone and taxonomic framework for prokaryotic systematics. In *Bergey's Manual* of Systematic Bacteriology, 2nd. ed, vol. I, D.R. Boone, R.W. Castenholz, and G.M. Garrity, eds. (New York: Springer), pp. 49–65.

Ludwig, W., and Schleifer, K.-H. (1994). Bacterial phylogeny based on 16S and 23S rRNA sequence analysis. FEMS Microbiol. Rev. *15*, 155–173.

Ludwig, W., Bauer, S.H., Bauer, M., Held, I., Kirchhof, G., Schulze, R., Huber, I., Spring, S., Hartmann, A., and Schleifer, K.H. (1997). Detection and *in situ* identification of representatives of a widely distributed new bacterial phylum. FEMS Microbiol. Lett. *153*, 181–190.

Macy, J.M., Nunan, K., Hagen, K.D., Dixon, D.R., Harbour, P.J., Cahill, M., and Sly, L.I. (1996). *Chrysiogenes arsenatis* gen. nov., sp. nov., a new arsenate-respiring bacterium isolated from gold mine wastewater. Int. J. Syst. Bacteriol. *46*, 1153–1157.

Madigan, M.T., Martinko, J.M., Dunlap, P.V, and Clark, D.P. (2009). Brock Biology of Microorganisms. 12th ed. (San Francisco: Pearson-Benjamin Cummings).

Meng, J., Wang, F., Wang, F., Zheng, Y., Peng, X., Zhou, H., and Xiao, X. (2009). An uncultivated crenarchaeota contains functional bacterochlorophyll a synthase. ISME J. 3: 106–116.

Mongodin, M.E.F., Nelson, K.E., Duagherty, S., DeBoy, R.T., Wister, J., Khouri, H., Weidman, J., Balsh, D.A., Papke, R.T., Sanchez Perez, G., Sharma, A.K., Nesbø, C.L., MacLeod, D., Bapteste, E., Doolittle, W.F., Charlebois, R.L., Legault, B., and Rodríguez-Valera, F. (2005). The genome of *Salinibacter ruber*: convergence and gene exchange among hyperhalophilic bacteria and archaea. Proc. Nat. Acad. Sci. USA *102*, 18147–18152.

Mori, K., Yamaguchi, K., Sakiyama, Y., Urabe, T., and Suzuki, K. (2009). *Caldisericum exile* gen. nov., sp. nov., an anaerobic, thermophilic, filamentous bacterium of a novel bacterial phylum, *Caldiserica* phyl. nov., originally called the candidate phylum OP5, and description of *Caldisericaceae* fam. nov., *Caldisericales* ord. nov. and *Caldisericia* classis nov. Int. J. Syst. Evol. Microbiol. *59*, 2894–2898.

Murray, R.G.E. (1974). A place for bacteria in the living world. In *Bergey's Manual* of Determinative Bacteriology, 8th Edition, R.E. Buchanan and N.E. Gibbons, eds. (Baltimore: Williams & Wilkins), pp. 4–9.

Murray, R.G.E. (1984). The higher taxa, or, A place for everything ...? In *Bergey's Manual* of Systematic Bacteriology, 1st Edition, vol. 1, N.R. Krieg and J.G. Holt, eds. (Baltimore: Williams & Wilkins), pp. 31–34.

Nesbø, C.L., Dlutek, M., Zhaxybayeva, O, and Doolittle, W.F. (2006). Evidence for existence of "Mesotogas," members of the order Thermotogales adapted to low-temperature environments. Appl. Environ. Microbiol. *72*, 5061–5068.

Oren, A. (2004a). Prokaryote diversity and taxonomy: present status and future challenges. Phil. Trans. R. Soc. London B *359*, 623–638.

Oren, A. (2004b). A proposal for further integration of the Cyanobacteria under the Bacteriological Code. Int. J. Syst. Evol. Microbiol. *54*, 1895–1902.

Oren, A. (2008). Nomenclature and taxonomy of halophilic archaea – Comments on the proposal by DasSarma & DasSarma [Saline Systems 4 (2008), 5] for nomenclatural changes within the order Halobacteriales. Int. J. Syst. Evol. Microbiol. *58*, 2245–2246.

Oren, A., and Tindall, B.J. (2005). Nomenclature of the cyanophyta/cyanobacteria/cyanoprokaryotes under the International Code of Nomenclature of Prokaryotes. Algological Studies *117*, 39–52.

Preston, C.M., Wu, K.Y., Molinski, T.F., and DeLong, E.F. (1996). A psychrophilic crenarchaeon inhabits a marine sponge: *Cenarchaeum symbiosum* gen. nov., sp. nov. Proc. Natl. Acad. Sci. U.S.A. *93*, 6241–6246.

Prokofeva, M.I., Kostrikina, N.A., Kolganova, T.V., Tourova, T.P., Lysenko, A.M., Lebedinsky, A.V., and Bonch-Osmolovskaya, E.A. (2009). Isolation of the anaerobic thermoacidophilic crenarchaeote *Acidiolobus saccharovorans* sp. nov. and proposal of *Acidilobales* ord. nov., including *Acidilobaceae* fam. nov. and *Caldisphaeraceae* fam. nov. Int. J. Syst. Evol. Microbiol. *59*, 3116–3122.

Rosselló-Mora, R., and Amann, R. (2001). The species concept for prokaryotes. FEMS Microbiol. Rev. *25*, 39–67.

Sakai, S., Imachi, H., Hanada, S., Ohashi, A., Harada, H., and Kamagata, Y. (2008). *Methanocella paludicola* gen. nov., sp. nov., a methane-producing archaeon, the first isolate of the lineage 'Rice Cluster I', and proposal of the new archaeal order *Methanocellales* ord. nov. Int. J. Syst. Evol. Microbiol. *58*, 929–936.

Skerman, V.B.D., McGowan, V., and Sneath, P.H.A. (1980). Approved lists of bacterial names. Int. J. Syst. Bacteriol. *30*, 225–420.

Sokolova, T., Hanel, J., Onyenwoke, R.U., Reysenbach, A.-L., Banta, A., Geyer, R., González, J.M., Whitman, W.B., and Wiegel, J. (2007). Novel chemolithotrophic, thermophilic, anaerobic bacteria *Thermolithobacter ferrireducens* gen. nov., sp. nov. and *Thermolithobacter carboxydivorans* sp. nov. Extremophiles *11*, 145–157.

Stackebrandt, E. (2006). Defining taxonomic ranks. In The Prokaryotes. A Handbook on the Biology of Bacteria: Ecophysiology and Biochemistry, vol. 1, M. Dworkin, S. Falkow, E. Rosenberg, K.-H. Schleifer, and E. Stackebrandt, eds. (New York: Springer), pp. 29–57.

Stackebrandt, E., Frederiksen, W., Garrity, G.M., Grimont, P.A.D., Kämpfer, P., Maiden, M.C.J., Nesme, X., Rosseló-Mora, R., Swings, J., Trüper, H.G., Vauterin, L., Ward, A.C., and Whitman, W.B. (2002). Report of the ad hoc committee for the re-evaluation of the species definition in bacteriology. Int. J. Syst. Evol. Microbiol. *52*, 1043–1047

Strous, M., Pelletier, E., Mangenot, S., Rattei, T., Lehner, A., Taylor, M.W., Horn, M., Daims, H., Bartol-Mavel, D., Wincker, P., Barbe, V., Fonknechten, N., Vallenet,

D., Segurens, B., Schenowitz-Truong, C., Médigue, C., Collingro, A., Snel, B., Dutilh, B.E., Op den Camp, H.J., van der Drift. C., Cirpus, I., van de Pas-Schoonen, K.T., Harhangi, H.R., van Niftrik, L., Schmid, M., Keltjens, J., van de Vossenberg, J., Kartal, B., Meier, H., Frishman, D., Huynen, M.A., Mewes, H.W., Weissenbach, J., Jetten, M.S., Wagner, M., and Le Paslier, D. (2006). Deciphering the evolution and metabolism of an anammox bacterium from a community genome. Nature *440*, 790–794.

Vandamme, P., Pot, B., Gillis, M., de Vos, P., Kersters, K., and Swings, J. (1996). Polyphasic taxonomy, a consensus approach to bacterial systematics. Microbiol. Rev. *60*, 407–438.

Wayne, L.G., Brenner, D.J., Colwell, R.R., Grimont, P.A.D., Kandler, O., Krichevsky, M.I., Moore, L.H., Moore, W.E.C., Murray, R.G.E., Stackebrandt, E., Starr, M.P., and Trüper, H.G. (1987). Report of the ad hoc committee on reconciliation of approaches to bacterial systematics. Int. J. Syst. Bacteriol. *37*, 463–464.

Woese, C.R. (1987). Bacterial evolution. Microbiol. Rev. *51*, 221–271.

Woese, C.R., and Fox, G.E. (1977). Phylogenetic structure of the prokaryotic domain: The primary kingdoms. Proc. Natl. Acad. Sci. U.S.A. *74*, 5088–5090.

Woese, C.R., Kandler, O., and Wheelis, M.L. (1990). Towards a natural system of organisms: proposal of the domains Archaea, Bacteria, and Eucarya. Proc. Natl. Acad. Sci. U.S.A. *87*, 4576–4579.

Yamada, T., Sekiguchi, Y., Hanada, S., Imachi, H., Ohashi, A., Harada, H., and Kamagata, Y. (2006). *Anaerolinea thermolimosa* sp. nov., *Levilinea saccharolytica* gen. nov., sp. nov. and *Leptolinea tardivitalis* gen. nov., sp. nov., novel filamentous anaerobes, and description of the new classes *Anaerolineae* classis nov. and *Caldilineae* classis nov. in the bacterial phylum *Chloroflexi*. Int. J. Syst. Evol. Microbiol. *56*, 1331–1340.

Yarza, P., Richter, M., Peplies, J., Euzéby, J., Amann, R., Schleifer, K.-H., Ludwig, W., Glöckner, F.O., and Rosselló-Mora, R. (2008). The All-Species Living Tree Project: a 16S rRNA-based phylogenetic tree of all sequenced type strains. Syst. Appl. Microbiol. *31*, 241–250.

Zhang, H., Sekiguchi, Y., Hanada, S., Hugenholtz, P., Kim, H., Kamagata, Y., and Nakamura, K. (2003). *Gemmatimonas aurantiaca* gen. nov., sp. nov., a Gram-negative, aerobic, polyphosphate-accumulating microorganism, the first cultured representative of the new bacterial phylum *Gemmatimonadetes* phyl. nov. Int. J. Syst. Evol. Microbiol. *53*, 1155–1163.

Zinder, S.H., and Dworkin, M. (2006). Morphological and physiological diversity. In The Prokaryotes. A Handbook on the Biology of Bacteria: Ecophysiology and Biochemistry, vol. 1, M. Dworkin, S. Falkow, E. Rosenberg, K.-H. Schleifer, and E. Stackebrandt, eds. (New York: Springer), pp. 185–220.

# Rooting the Tree of Life

Greg Fournier

Abstract

Defining the evolutionary relationships between groups of organisms is a major part of modern-day microbiology. With the continuing dramatic increase in the availability of genomic data, these techniques have been extended to describing an all-encompassing 'tree of life'. However, identifying the location of the root of this tree corresponding to the most recent common ancestor is a challenging and distinct problem that has yet to be solved. To date, many investigations have proposed various roots, using a wide diversity of biological data and techniques. A survey of the most promising of these models illustrates the difficulty faced in reaching a scientific consensus on the issue, as well as the additional philosophical complications posed by our emerging understanding of the role of horizontal gene transfer in genome evolution.

## Introduction

All phylogenies depict relationships between sets of objects as a furcating tree, based on a measurement of their similarity using defined criteria. In many cases, a phylogeny is 'natural', in that it not only organizes and classifies objects, but also relates to the process by which these objects became distinct entities. In these cases, it is possible to root the tree by adding a temporal dimension. This shows which splits within the tree occurred before others (along the same branch-path), and identifies the specific branch containing the actual root. Since all organisms on Earth trace their ancestries back to a common origin, the history of life can be depicted phylogenetically based on sets of characters inherited through evolutionary processes. Consequently, the rooting of such a 'Tree of Life' (ToL) orders the evolutionary history of life on Earth with respect to time, and can potentially provide important information about the nature of the earliest life forms. However, this is a challenging goal, as rooting universal trees requires unconventional methods frequently based on obscure external criteria, both often proving controversial. Furthermore, the very concept itself relies on many assumptions about the tree-like nature of organismal evolution that must be addressed.

## What is the 'Tree of Life'?

All microbial individuals reproduce by division of a parent individual, and all sexually reproducing and/or metazoan organisms arise directly from the genetic and somatic material of one or two parent organisms. Invoking theoretical omniscience, this map of 'cellular descent' would coalesce into a furcating 'tree of cellular divisions' (ToCD) describing the phylogenetic relationships between all species (Cavalier-Smith, 2006b). Therefore, a vertical line of organismal descent tracing all cellular ancestry directly back to a universal, most recent common ancestor (MRCA) is a biological reality. However, evolution is a complex phenomenon that involves not only replication and descent but the inheritance of genetic information, which can be transferred in a variety of ways that do not conform to defined lineages. To what, then, does the ToL refer to, and how does this definition relate to the emerging picture of complex, reticulate evolution in the biological (and especially prokaryotic) world?

The replication and inheritance of genetic material is not theoretically essential to reproduction and a cellular line of descent; one can imagine a membrane-enclosed, autocatalytic 'organism' which grows and replicates without genes. However, as such an entity lacks heritable information it would be incapable of Darwinian evolution. While the earliest self-replicating entities on Earth may well have resembled such a 'precellular' autocatalytic anabolic system (Wächtershäuser, 1990, 1992, 2000), the development of heritable information was of such enormous consequence that its existence is generally considered essential to any useful definition of life. This heritable information in the form of genes and genomes is also by far the most substantial evidence we have for retracing evolutionary history; it is therefore no surprise that the concepts of organismal and genomic/molecular evolution have become so intimately intertwined.

Ironically, as the study of genetics has moved from studying individual genes to entire genomes, and most recently to the metagenomes of entire communities and ecosystems, evolutionary thinking has become increasingly gene-centric (Doolittle, 1999; Koonin *et al.*, 2009; Puigbo *et al.*, 2009). This shift has largely been driven by the discovery of surprising levels of genomic variability within and across prokaryotic species, revealing the substantial role of horizontal gene transfer. It is increasingly popular to speak of 'selfish genes' and even of the obsolescence of the organism as a true biological entity from an evolutionary perspective, with individual genetic elements being the 'fundamental units of evolution' (Dawkins, 1976; Koonin *et al.*, 2009).

This gene-centric view of evolution leads to a mereological essentialism (or an even more extreme mereological nihilism) which proposes that since genes are the basic unit of evolution and frequently rearranged between lineages via transfer, then the wholes comprised of and defined by these parts (genomes and organisms, respectively) do not exist in an evolutionary meaningful context, and the concept of 'organismal' lineages (and even species) is biologically invalid (Koonin *et al.*, 2009). Instead, what remains are the many disparate evolutionary histories of individual genes, which may incidentally share trajectories indicative of a 'central tendency' of limited evolutionary relevance. Clearly, this also invalidates the biological concept of a 'tree of life', and precludes any discussion or characterization of its rooting.

It is true that most (if not all) genes have histories that are incompatible with a singular, monistic evolutionary history (Bapteste *et al.*, 2009). However, while evolution presupposes and requires a vertical line of descent via reproduction, the two are distinct processes, and the complexity of the former should not be used to deny the reality and significance of the latter. Evolution is, by definition, the change in the genetic material within a population of organisms across generations; therefore, any process by which genetic material within a population changes that is unrelated to reproduction will show a history that is unrelated to the organismal vertical line of descent. Most notably, these reticulate processes include hybridization (introgression), the duplication and selective loss of genes within lineages (lineage sorting) and horizontal gene transfer. Awareness of this complexity has led to the pursuit of a genome-based ToL, a middle ground between gene-centric and organismal paradigms.

Two common approaches in constructing a genome-based ToL are supermatrix analyses, in which sequence alignments for individual gene families are concatenated into a single dataset which is then used to construct a tree (Delsuc *et al.*, 2005), and supertree analyses, in which a consensus phylogeny is constructed from multiple gene trees (Bininda-Emonds, 2004). Regardless of the particular approach used, these methods result in an 'averaging' of phylogenetic signal across the genome, which approximates the underlying organismal history. While these approaches acknowledge that individual gene histories within a genome must be taken into account if attempting to reconstruct an organism-based ToL, there is no guarantee that an average or median phylogeny is a good proxy for the signal of vertical inheritance. Often, these approaches will give equal weight to genes with different rates of HGT at different taxonomic levels. This is especially a problem if there are 'highways' of gene sharing- that is, large numbers of genes shared between specific groups of otherwise phylogenetically distinct organisms (Beiko *et al.*, 2005). This can easily be mistaken for a consistent signal supporting an organismal tree. For example these types of analyses often group members of the order Thermotogales with the Firmicutes, and the members of the Aquificales with the Epsilonproteobacteria. In contrast, 16S rRNA gene phylogenies and concatenated ribosomal protein phylogenies strongly support a very different phylogeny, with both Thermotogales and Aquificales being deeply branching bacterial lineages (Boussau *et al.*, 2008; Zhaxybayeva *et al.*, 2009). Perhaps even more problematic, the existence of more than one strong, consistent phylogenetic signal (caused by any combination of vertical inheritance and biased HGT) can cause the final consensus phylogeny to contain groups not supported or even present in any individual gene history (Wilkinson *et al.*, 2007).

Various tree-based and surrogate methods (such as detecting compositional or codon usage biases) exist for the detection of horizontal gene transfers within genomic datasets (Beiko and Hamilton, 2006; Cortez *et al.*, 2009; Ragan, 2001). With a confidence in these methods, it is possible to avoid (or at least diminish) the problem of averaging in genomic datasets by removing genes with evidence of HGT. It is then hoped that the remaining 'core' genes are more reliable for generating a consensus phylogeny indicative of vertical descent (Ciccarelli *et al.*, 2006; Galtier and Daubin, 2008; Wu and Eisen, 2008). While this approach depends entirely on the ability to accurately identify horizontally transferred genes, it can be victim to both false positives due to phylogenetic reconstruction and compositional artifacts, and false negatives due to a lack of phylogenetic information. The latter are especially problematic, as their cumulative effect can preserve the 'fallacious averaging' which plagues whole-genome methods. However, the more stringent the criteria for excluding genes which may have been transferred, the fewer and fewer genes remain with which to infer a vertical ToL. This is the major philosophical challenge to what has been labelled the 'tree of one percent' as it is clear that the resulting phylogeny only takes into account a small minority of the evolution occurring in genomes. As such, '... a single bifurcating tree is an insufficient model to describe the microbial evolutionary process' (Dagan and Martin, 2006).

This objection is reinforced by the observation that these methods tend to treat HGT as 'noise' to be removed from the true vertical signal, as opposed to a major evolutionary process in their own right. However, as discussed in Bapteste *et al.* (2009) and the accompanying commentary, this problem can largely be resolved by the adoption of a 'pluralistic' view which explicitly acknowledges that the ToL is a distinct concept from the complete evolutionary history of life. Furthermore, and for similar reasons, the concepts of 'ancestry' and 'relatedness' must be treated more distinctly, as the former is more closely associated with the organismal tree, while the latter takes into account the substantial contribution of HGT to the gene content and phenotypic character expressed by genomes. In such a pluralistic framework, a set of highly conserved genes, which are likely vertically inherited is free to be used as a proxy for organismal descent, and a more narrowly defined ToL. This tree, even if it comprises the minority of the phylogenetic signal within genomes, can then be used as the scaffold onto which the network-like relationships of most other genes can be mapped, in order to produce a more comprehensive view of evolution (especially among prokaryotes). Thus, albeit somewhat diminished in evolutionary scope and significance, a biologically relevant organismal tree of life can still be inferred, and its rooting provides insights into the nature of early life and its diversification.

## Rooting and topology

Given the above-stated technical and philosophical complexities, rooting the ToL may seem like an impossible task, or at least conceptually premature. However, rooting the ToL is not necessarily dependent on the phylogenetic accuracy of the entire tree, or even that the evolution of all groups be primarily 'tree-like'. Rather, it is only essential that the root resides within a well-supported branch, and that the method of identifying the root does not implicitly assume or depend on the accuracy of the rest of the tree. For this reason, many of the issues in constructing a ToL, while important, can often be sidestepped. Even if there is disagreement between conflicting ToL topologies, these only substantially impact analyses if the conflicting branches coincide with potential locations of the root.

The initial phylogenetic analysis of 16S rRNA sequences that identified the Archaea as a distinct clade and led to the current three domain system also identified two major monophyletic groups within Archaea (Woese and Fox, 1977; Woese and Olsen, 1986). However, this 'archaebacterial' model was challenged by subsequent analyses of structural ribosomal features (Lake *et al.*, 1984), elongation factor EF-1 structure and phylogeny (Rivera and Lake, 1992), and a re-analysis of 16S rRNA phylogeny (Lake, 1987), supporting an alternative model in which the Archaea were paraphyletic. In this so-called 'eocyte' model, Euryarchaeota groups with Bacteria, and Crenarchaeota groups with Eukarya. This produces a novel deep branch within the 16S rRNA phylogeny which also provides a convenient location for the root of the ToL. In this model, eukaryotes are derived from crenarchaeal ancestors, while the euryarchaeotes are a sister group to the bacteria.

Continuing the debate, subsequent investigations identified that the 16S analysis supporting the eocyte model was likely the result of an artefact, and that other ribosomal support agreed with an archaebacterial model (Gouy and Li, 1989; Olsen and Woese, 1989). Phylogenetic analyses of genes encoding elongation factor EF-G(2) and components of the protein targeting machinery also supported the monophyly of the Archaea and Eukarya, which is incompatible with the division of the Archaea proposed under the eocyte hypothesis (Cammarano *et al.*, 1999; Gribaldo and Cammarano, 1998a). While additional phylogenetic analyses providing support to the eocyte model have recently been presented (Cox *et al.*, 2008), the balance of evidence is generally currently accepted as being in favour of the archaebacterial model. In any case, a paraphyletic archaeal clade is compatible with many of the proposed ToL rootings.

## The nature of the MRCA

The root of the ToL is not the same as the origin of life. By definition, the root is the shared common ancestor of all extant lineages; it does not rely on or encompass any biological evolution that precedes this point in time. The distance between the root and the actual origin of life is an interesting problem, however, with direct relevance to the nature of the MRCA. In 'root early' models, the root closely coincides with the origin, with Darwinian evolution, cellularization, and translation only evolving later out of this progenote state (Koonin and Martin, 2005; Woese and Fox, 1977). In 'root late' models, the MRCA was a fully functioning, if primitive, prokaryotic cell (Fig. 6.1). This presupposes a substantial period of Darwinian evolution between the origin of life and the MRCA, and is supported by the universal conservation of many genes, including those encoding ribosomal proteins and RNA, aminoacyl-tRNA synthetases, ATP synthases, and other proteins involved in translation. As such, it also predicts that the MRCA (or the lineage leading directly to it) was not alone, and that substantial diversification and extinction could have taken place by this time. Furthermore, due to the pervasive role of horizontal gene transfer, this also predicts that the MRCA likely contained genetic material that had originally evolved in other lineages that did not survive to leave existing descendants. As such, extant life likely inherited many genes from this diverse primitive extinct community, even if the MRCA was a single organismal population (Fournier *et al.*, 2009; Gogarten *et al.*, 2007).

Regardless of the specific nature of life at the time of the MRCA, it should not be thought of as a single individual. Cellular entities, like organisms today, would have existed as populations, likely within an ecosystem together with other (subsequently extinct) organisms with which it could share genes. If a progenote existed, it would have evolved as a community, without organismal genealogy or intralineage variation (Woese, 2002). In either case, the concept of a root retains its validity, even if the direct inference to an organismal line of descent breaks down along the deepest branch in the latter scenario. It is also important to note that many of the most powerful techniques for rooting ToL phylogenies depend upon the existence of paralogous genes diverging before the time of the MRCA. As such, even in the case of 'root early' scenarios, a substantial amount of evolution must have taken place within these primitive systems.

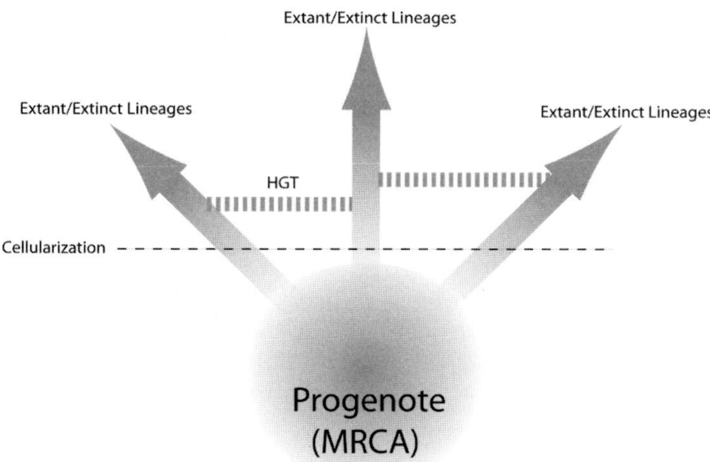


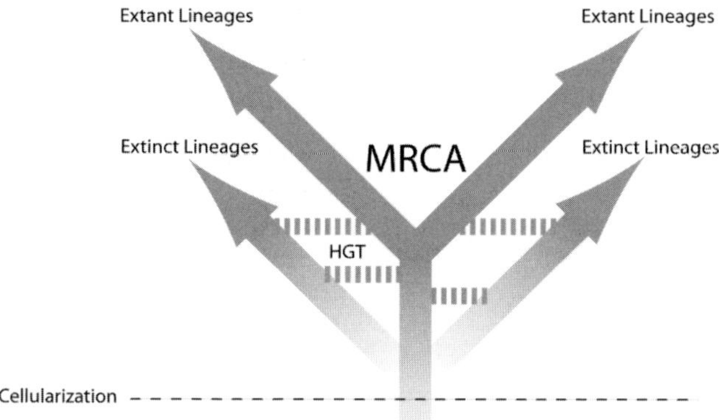


**Figure 6.1** Models illustrating the emergence of organismal lineages from the MRCA. The dotted line representing time of cellularization coincides with the emergence of discrete lineages subject to Darwinian evolution. The larger progenote cloud in the MRCA Early model represents a greater amount of molecular evolution occurring within this state.

### Methods for rooting

Typically, a phylogenetic tree is rooted by the inclusion of an outgroup, a related group which is less similar to other members of the tree than they are to one another. Ideally, this outgroup joins the tree at their point of common origin, rooting it. However, by definition, no outgroup exists for a universal tree containing representatives from all domains of life. Therefore, more creative approaches must be used. Methods to root the ToL are numerous, but can be broadly categorized based on two primary distinctions: (1) inclusive vs. exclusive; and (2) sequence vs. non-sequence characters. Inclusive rooting methods attempt to identify the specific branch within the tree that contains the root, while exclusive methods identify phylogenetic groups containing states which are shown to be derived, therefore excluding the root from any location within these groups or their ancestral branches. Sequence-based approaches utilize characters defined by protein, RNA, or DNA sequences, such as signature residues, insertion/deletions (indels) or phylogenetic trees derived from multiple sequence alignments. Sequence independent approaches depend upon physiological characters that are not directly based on sequence (e.g. growth conditions or membrane structure), yet nonetheless can be compared across universal trees. However, these distinctions are not entirely clear-cut, and can overlap conceptually, as sequence and physiology are frequently intimately related (Singer and Hickey, 2003; Tekaia *et al.*, 2002). Different combinations of these approaches and types of characters are often used, each with their own advantages and shortcomings.

## Paralogue rooting

Many protein families contain ancestral gene duplications, where divergent, paralogous copies of a specific gene existed at the time of the most recent common ancestor (MRCA). In effect, this allows each paralogue to act as outgroup for the other within a phylogenetic reconstruction, producing a reciprocal rooting of the tree. Ancestral gene duplications have been frequently used to root the tree of life, an approach pioneered using protein sequences of the paralogous catalytic and non-catalytic subunits of the ATP synthase complex, and paralogous elongation factors (Gogarten *et al.*, 1989; Iwabe *et al.*, 1989). This approach was subsequently expanded to many other gene families containing ancestral duplications, including aminoacyl-tRNA synthetases (Brown and Doolittle, 1995), protein targeting machinery components (Gribaldo and Cammarano, 1998b), and a revisitation of elongation factors (Baldauf *et al.*, 1996). Aminoacyl-tRNA synthetases (aaRS) are an especially rich resource, containing many pairs or triplets of ancient paralogues with clear homology that allows reliable alignment and tree reconstruction (TyrRS/TrpRS, AsnRS/AspRS, GlnRS/GluRS, ValRS/IleRS/LeuRS, MetRS/CysRS). While other sets of aaRS show clear structural and sequence homology, they are too divergent to have a clear phylogenetic signal useful for paralogue rooting. Additionally, this approach has been applied in at least one case to ancient internal gene duplications, using the large subunit of carbamoyl phosphate synthetase (Islas *et al.*, 2007; Lawson *et al.*, 1996; Olendzenski and Gogarten, 1998; Schofield, 1993). All of these analyses support placing the root either on the branch leading to the bacteria, or, additionally in the case of the carbamoyl phosphate synthetase, possibly within the bacterial domain itself. However, phylogenetic analysis is complicated by the possibility of ancient horizontal gene transfer events, differential paralogue loss, as well as artifacts of tree reconstruction, especially those caused by long branches frequently associated with interdomain relationships (Zhaxybayeva *et al.*, 2005).

The origin of the deep long branches observed between paralogues is especially relevant to these arguments. It has been suggested that long, bare branches are an artefact of divergence between paralogues, driven by the acquisition of novel function in one or both gene copies (Cavalier-Smith, 2006b). As such, the location of the paralogue root will gravitate to the longest interdomain branch (typically that leading to the bacteria), which is the observed result. However, it is unclear that this is the case; many aminoacyl-tRNA synthetases show branches between paralogues of only modest length, which still place the root on the branch leading to the bacteria. Since the divergence of these proteins was driven almost entirely by substrate specificity at limited binding sites, strong positive selection (or relaxing of purifying selection) resulting in an increase in the substitution rate across the entire protein is unexpected. Therefore, it seems that paralogue rooting is most reliable in proteins with limited functional changes between paralogues, and may be less reliable due to the aforementioned artifact only in cases of radical functional change. Clearly, this potential does not affect any paralogue rootings within the bacterial domain, due to the absence of substantially long branches (Pereto *et al.*, 2004). Interestingly, however, excessively short branches between paralogues may also be an indication of lack of reliability, as this pattern can also be caused by recent and pervasive HGT. Variation of branch length between paralogues can also simply reflect their divergence times, especially under the 'MCRA late' model. Paralogues separated by longer branches lengths would correlate to earlier divergence times, as more evolution had an opportunity to occur before the time of the MRCA. It is also important to note that the length of bare branches between domains may not necessarily indicate an increased rate of evolution, as evidenced by the absence of deeply branching lineages. Rather, these bare branches also appear in simulations which assume steady states of extinction and speciation: older lineages become less likely to have any surviving sister taxa, producing 'sparser' deep branches (Zhaxybayeva and Gogarten, 2004). For these reasons, while reciprocal rooting of deep paralogues is a conceptually sound method that provides largely consistent results, caution is required in discerning sources of possible bias given the particular dataset used, and corroboration by additional methods less dependent on tree reconstruction are required to make any definitive claims.

## Archaeal rooting and tRNA

Rooting of ancient paralogues is not restricted to proteins. Structural RNAs such as tRNA can also duplicate and diverge, and the large number of specific tRNA species found within organisms attests that this process is a pervasive force in their evolution. As multiple, specific tRNAs are required for translation, these duplications must also have occurred very early in the history of life, and therefore are suitable for use in determining the root of the ToL. The earliest attempt at rooting the ToL using tRNA sequences relied on direct phylogenetic reconstruction (Fitch and Upper, 1987). With tRNA sequences from only a few genomes available at the time, this analysis did not have a high resolution, and produced conflicting results for different sets of tRNAs, each locating the root on a branch leading to a different domain. A similar and more comprehensive analysis using tRNA was later performed with the benefit of extensive genomic data availability (Xue *et al.*, 2003). This investigation identified alloacceptor pairs of tRNA within genomes that contain the most recognizable evidence of shared ancestry due to sequence similarity. Then, the overall sequence divergence of tRNA was quantified from a sample of genomes across all three domains. The two genomes with the strongest clustering of tRNAs were *Methanopyrus kandleri* and *Aeropyrum pernix*, implying that their tRNA complements are among the most primitive and therefore closest to the root, which was subsequently identified as existing somewhere between the branches leading to these groups. In the case of either the archaebacterial or eocyte model, this places the root very close to the origin of both major archaeal groups. However, an archaeal rooting using tRNA similarity is also subject to several artifacts to which this result could be attributed (Cejchan, 2004). More recently, phylogenetic analysis of tRNA sequences with the inclusion of secondary structure characters has produced a similar result (Sun and Caetano-Anolles, 2008a,b). Adding a monophyly constraint to the three domains and a fourth group for viral sequences, this analysis finds the most parsimonious rooting of the ToL to be between Bacteria/Eukarya and Archaea/viruses. Furthermore, it supports a very ancient divergence of viruses and Archaea, with a more recent divergence of Bacteria and Eukarya.

An alternative rooting of the ToL using tRNA sequences has also been proposed (Di Giulio, 2006), given the discovery of split tRNA genes within *Nanoarchaeum equitans* (Randau *et al.*, 2005a,b). Assuming that the process of short RNAs combining via *trans*-splicing into a functional tRNA recapitulates deep evolutionary history, this would be a primitive character unique to this organism. As such, the root of the ToL would be located along the branch leading to the Nanoarchaeota. Interestingly, this approach agrees with other tRNA-based analyses in rooting the ToL within the archaeal domain, albeit on a branch leading to a distinct group. However, given the unique intracellular parasite physiology of the Nanoarchaeota and great amount of sequence divergence observed between its proteins and those of other Archaea, it is likely that this group has undergone extensive evolution, and that any unusual traits are more likely to be derived than primitive.

Ten species of tRNA within the genome of *Caldivirga maquilingensis* have also been discovered to be encoded by split genes, although, in this instance, the tRNA appears to be in three fragments instead of two (Fujishima *et al.*, 2009). In comparison with *N. equitans*, different tRNAs are involved, and the locations of the splits are not conserved. Furthermore, the intervening regions of these fragmented tRNAs are homologous to tRNA introns found in other *Thermoproteales*. This evidence agrees with an alternative hypothesis in which split tRNAs are descended from intron-containing tRNAs, and do not represent a plesiomorphic (ancestral) character. Also central to the debate is the cladistic status of the Nanoarchaea; originally suspected to be a 'deep branching' archaeal lineage, more recent phylogenetic investigations have supported that this group is actually a derived euryarchaeote (Brochier *et al.*, 2005), further diminishing support for a nanoarchaeal rooting of the ToL.

## Paralogues and indels

The technique of paralogue rooting can also be used for exclusive methods, avoiding many of the artifacts of tree reconstruction inherent to sequence-based reciprocal rooting. Regions of sequence having undergone insertion/deletions (indels) are used as reliable larger-scale characters in these analyses. In many cases ancient paralogues will differ in their inclusion of an indel; if these differences are isolated to a subset of sequences within one paralogue, this can be inferred to be the derived state, and therefore exclude the root from within the group (Baldauf and Palmer, 1993; Gupta, 1998; Rivera and Lake, 1992). This approach also has the advantage of not being restricted to the use of universally distributed proteins, or paralogues that diverged before the MRCA. This is because exclusive methods are additive, and while individual datasets may only exclude the root from regions of the ToL, a combination of them will potentially exclude all but one branch, which therefore must be the location of the root. However, while theoretically powerful, this approach has two major limitations. First, as indels are single characters, the presence of convergent evolution (homoplasy) can completely invalidate individual analyses. This is especially true for small indels which could more easily be gained and lost with little or no impact on fitness, or for regions of sequence corresponding to protein surfaces, which are more likely to contain structures such as loops that can be easily lost. Second, in order to produce a convincing result HGT must be excluded as a possibility. As HGT detection usually relies on the same tree reconstruction methods as inclusive rooting, the potential for artifacts re-enters the analysis at this point.

Combining several of these polarizing indel analyses (Gao and Gupta, 2005; Gupta, 1998; Servin *et al.*, 2008; Skophammer *et al.*, 2006, 2007), it has been proposed that the location of the root of the ToL can be excluded from every location except for a branch within the bacteria, between the Firmicutes and the Actinobacteria/double-membrane prokaryotes (Lake *et al.*, 2009; Skophammer *et al.*, 2007). Interestingly, since this model includes an analysis of protein Hsp70 that groups together double-membrane bacteria and eukaryotes as derived, and another analysis of IF2/EFG paralogues that excludes the root from the combined grouping of eukaryotes and Archaea, the only consistent way to portray this particular schema of indel rooting is with a 'ring of life' presuming a fusion of lineages, possibly resulting in the formation of the double-membrane bacteria (Lake, 2009) (Fig. 6.2). However, this model is only as strong as the individual analyses that lead to it, which are not without criticism (Valas and Bourne, 2009).

One clear trend in indel analysis is that the quality of the signal used to exclude the root from specific groups seems to vary inversely with the size and sequence diversity of the group implied to be derived. For example, polarized GyrA/ParC and IF2/EFG indels excluding the root from the Actinobacteria and eukaryal/archaeal groups, respectively, are well supported and unlikely to

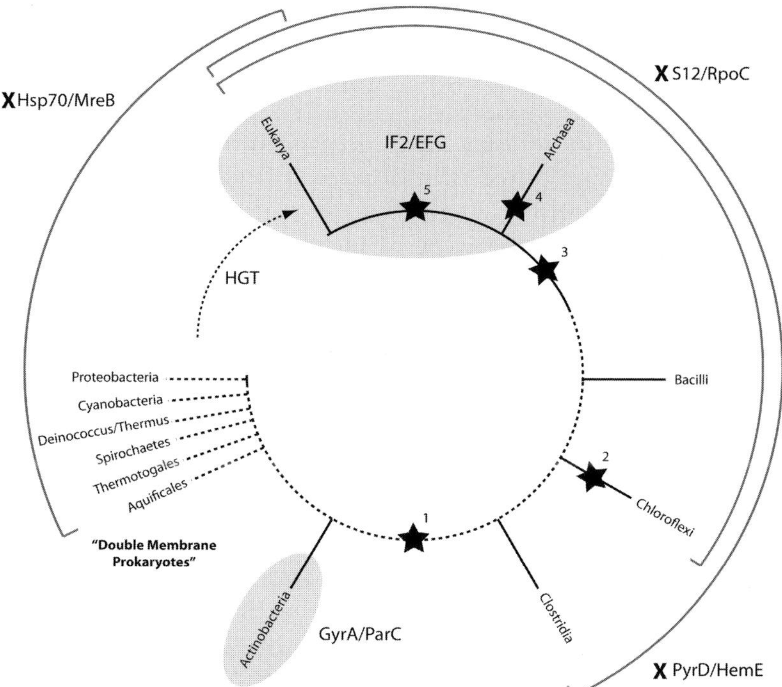


**Figure 6.2** The broken 'Ring of Life'. Shaded regions indicate where reliable indels associated with pairs of paralogues exclude the location of the root. Proposed indels which are found to be unreliable (as explained in the text) are indicated with brackets and 'X' marks next to the associated putative pairs of paralogues. Stars indicate the locations of proposed rootings of the ToL based on the following methods and models: (1) Indel-based; (2) transition analysis; (3) reciprocal rooting of paralogues/compositional analysis; (4) tRNA analysis; (5) Protoeukaryotic rooting. As the topology of the original 'ring of life' (Rivera and Lake, 2004) was defined by the groups excluded due to various indels, these otherwise unsupported relationships are now represented by a dotted line. Furthermore, double-membrane prokaryotes are broken up to represent the unknown phyletic status of this group. The possibility of extensive horizontal gene transfer between bacteria and eukaryotes (largely via endosymbionts) is indicated as 'closure' of the ring, preserving the spirit of the original model, if only in metaphor.

be due to HGT or homoplasy. Additional support for excluding the root from the archaeal/eukaryal group is provided by a polarized indel within the vacuolar ATPase (Gogarten *et al.*, 2007, 1989b; Gogarten and Taiz, 1992). However, other polarized indels which are shown to exclude the root from wider regions of the ToL are less convincing, due to various shortcomings in data and analysis. Ribosomal protein S12 has a 13 aa insert present exclusively within Firmicutes (excluding some Clostridia), Archaea, and Eukarya (Gupta, 1998). Given an analysis detecting homology between S12 and RpoC, the corresponding region between paralogues was observed to lack this insert, thus polarizing the indel supporting these three clades as a derived group, excluding the root (Skophammer *et al.*, 2007). While the inserted region shows clear sequence homology between the excluded groups (disproving homoplasy), and does not have any phylogenetic signature of HGT, independent analysis of sequence and structure of S12 and RpoC support that it is unlikely these proteins are paralogous, and thus the indel in S12 cannot be reliably polarized (Valas and Bourne, 2009). Furthermore, several Chloroflexi genomes apparently contain partial indels, with shorter, variable sequence regions corresponding to the insertion. These sequences were not included in the original analysis. In the absence of a reliable paralogue for polarization, this indel can be explained using a more traditional rooting, being an ancestral character inherited by Archaea,

Eukarya, and Bacteria, and subsequently lost in lineages leading to Actinobacteria, a subset of the Clostridia, Proteobacteria, and other groups, including a partial loss within the Chloroflexi. As the deep relationships between many of these groups are unclear, it cannot be precisely determined how many individual deletion events this would require, although only a few seem necessary in any case. However, as an unpolarized plesiomorphy, this character cannot provide any compelling additional evidence for any particular rooting.

This model also excludes the root from within the double membrane prokaryotes and Eukarya by a polarized indel in HSP70/MreB paralogues (Gupta, 1998; Lake *et al.*, 2007). However, while subsequent structural and sequence analysis support that these two proteins are in fact paralogues, the region of the indel is not clearly aligned between the two, and multiple insertion–deletion events are required in order to explain the evolution of the region, preventing polarization (Valas and Bourne, 2009). An additional problem with this analysis is the *a priori* assumption that double membrane prokaryotes constitute a valid monophyletic group. While the vast majority of known prokaryotes with a double membrane are members of the Proteobacteria, this group also includes Spirochaetes, Thermus/Deinococcus, Cyanobacteria, Thermotogales, and Aquificales, which have not been shown to be monophyletic by any sequence-based phylogenetic analyses except for HSP70. Closer investigation reveals evidence that this protein may have been subject to HGT between major groups, which would invalidate the assumption of monophyly essential to the previous indel-based analysis (Gribaldo *et al.*, 1999). As HSP70 and MreB are valid paralogues with a clear indel signature, however, this dataset could still potentially be used for more accurate root exclusion. Specifically, such an analysis should attempt to resolve the pattern of multiple indels within these proteins, and treat the aforementioned constituent phylogenetic groups as distinct entities that may have undergone unique insertion or deletion events. Additionally, the complicating presence of HGT must be taken into account.

Finally, this model excludes the root from Eukarya, Archaea, and the Firmicutes (Bacilli + Clostridia in their analysis) using short one-, two-, or three-amino-acid indels in PyrD/HemE paralogues. While these proteins do possess detectable homology, structural features arising from internal duplications make reliable alignment difficult. Additionally, some variants of these proteins contain additional small indels near those used for exclusive rooting. This, combined with the small size of the indels, provides support for the alternative hypotheses that these features could have arisen by convergent evolution, making these paralogues unreliable for root exclusion (Valas and Bourne, 2009).

One major problem with a composite approach to exclusive rooting via indels is that by necessity genes are selected based on their suitability for analysis; genes must have recognizable deep paralogues, as well as a robust indel signature. As several such genes are required to provide enough coverage to exclude the root from most of the tree, this means that there is little room for other criteria to come into play for gene selection. This is problematic, as ToL rooting must assume that sets of genes used are a suitable proxy for vertical inheritance, or, at the very least, represent the rooting of some distinct biological component within cells. For example, many unrooted universal trees representing the ToL are based on components of the ribosomal machinery (ribosomal proteins, RNA, and associated factors). These components, while distinct, are part of a functional whole and therefore must have co-evolved, justifying their inclusion within a dataset presumed to share a cohesive phylogenetic signature. In the strong sense, given the additional information that these genes show minimal phylogenetic incongruence due to HGT, they can be considered a proxy for vertical organismal evolution. Even in the weak sense, these genes still reflect the evolution of a singular, complex biological system, and could be called the 'ribosomal tree of life'. However, the genes used for exclusive indel rooting (GyrA, IF2, HSP70, PyrD, S12, V-ATPase) show no such biological cohesion which would support their being used to root anything but a 'tree of indel-containing proteins'. Such a tree could, in theory, be a proxy for an organismal ToL (and thus also identify the root) if most genes could generally be relied upon to provide similar phylogenetic patterns reflecting vertical inheritance, but this seems less and less likely to be the case. That being said, exclusive methods such as indel rooting are still of interest in rooting the ToL, albeit in a less direct manner. The discussed analyses strongly support that the initiation/elongation protein systems (IF2/EFG) did not evolve within the Archaea, Eukarya, or any other common ancestor descendants, and that DNA gyrase/topoisomerases (GyrA/ParC) did not originally evolve within the Actinobacteria. Any rooting of the ToL must therefore be compatible with these evolutionary scenarios, even if in each instance these polarized indels cannot, in themselves, provide direct evidence for an organismal rooting.

## Transition analysis

Interestingly, another distinct rooting method has been proposed using the variable tertiary structure of PyrD, which can exist as a monomer, homodimer, or heterotetramer with other proteins in different lineages (Valas and Bourne, 2009). Specifically, PyrD2 is a membrane-bound monomer found within Gram-negatives, Actinobacteria and *Staphylococcus*; PyrD1A is a homodimer found within Lactobacillales; PyrD1B is a heterotetramer together with the non-homologous PyrK subunit, and is found across Archaea and Firmicutes (except *Staphylococcus*). Arguing from the logic of increasing complexity from monomer to heterotetramer, this would polarize the PyrD tree, excluding the root from within Archaea, Firmicutes, or their common ancestor, similar to the proposed polarizing indel within S12/RpoC (Skophammer *et al.*, 2007).

This type of 'transition analysis' is compelling in that it makes use of an apparently natural tendency towards complexity over time. In practice, it is very difficult to prove that such trends are irreversible, and therefore useful in tree polarization. If parsimony is accepted as a major philosophical guide in determining historical events, then much weight will be placed on analyses making arguments of this type, and their validity can appear self-evident. However, in an evolutionary history dominated by gratuitous complexity driven by many neutral processes in addition to selection, parsimony may often provide a poor guide to biological reality.

A phylogenetic analysis of PyrD homologues (Fig. 6.3) illustrates that in the absence of a dependence on arguments from parsimony, plausible explanations remain given the conventional rooting scheme. The two major PyrD groups (1 and 2, or cytosolic and membrane bound), are phylogenetically distinct, and separated by a long branch (about 1.5 substitutions/site). Sequence analysis also supports this classification, as in addition to an N-terminal helix for binding to the membrane, PyrD2 contains a seven-aa indel within a highly conserved region. Additionally, PyrD1 sequences contain a completely conserved Glu residue signature in another conserved region near the middle of the protein. The same position in PyrD2 is occupied by Thr, Ala, or Val, physiochemically distinct residues. If rooted within the monomeric PyrD2-containing species (mostly Proteobacteria), this tree supports the Archaea and Firmicutes (as well as several other disparate bacterial groups) as being two major derived subgroups each containing the heterodimeric PyrD1B. Interestingly, PyrD1A appears to constitute an equally distinct, albeit smaller group, rather than being clearly 'intermediate' between PyrD2 and PyrD1B, or being clearly derived within the Firmicutes, as would be expected in a 16S phylogeny. This group also includes a 1A variant found within *Frankia* species. As Actinobacteria are not closely related to Lactobacillales, this is likely evidence of HGT. The fact that PyrD1B variants within Lactobacillales group together with Bacilli and other Firmucutes, as would be expected from vertical descent, further supports that PyrD1A homologues within these genomes have undergone transfer and are unreliable for rooting. Three other distant, duplicate PyrD1 variants are also found within Gammaproteobacteria, Bacteroidetes and Thermoproteales, supporting that this gene family may contain extensive diversity present within a subset of genes undergoing HGT. Other HGT events appear to take the role of replacements, as indicated by a bacterial PyrD2 homologue found within haloarchaea, and a Firmicute-like PyrD1B homologue found within *Magnetococcus* (Alphaproteobacteria).

It is possible that PyrD2 homologues are on a long branch due to a large amount of positive selection (or relaxation of purifying selection after losing the requirement for protein-protein

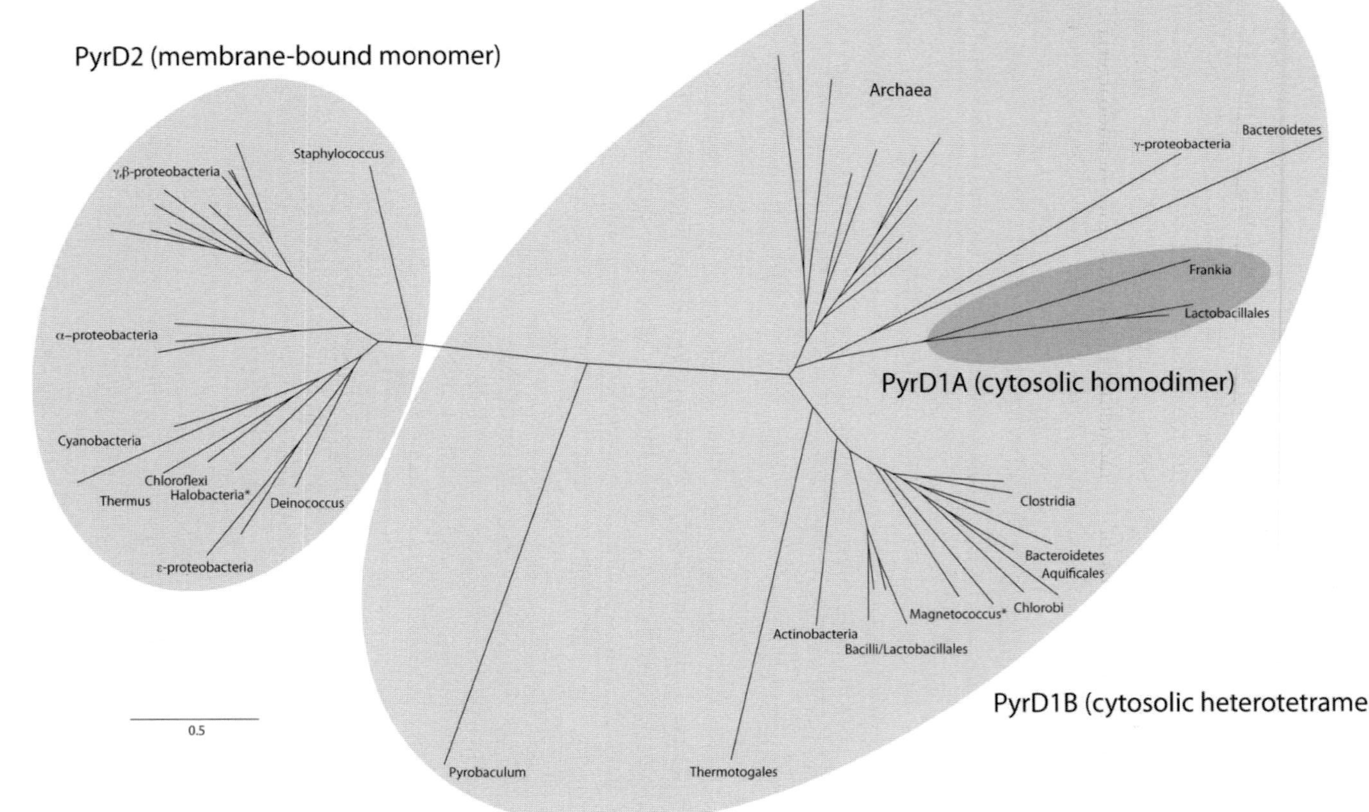


**Figure 6.3** Maximum-likelihood tree indicating phylogenetic relationships of PyrD homologues. Circled regions indicate the various PyrD types. Note that all deep-branching bacterial representatives (PyrD1A and the adjacent two sequences) are from groups that are also represented elsewhere in the tree at phylogenetically accepted positions, supporting that an extensive diversity reservoir exists for this protein, which is likely being shared via HGT. The deep branching archaeal sequence representing *Pyrobaculum* is likely a highly divergent homologue subject to long branch attraction. The fact that other deeply branching bacterial sequences do not also join the tree in this vicinity supports that their deeper rooting within the tree is not the result of a similar artifact.*Additional groups that probably underwent HGT across domains or phyla.

interactions) resulting from their transition to a membrane-bound state from an ancestral PyrD1B system. In this case, long-branch attraction would tend to root these sequences to the base of the PyrD1B homologues between other bacterial and archaeal sequences, as is observed. This explanation excludes the root from within those groups containing PyrD2, compatible with more traditional rooting scenarios. Presumably, a similar long-branch artifact in PyrD1A could also have resulted from structural divergence after losing its PyrK interactions. These two long branch artefacts would obfuscate the origins of the PyrD2 and PyrD1A homologues, making any phylogenetic inference to support or refute the parsimony model extremely difficult. These artifacts also make it equally likely that these sequences both root within the bacterial PyrD1B homologues in locations compatible with the traditional bacterial phylogeny and rooting of the ToL. The argument is largely academic, however, as such a plastic and easily modified biological component makes a poor proxy character for polarizing an organismal ToL.

Eukaryotes also contain a PyrD2 homologue derived from mitochondria (Valas and Bourne, 2008); presumably this version either directly replaced the 'archaeal-type' PyrD1B homologue, or the eukaryotes had already lost them before mitochondrial transfer took place. While admittedly a special case, this does show that reversals in subunit complexity within organisms can happen by a variety of means.

## The negibacterial root

Transition analyses more typically rely on sequence-independent characters that are presumably more robust, and more useful in polarizing the tree. These analyses are also purported to be a more traditional evolutionary approach, making use of well-established principles in paleontology and cladistics, and eschewing more computationally intensive analyses of molecular sequences with their associated propensity for artifacts and error. It is difficult to say if these represent inclusive or exclusive methods, as they attempt to identify a necessary linear ordering to characters that simultaneously define the ancestral and derived states, and therefore the location of the root as well. The definitive attempt at rooting the ToL via transition analysis is the extensive work leading to the 'Negibacterial Rooting' model (Cavalier-Smith, 2006b). In short, this work identifies and orders a wide array of so-called transitions which serve to polarize and order the emergence of different types of cells, with four major kinds being apparent (in order of emergence): (1) Negibacteria (Chloroflexi, Thermus/Deinococcus, Cyanobacteria, Spirochaetes, Proteobacteria, Aquificales, Chlamydiae/Verrucomicrobia, presumably Thermotogales, and others); (2) Posibacteria (all other bacteria, including Firmicutes and Actinobacteria); (3) Archaebacteria; and (4) Eukaryota. Under this system, Eukarya and Archaebacteria are referred to as 'Neomura'. There are additional fine transitions within each of these major groups, nearly completely structuring a linear succession of phenotypes, superficially reminiscent of the 'great chain of being' popular in pre-Darwinian natural philosophy. The most important of these transitions occurs within Negibacteria, locating the root within or adjacent to the Chloroflexi (Fig. 6.4).

The key initial assumption in the Negibacterial Root hypothesis is that paleontological evidence for the early domains of life is inconsistent with a rooting between the Bacteria and Archaea, with eukaryotes as an archaeal sister group. Rather, these data are interpreted to support an earlier origin for Bacteria at 2.8–3.5 Gya, and a much more recent emergence of Archaea and eukaryotes at 1.1–0.8 Gya (Cavalier-Smith, 2006a; Schopf, 1999). Since several additional biochemical and paleontological analyses support a much more ancient origin of Eukaryota and Archaea, these assumptions are controversial in light of other evidence. However, even accepting the claims of evidence for an early origin of Bacteria, this chronology does not preclude a traditional rooting of the ToL. The earliest descendants of the MRCA (or lineages at the time of the MRCA itself) could have left bacterial-like biosignatures. Additionally, the dependence of mitochondrial origin on the emergence of Alphaproteobacteria (a derived group of Bacteria), is only problematic if one assumes that the mitochondrial ancestor invaded the proto-euykaryote host early on, close to the point of divergence of eukaryotes and Archaea. A later invasion closer to the most recent eukaryotic

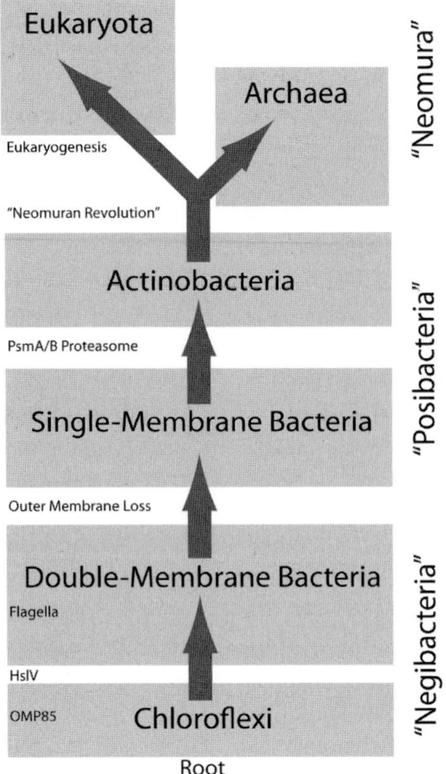


**Figure 6.4** Negibacterial rooting of the ToL, based on transition analysis. Major character transitions are labelled on the left-hand margin. 'Eukaryogenesis' and 'Neomuran Revolution' are defined by many character transitions that are extensively discussed in the literature and not peculiar to this model. Phylogenetic nomenclature specific to this model is labelled on the right-hand margin in quotations. Figure adapted from (Cavalier-Smith, 2006).

ancestor is not chronologically problematic, given the times assumed for the origin of Eukaryota (Cavalier-Smith, 2006b). Even as a sister group to the eukaryotes, since the Archaea typically are separated from the neomuran ancestor by a much shorter branch, biochemical signatures of archaeal metabolism could be expected to be found much earlier in the fossil record. It is clear that given the numerous interpretations of biochemical and microfossil evidence, combined with the plasticity of relative chronologies inherent in most rootings of the ToL, current evidence of this type is not particularly useful in determining the major organization of the domains of life, or their origin. However, this being said, until this line of evidence is more clearly established it does not preclude the proposed direct ancestry of bacterial groups to Archaea and eukaryotes, and the model is better evaluated based on the evidence for the individual transitions predicated upon this assumption.

This model is also supported by the description of many polarizing characters which exclude the root from any position within the Neomura, and most specifically from within Eukaryota or Archaea; however, as many of these characters are well-described and assumed to be derived in the general literature, these do not have a central role in locating the root of the ToL in this analysis.

### Outer membrane biogenesis

The strongest polarizing transition in this model concerns the origin of the outer membrane, which is found in all double-membrane bacteria (Negibacteria). In single-membrane bacteria (Posibacteria), the peptidoglycan layer is thick and attached directly to the cytoplasmic membrane. Negibacteria also have a peptidoglycan layer, although it is thinner and attached to the outer membrane. Proteins present in the outer membrane are transported from the cytoplasmic membrane across the periplasmic space using complex protein transport machinery. Since these are homologous proteins within bacteria with double membranes, the most parsimonious explanation would place the bacterial root within this group, with singular or multiple independent losses of the outer membrane machinery (and the outer membrane itself) leading to the Posibacteria. This loss would be compensated for by a thickening of the peptidoglycan layer. In this model the reverse polarity is considered highly unlikely, as at that point the single membrane + peptidoglycan is an embedded character, with no clear evolutionary mechanism of expansion to a double membrane without complete disruption of the cell. Of course, an ancestral cell would still have developed from a single to a double-membrane system; however, this process could have been gradual and need not have been complicated by the existence of lipopolysaccharides, a peptidoglycan wall, or other molecular components which evolved at a later time.

There are two major objections to this argument based on microbial comparative physiology. First, it subjectively assumes that developing an outer membrane as a derived character is prohibitively difficult, while taking for granted the feasibility of the other major membrane based transition of life, the different stereochemistry of bacterial and archaeal membranes. While Archaea use ether lipids derived from glycerol-1-phosphate, bacteria use ester acyl-lipids based on glycerol-3-phosphate (Pereto *et al.*, 2004). In many ways this transition seems to be even more profound, and would be strong evidence for placing the root between Bacteria and Archaea. Second, it ignores that in one demonstrated case, double-membrane microbes (the archaeon *Ignicoccus*) evolved from a single-membrane ancestor (Nather and Rachel, 2004). While not directly homologous to double-membrane physiology observed within bacteria, the existence of this system within a clearly derived group demonstrates that even if the precise evolutionary mechanism is unknown, such complex embedded characters can and do evolve. This example also serves to illustrate a wider philosophical weakness of transition analysis-based arguments, which amounts to a kind of negative ontological argument: if a mechanism for a transition polarity is not apparent, then it is assumed to be impossible, and the inverse transition polarity correct. The critical flaw in this logic occurs when envisioning a transition where neither transition polarity has a readily apparent evolutionary mechanism; in such a case, to apply the same logic leads to the argument from irreducible complexity often used by opponents of evolutionary theory. Given such a case, it would be more appropriate to assume that current information was simply insufficient to infer the evolutionary mechanism at work. This same conclusion should suffice for any single transition polarity. While parsimony can, and should, be a reliable indicator of polarity, it can only be correctly employed between two alternative explanations with quantifiable probabilities of occurrence; an absence of a viable explanation does not equate to a non-parsimonious explanation. In this way, a certain agnosticism is required in investigating the evolution of complex biological systems, as history has proven that new discoveries can radically alter the assumptions as to what evolutionary mechanisms are biologically plausible, parsimonious, or even necessary in their occurrence.

### Origin of the flagella

Another argument for placing the root within the Negibacteria concerns the co-evolution of the bacterial flagella and the double-membrane system. Bacterial flagella are complex macromolecular machines, with central components showing homology to the type III secretion system, as well as the slime secretion nozzle used by some bacteria for gliding motility (Pallen *et al.*, 2005). While the flagellar motor is fixed within the cytoplasmic membrane, it requires specific protein structures for penetrating through the outer membrane (the P-ring, the L-ring, or, in some cases such as Thermotogales, both). As such, it is argued that it would be very difficult for this system to be independently gained within different bacterial lineages. Additionally, it is argued that as the hook structure at the flagellar base is more similar to the slime-secretion nozzle found within some species of Negibacteria, this is likely the ancestral state of the flagellar system, rather than the type-III secretion system that has a phylogenetic distribution incompatible with this model. This rooting is even further supported by the assertion that the ancestors of the flagellar motor proteins (TonB complex proteins) are also exclusively found within Negibacteria.

Additional transitions within the Negibacteria further identify the root to be either within or adjacent to the Chlorobacteria, a group comprising of non-sulfur green bacteria (e.g. *Chloroflexus*), as well as other heterotrophic species (e.g. *Dehalococcoides*). The strongest argument supporting this rooting is the conspicuous absence of the OMP85 protein within these lineages, which is necessary for the insertion of beta-barrel porin complexes found in outer membranes (Gentle *et al.*, 2004). This absence also coincides with the lack of LolC and LolE proteins required to move lipoproteins into the periplasm, and the entire flagellar machinery present in other presumably more derived Negibacteria. In short, the Chlorobacteria have an extremely simplified double-membrane system, indicative of being near the location of the root of the ToL in this model. While all of these features could also be explained by gene losses and reduction of membrane complexity within this clade, the polarization placing Negibacteria as a basal lineage based on the other discussed transitions decreases the parsimony of this explanation. Furthermore, OMP85 and the Lol proteins apparently have not been lost in any other bacterial lineages, making their loss within Chlorobacteria a less parsimonious special case. Obviously, these arguments are contingent upon the other transitions placing the Negibacteria at the base; a rooting in any other location necessitates gene loss as an explanation, which is not problematic from any apparent arguments of complexity or embedded characters.

## Evolution of the proteasome

The second major component of the negibacterial rooting via transition analysis is provided by the evolution of the 20S Proteasome and its two components, PsmA and PsmB. These are restricted in their distribution to the Neomura and Actinobacteria. Additionally, some members of the Chlamydiae/Verrucomicrobia appear to have PsmA. Another proteasome complex consisting only of homologous HslV proteins is found within a wide and scattered group of bacteria, including members of the Proteobacteria, Firmicutes, Aquificales, Thermotogales, and Spirochaetales. Transition analysis infers that this simpler HslV system is ancestral to PsmA/PsmB, which later diverged within the lineage leading to the therefore monophyletic Neomura + Actinobacteria. While it has been assumed that HslV is too divergent to include in a phylogenetic analysis with PsmA and PsmB (Cavalier-Smith, 2006b), multiple sequence alignment suggests otherwise, with each paralogue being aligned along its full length with obvious sequence homology. A maximum-likelihood tree based on this alignment reveals that proteasome subunits are divided into seven major coherent groups (Fig. 6.5). This tree reveals several key features that allow us to evaluate the claim that the 20S proteasome is a polarizing transition grouping the Neomura together with Actinobacteria as the most derived members. First, for both PsmA and PsmB, it is clear that eukaryotic and archaeal homologues are more similar to each other than they are to Actinobacteria. Furthermore, reciprocal rooting of each paralogue places the root between bacterial and neomural sequences. Perhaps most importantly, HslV is not a deep branching group, but exists as the sister group to the Actinobacterial PsmB subunit. Surprisingly, the longest branch in the tree actually leads to actinobacterial PsmA homologues. This is also supported by the presence of a conserved signature 'Gly-Ser-Gly' motif found within all PsmB sequences, archaeal PsmA, and HslV. This same region contains a conserved 'Gly-Gly' motif within actinobacterial PsmA sequences, and a less conserved single 'Gly' residue within eukaryotic and Chlamydiae/Verrucomicrobia sequences. Given a negibacterial rooting within or adjacent to the HslV group produces a highly unlikely pattern of evolution, with an initial divergence to actinobacterial PsmB subunits, followed by a divergence between all PsmA subunits and neomuran PsmB subunits. In other words, this rooting renders PsmB paraphyletic, which is very unlikely to be correct. Additionally, while HslV is present in a very broad sampling of bacterial species, these sequences appear to be less divergent

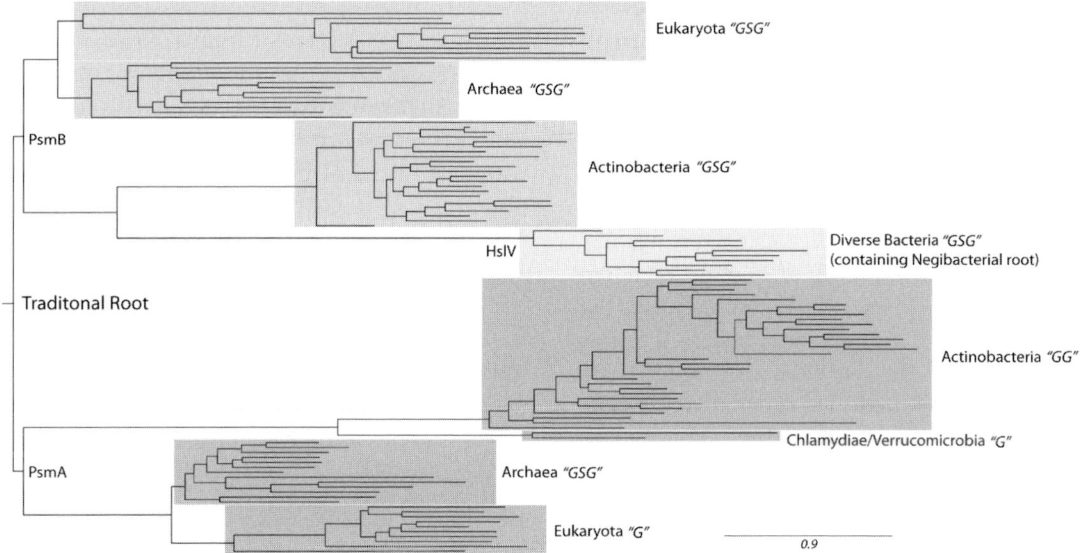


**Figure 6.5** Maximum-likelihood tree of proteasomal protein paralogues. Protein paralogues (PsmA, PsmB, HslV) are labelled and indicated by shared colour. The tree is rooted as to place the root of PsmA and PsmB paralogues at the traditional position (along the bacterial branch), maintaining the monophyly of all three paralogues and placing HslV as a derived group within PsmB. Signature motifs at a conserved position are indicated after the labelled clades within each paralogue. Note that the distribution of the conserved signature motif '*GSG*' supports the chosen location for the root, as well as for HslV being a derived subset within the bacterial domain. The observed high sequence similarity for such a phylogenetically diverse group also supports relatively recent, extensive HGT of HslV homologues.

than those observed within Actinobacteria, or other homologues within each domain. This can be resolved by a traditional rooting (as indicated by the reciprocal paralogue rooting observed in this case), complicated by rampant HGT within the bacterial domain. In this scenario, PsmA and PsmB are ancient paralogues inherited from the cenacestor. However, within the bacterial domain rapid diversification and transfer occurs between different proteasome subunits, resulting in the unusual distributions observed within the bacterial branches of HslV. In this model, HslV may be a particularly recent and successful proteasome system that underwent simplification, being transferred widely across the bacterial domain. This also illuminates why the branch lengths between proteobacterial and firmicute HslV sequences are shorter than those observed between actinobacterial species, which retained the organization of the ancestral system. This model also predicts that additional homologous proteosomal subunits have yet to be discovered within bacteria, which should show similar scattered distributions. In fact, at least two such proteins have recently been identified, although these have not been subjected to phylogenetic analysis (Valas and Bourne, 2008).

If these investigations provide evidence against the Neomura being rooted within the Actinobacteria and/or Posibacteria, this does not necessarily invalidate other arguments that the double-membranes of Negibacteria are an ancestral bacterial character. In fact, this later claim would still be compatible with a traditionally rooted ToL. The lack of phylogenetic resolution at the base of the bacterial domain may indicate a very rapid radiation and diversification, possibly following a cataclysmic event (Gogarten-Boekels *et al.*, 1995), which, combined with HGT, would mask the order of emergence of complex traits within an 'evolutionary singularity'. In such a scenario, transition analysis is useful for determining the order of events; the catch is that these events could also have occurred so rapidly that the ordering is largely irrelevant within the broader context of the tree of life, or its rooting.

Transition analysis is correct in its insight that complex biological characters are more important than individual genes in establishing polarity within unrooted organismal phylogenies. Furthermore, this approach highlights that inheritance occurs not just for genetic information, but for the structures and systems replicated within each cell division, as well. Therefore, evolutionary events must take the plausibility of such 'transitions' into account, an aspect that is often lacking in studies of evolution that focus primarily on events within genes and genomes. However, in testing these hypotheses there is surely a role for the construction of phylogenetic trees, especially for determining the evolutionary relationships between the many proteins found within the complexes that comprise these characters. These studies need to be undertaken for the above model to be confirmed, despite the possibility of artifacts and HGT, as such trees are useful in either verifying or falsifying any arguments based on transition analysis.

## A eukaryotic root

It is generally taken for granted that the MRCA must resemble a prokaryote, and that eukaryotes represent a highly derived group. This derivation is often attributed to the accumulation of several complex traits through vertical evolution and innovation, as well as endosymbiotic events involving some combination of bacterial and archaeal cells, or even viruses (Bell, 2009). By extension, rooting the ToL has typically been a prokaryotic endeavour. However, there are inclusive arguments for placing the root of the ToL on the branch leading to the eukaryotes, relying on similar non-sequence characters to those employed in transition analysis.

In the ToL based on the SSU ribosomal RNA (Woese and Fox, 1977), eukaryotic sequences occupy the end of a long branch; in most rooting schemes, this branch is the sister group to the Archaea, which are separated from their common ancestor by a shorter branch. This arrangement has led to an ongoing debate about eukaryogenesis, and the nature of the eukaryotic ancestor. It is well established that the ancestor to all known eukaryotes was descended from a lineage that had somehow 'enslaved' an endosymbiotic member of the Alphaproteobacteria, giving rise to mitochondria. However, did this event occur early, within an archaeal lineage destined to become eukaryotes (Horiike *et al.*, 2001; Margulis, 1996; Martin and Muller, 1998; Rivera and Lake, 2004; Vellai and Vida, 1999)? Or, had the 'protoeukaryote' lineage already substantially diversified and acquired many of the other eukaryotic characters, such as a nucleus, cytoskeleton, and spliceosome (Glansdorff, 2000; Glansdorff *et al.*, 2008)? Or were multiple endosymbiotic events involved, with both archaeal and bacterial cells contributing to a third 'chronocyte' lineage (Hartman and Fedorov, 2002)?

While these different observations and hypotheses (as well as many others relating to eukaryogenesis) are complex and beyond the scope of this survey, in some cases they dovetail with arguments for rooting the ToL in the branch leading to the existing eukaryotic lineages (i.e. within the 'protoeukaryotes'). The primary arguments to this effect include the possible role of RNA molecules and the spliceosome in early life, as well as the phylogenetic distribution of major genes families between domains (Glansdorff, 2000; Glansdorff *et al.*, 2008).

It is frequently assumed that before the existence of protein-based cells, primitive life existed in a state where both catalytic and informational roles were performed by RNA, in an 'RNA world' (Glibert, 1986). Evidence supporting this model still exists across all three domains of life, as many of the central components of protein synthesis rely on RNA for catalytic and structural function, such as ribosomal RNA (rRNA) and transfer RNA (tRNA). However, catalytic and small RNAs are especially abundant within eukaryotes, where extensive RNA processing also occurs. This supports an 'RNA continuity' model, in which a substantial amount of primordial RNA-based physiology is preserved within the eukaryotic lineage, supporting that early eukaryotes are more similar to the state of the MRCA than either Bacteria or Archaea, in which many RNA-based systems were lost (Collins *et al.*, 2009; Collins and Penny, 2009; Penny and Poole, 1999). Taken a step further, this also provides support for rooting the ToL within the branch leading to modern eukaryotes (Glansdorff *et al.*, 2008), treating the extensive role of RNA as a complex polarizing character.

Related to their extensive use of RNAs, eukaryotes are also unique in their possession of a spliceosome, a complex machine comprised of numerous proteins and RNA that removes non-coding introns from RNA transcripts, rejoining the remaining exons into a precursor mRNA. The spliceosome has no counterpart within either Archaea or Bacteria; as such, it has typically been assumed that it is a derived eukaryotic character, evolving over time within the protoeukaryote lineage. However, since the spliceosome is essential in the presence of (aptly named) spliceosomal introns, evidence for the early presence of these introns would presuppose that the spliceosome is ancient as well, providing an additional character in support of a protoeukaryotic rooting. The 'introns early' hypothesis states that intervening introns facilitated the recombination of protein domains via exon shuffling as a crucial step in the evolution of major protein folds and families (Fedorov and Fedorova, 2004; Penny *et al.*, 2009; Roy and Gilbert, 2006). This would suggest that the spliceosomal machinery existed at a time before the MRCA, which would then most parsimoniously place the root within protoeukaryotes, with Archaea and Bacteria losing this system (although retaining self-splicing and endonuclease-dependent group I and II introns). This hypothesis also predicts that exons should correlate with distinct protein secondary structure domains (presumably the units of intragenic recombination). While this has been reported to be the case (Fedorov *et al.*, 2001; Fedorov *et al.*, 2003), other work has suggested that the correlation is substantially weakened by the inclusion of additional data, and can likely be explained other selective effects (Logsdon and Palmer, 1994; Logsdon *et al.*, 1995; Stoltzfus *et al.*, 1994, 1995).

The concept of a large, complex eukaryote-like ancestor giving rise to Archaea and Bacteria via 'streamlining' is also supported by studies of gene family content and phylogenies between the three domains (Kurland *et al.*, 2007; Ouzounis *et al.*, 2006). These show that the majority of gene families exist within the eukaryotic lineage (over 1000), while genes families specific within either Archaea or Bacteria are many fewer in number (less than 150). Assuming that horizontal gene transfer plays a minor role in the distribution of genes, the alternative hypothesis of differential loss within lineages will therefore root the ToL within the branch leading to modern eukaryotes (Kurland *et al.*, 2003). The fact that eukaryotes are generally mesophilic, while many deep-branching Archaea and bacteria are thermophiles additionally supports that these differential losses may have taken place via 'thermoreduction' (Forterre, 1995). Given this model, the fact that many bacterial and archaeal systems are analogous yet unrelated additionally implies that the protoeukaryotic ancestor contained a large amount of genetic and physiological redundancy, possibly a consequence of a particularly diverse protoeukaryotic population, or to compensate for a less than accurate cell division mechanism (Glansdorff *et al.*, 2008).

These claims become much weaker if one is more sympathetic to substantial ancient (and continuing) horizontal gene transfer, which has two effects; it doesn't necessitate massive differential gene loss, and it makes a protoeukaryotic rooting less parsimonious. The observed distribution of gene families is also compatible with a large amount of evolutionary innovation on the branch leading to the eukaryotes. As is the case with the negibacterial rooting proposed by Cavalier-Smith (2002), rapid character losses are easier to explain than gradual character gains. This causes bias in favour of reductive models, despite the implication of a complex ancestral state that is not otherwise considered likely to have existed. The arguments for 'thermoreduction' are interesting in that they provide independent justification of the model. However, the causal relationship between the streamlined nature of prokaryotic cells and ancestral thermophily is unclear. One proposed explanation for a mesophilic origin of life (and MRCA) coinciding with thermophilic ancestry of both Archaea and Bacteria is the 'bottleneck hypothesis', in which a cataclysmic event such as a major impact destroyed all early life except that within protected deep-sea environments, largely populated by thermophiles at hot vents (Gogarten-Boekels *et al.*, 1995). While this scenario is not compatible with a protoeukaryotic rooting (as eukaryotes do not demonstrate any thermophilic adaptations that could have ensured impact survival), it does illustrate that strong purifying selection could have the same apparent result as positive selection driving a lineage towards its shared derived state via a reduced physiology and genome size.

Finally, it is important to note that many of these arguments can be read to support a weaker model that the MRCA was eukaryote-like in its physiology, regardless of the location of the root of the ToL. For example, the traditional rooting could be associated with a protoeukaryotic ancestor; in this scenario, an early divergence and reduction would result in the Bacteria, while a later divergence and reduction would result in the Archaea, with the ancestral lineage to modern eukaryotes still existing as a sister group (Glansdorff *et al.*, 2008). However, the required convergence of multiple independent reductions, and the paralogue partitioning such a scenario would implicitly assume for a eukaryote-like MRCA seems most at home within the protoeukaryotic branch, with a single reductive lineage undergoing additional partitioning and refinement to give rise to the other two domains of life.

## Genetic code evolution and composition analysis

As previously described, inclusive methods for rooting the ToL are frequently complicated by tree reconstruction artifacts in the case of sequence-based analyses, and subjective assumptions about the nature of early life in the case of non-sequence character-based analyses. Furthermore, many analyses have depended either directly or indirectly upon specific genes or other biological characters which show poor conservation and/or a history of frequent interdomain HGT, making them poor proxies for an organismal ToL. It seems likely that the best genetic proxy for an organismal ToL are phylogenetic trees generated using the ribosomal machinery, specifically the three rRNAs and about 29 universal ribosomal proteins that comprise the 'core ribosome'. While ribosomal protein and RNA encoding genes have been transferred in the past (Gogarten *et al.*, 2002), these genes are resistant to transfer across large phylogenetic distances, with most transfers occurring between close relatives (Sorek *et al.*, 2007; Zhaxybayeva, 2009). This has little effect on tree topology among deep branches where the root of the ToL may reside. Additionally, the high level of structural, sequence, and functional conservation in ribosomal proteins promotes higher confidence in tree topologies being free of artifacts, especially those producing long branches via radical functional divergence following duplication, as proposed for some deep paralogues (Cavalier-Smith, 2006b). For these reasons, a ribosomal ToL has been proposed to be an ideal scaffold upon which to map horizontal gene transfers, clearly depicting their distinct contribution to genomic (and organismal) evolution (Dagan *et al.*, 2008; Gogarten, 1995). Unfortunately, since there are no confirmed paralogues for ribosomal proteins or RNA, and their strong functional and structural conservation precludes any use of transition analysis, it has seemed that while ideal for constructing a ToL, the ribosome itself does not contain any information that could be used to root it.

The ribosomal machinery is intimately associated with one unambiguously primitive character; however, the genetic code shared by all cellular life, likely present in its complete form at the time of the MRCA (Fournier and Gogarten, 2007; Knight *et al.*, 2001; Miranda *et al.*, 2006). Although by definition the genetic code and its requisite translation machinery must have evolved in a pre-protein (and likely RNA-based) world, such a complex system could only have evolved incrementally. As such, it is likely that specific genetically encoded amino acids were incorporated into protein synthesis gradually, until the code reached its current 20 amino acid retinue. At least two distinct approaches have used amino acid usage at ancient positions to infer the evolutionary history of the genetic code, albeit with somewhat differing results (Brooks *et al.*, 2004; Brooks and Fresco, 2002; Brooks *et al.*, 2002; Fournier and Gogarten, 2007). Amino acids could not be fixed at specific positions within proteins until their establishment in the code; therefore at the time of the MRCA, the most recent amino acids would have had less time to become fixed, and should have been under-represented. Conversely, more ancient amino acids should be over-represented (Fournier and Gogarten, 2007).

In each case, those analyses assumed the location of the root was a known quantity. However, an inversion of this approach can be used to empirically determine the root of the ToL, by calculating the compositional bias in amino acid usage along each branch of a universal phylogeny. The branch containing the root should contain the strongest overall bias, retaining a distinct and unique signature of amino acid composition imposed by the echoes of earlier stages of code evolution (Fournier and Gogarten, 2010). This rooting analysis was performed on a dataset of 29 concatenated universal ribosomal proteins, with 9,258 aligned positions. For each branch within a generated phylogenetic tree, conserved positions with 90% posterior probability of being fixed for a specific amino acid were identified. Amino acid usage at conserved positions along each branch were then compared to those generated from simulated datasets, in order to identify usage bias independent of dataset composition, or artifacts arising from substitution models acting over long branches. The branch leading to the bacterial domain was identified as containing the largest number of amino acids with strong compositional bias, a significant over-representation of Gly, Ala, and Asn, and a significant under-representation of Gln, Phe, Ser, Trp, Tyr, and Cys (Fig. 6.6). Reassuringly, these results agree well with recently published work (Higgs, 2009) which reconstructs the organization and evolution of the genetic code using a selection-based model, evaluating the fitness effect of code expansion via the addition of new amino acids via stepwise codon space partitioning. These amino acid biases are also similar to a consensus ordering of code evolution based on several independent analyses (Trifonov, 2000).

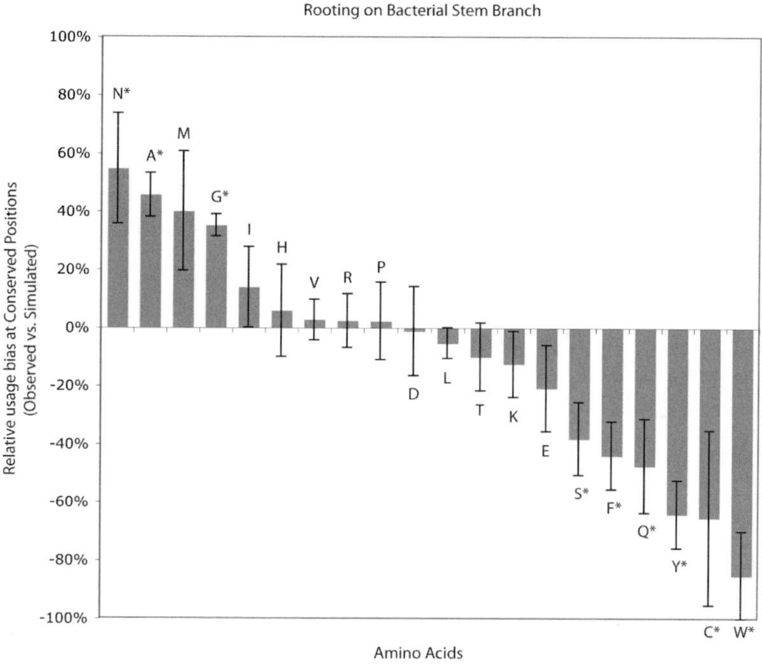


**Figure 6.6** Compositional bias at conserved positions on the bacterial branch within a reconstructed ribosomal protein phylogeny. The large number of amino acids with significant bias congruent with models of genetic code evolution identify this branch as containing the root of the ribosomal machinery, and, by proxy, the organismal ToL. Error bars represent the standard deviation of the distribution of results from model simulations. '*' denotes amino acids with statistically significant biases ($P<0.05$, two-sided z-test). Met was not evaluated as the low raw count of conserved positions for this aa prevented a reliable estimate of variance.

While significant compositional bias was detected on many other branches, they fit expectations given the physiological and compositional characters of the associated organisms. The over-representation of Asp and Glu in halophiles is an adaptation to the ionic, hypersaline environment of their intracellular space (Dennis and Shimmin, 1997). Thermophiles prefer charged residues and avoid polar residues due to the thermodynamics of protein folding, and tend to favour proline residues to promote rigidity in protein structures (Fukuchi and Nishikawa, 2001; Watanabe *et al.*, 1991; Zhou *et al.*, 2008). Finally, genomes with strong G+C biases tend to favour and avoid specific sets of amino acids based on the nucleotide content of their codons (Paila *et al.*, 2008). Importantly, each of these observed physiologically imposed compositional biases (or any combination thereof) is incongruent with the set of biases observed on the bacterial branch; this reinforces that this signature is not the reflection of an ancestral physiological state, but an echo of a more primitive genetic code at the deepest branch of the ribosomal ToL. While it is possible that the observed biases are physiologically derived based on a set of environmental criteria that are currently unknown, this would be extraordinarily coincidental, considering how closely the identified amino acid usages match those agreed upon by several models of genetic code evolution.

Amino acid fixation analysis is novel in that, while it is an inclusive approach based on gene sequences, the root is not inferred directly from the position of a given branch in a tree; rather, the tree topology is accepted *a priori*, which subsequently defines the observed compositional biases via ancestral sequence reconstruction. This makes the results very unlikely to be due to tree reconstruction artifacts. Even in the case of misplaced branches within the tree, these tend to be in terminal locations and will not strongly affect the overall signal. Deeper misplaced branches are likely very short within phylogenetically ambiguous regions (e.g. near the root of the bacteria) that show weak levels of support in this phylogenetic reconstruction. However, once again the result would not be largely affected, as by definition the ancestral sequences separated by such short branches are nearly identical. For all its merits, a cautious enthusiasm is warranted with this approach, however, as it is among the most recent proposed and has not yet been subjected to broad scientific criticism, an especially valuable resource in this field.

By empirically locating the root of the ToL on the branch leading to the bacteria, this analysis gives independent support to the results of many reciprocal paralogue rootings, increasing the likelihood that these reflect a biological reality. This is especially true for elongation factors and aminoacyl-tRNA synthetases, as they are part of the same translation machinery that co-evolved with the ribosome (and the genetic code). However, this empiricism is a double-edged sword: while it permits a rooting free of any controversial physiological assumptions about the nature of early life (other than it had an evolving genetic code), it also means the model provides few predictions about the nature of the MRCA. If the main goal of rooting the organismal ToL is to define and characterize the MRCA, then taken on its own this approach appears particularly sterile, especially compared to more predictive character-based models. However, this same agnosticism about the physiological nature of the MRCA permits a flexible integration with other hypotheses, as long as they are compatible with rooting along the bacterial branch. There is nothing in this model to prevent the MRCA from being a protoeukaryote, a double-membrane prokaryote, or even a pre-Darwinian progenote community. This serves to emphasize that rooting the organismal ToL is not necessarily the same task as characterizing the MRCA, although each must be compatible with the other.

## Conclusions and outlook

There are few evolutionary problems as challenging as rooting the ToL. Given the diverse lines of evidence supporting an equally diverse set of models, it seems that the dramatic increase in biological information over the past few decades has multiplied hypotheses rather than narrowed the possibilities, and resolved few of the conflicts within the field. Moreover, this explosion in genomic data has also revealed the true complexity of evolution, bringing into question the very definition of a Tree of Life. Clearly, lack of information is no longer a viable justification for our ignorance. Rather, what is needed are new quantitative methods for analysing sequence and non-sequence characters, detecting sources of artifacts and bias, and more critically re-evaluating previous claims. Even with these improvements, however, it is hard to imagine that a single breakthrough will ever conclusively root the tree to the full satisfaction of the scientific community. It is far more likely that, as in other historical fields, gradual accumulation of evidence and removal of conflicts will eventually result in a consensus. Finally, there exists the spectre of strict agnosticism, in that there simply may be no reliable, conclusive evidence left, and that the location of the organismal root is unknowable. As such a proposition is unverifiable, however, there seems no scientific merit to taking this position.

It is no surprise that each proposed ToL rooting relies upon information and methods best known to its adherents. Nor it is surprising that criticisms of each model are most casually dismissed or ignored when based on studies or observation outside of one's own field. Far from reflecting insincerity, this is likely caused by the vast amount of biological knowledge relevant to the problem, surely held in its entirety by no single individual, and a daunting obstacle to rigorous evaluation of claims. When synthesizing and evaluating competing models, these difficulties multiply even further. It is impossible to pull one thread without discovering the complex network of assumptions, models, and fact to which it is inextricably connected; to weave a cohesive theory out of these entwined threads is a staggering challenge. So much so, in fact, that it is tempting to simply remove those unmanageable strands, and replace missing pieces with custom-tailored patches. To do so too eagerly, however, may unravel the very threads that maintain the essential connections between theory and reality.

## References


Baldauf, S.L., and Palmer, J.D. (1993). Animals and fungi are each other's closest relatives: congruent evidence from multiple proteins. Proc. Natl. Acad. Sci. U.S.A. *90*, 11558–11562.

Baldauf, S.L., Palmer, J.D., and Doolittle, W.F. (1996). The root of the universal tree and the origin of eukaryotes based on elongation factor phylogeny. Proc. Natl. Acad. Sci. U.S.A. *93*, 7749–7754.

Bapteste, E., O'Malley, M.A., Beiko, R.G., Ereshefsky, M., Gogarten, J.P., Franklin-Hall, L., Lapointe, F.J., Dupré, J., Dagan, T., Boucher, Y., and Martin, W. (2009). Prokaryotic evolution and the tree of life are two different things. Biol. Direct *4*, 34.

Beiko, R.G., and Hamilton, N. (2006). Phylogenetic identification of lateral genetic transfer events. BMC Evol. Biol. *6*, 15.

Beiko, R.G., Harlow, T.J., and Ragan, M.A. (2005). Highways of gene sharing in prokaryotes. Proc. Natl. Acad. Sci. U.S.A. *102*, 14332–14337.

Bell, P.J. (2009). The viral eukaryogenesis hypothesis: a key role for viruses in the emergence of eukaryotes from a prokaryotic world environment. Ann. NY Acad. Sci. *1178*, 91–105.

Bininda-Emonds, O.R. (2004). The evolution of supertrees. Trends Ecol. Evol. *19*, 315–322.

Boussau, B., Gueguen, L., and Gouy, M. (2008). Accounting for horizontal gene transfers explains conflicting hypotheses regarding the position of Aquificales in the phylogeny of Bacteria. BMC Evol. Biol. *8*, 272.

Brochier, C., Gribaldo, S., Zivanovic, Y., Confalonieri, F., and Forterre, P. (2005). Nanoarchaea: representatives of a novel archaeal phylum or a fast-evolving euryarchaeal lineage related to Thermococcales? Genome Biol. *6*, R42.

Brocks, J.J., Logan, G.A., Buick, R., and Summons, R.E. (1999). Archean molecular fossils and the early rise of eukaryotes. Science *285*, 1033–1036.

Brooks, D.J., and Fresco, J.R. (2002). Increased frequency of cysteine, tyrosine, and phenylalanine residues since the last universal ancestor. Mol. Cell Proteomics *1*, 125–131.

Brooks, D., Fresco, J., and Singh, M. (2004). A novel method for estimating ancestral amino acid composition and its application to proteins of the Last Universal Ancestor. Bioinformatics *20*, 2251–2257.

Brooks, D.J., Fresco, J.R., Lesk, A.M., and Singh, M. (2002). Evolution of amino acid frequencies in proteins over deep time: inferred order of introduction of amino acids into the genetic code. Mol. Biol. Evol. *19*, 1645–1655.

Brown, J., and Doolittle, W. (1995). Root of the universal tree of life based on ancient aminoacyl-tRNA synthetase gene duplications. Proc. Natl. Acad. Sci. U.S.A. *92*, 2441–2445.

Cammarano, P., Creti, R., Sanangelantoni, A.M., and Palm, P. (1999). The archaea monophyly issue: A phylogeny of translational elongation factor G(2) sequences inferred from an optimized selection of alignment positions. J. Mol. Evol. *49*, 524–537.

Cavalier-Smith, T. (2002). The neomuran origin of archaebacteria, the negibacterial root of the universal tree and bacterial megaclassification. Int. J. Syst. Evol. Microbiol. *52*, 7–76.

Cavalier-Smith, T. (2006a). Cell evolution and Earth history: stasis and revolution. Phil. Trans. R. Soc. Lond. B *361*, 969–1006.

Cavalier-Smith, T. (2006b). Rooting the tree of life by transition analyses. Biol. Direct *1*, 19.

Cejchan, P.A. (2004). LUCA, or just a conserved Archaeon?: comments on Xue *et al.* Gene *333*, 47–50.

Ciccarelli, F.D., Doerks, T., von Mering, C., Creevey, C.J., Snel, B., and Bork, P. (2006). Toward automatic reconstruction of a highly resolved tree of life. Science *311*, 1283–1287.

Collins, L.J., Kurland, C.G., Biggs, P., and Penny, D. (2009). The modern RNP world of eukaryotes. J. Hered. *100*, 597–604.

Collins, L.J., and Penny, D. (2009). The RNA infrastructure: dark matter of the eukaryotic cell? Trends Genet. *25*, 120–128.

Cortez, D., Delaye, L., Lazcano, A., and Becerra, A. (2009). Composition-based methods to identify horizontal gene transfer. Meth. Mol. Biol. *532*, 215–225.

Cox, C.J., Foster, P.G., Hirt, R.P., Harris, S.R., and Embley, T.M. (2008). The archaebacterial origin of eukaryotes. Proc. Natl. Acad. Sci. U.S.A. *105*, 20356–20361.

Dagan, T., Artzy-Randrup, Y., and Martin, W. (2008). Modular networks and cumulative impact of lateral transfer in prokaryote genome evolution. Proc. Natl. Acad. Sci. U.S.A. *105*, 10039–10044.

Dagan, T., and Martin, W. (2006). The tree of one percent. Genome Biol. *7*, 118.

Dawkins, R. (1976). The Selfish Gene (Oxford: Oxford University Press).

Delsuc, F., Brinkmann, H., and Philippe, H. (2005). Phylogenomics and the reconstruction of the tree of life. Nature Rev. Genet. *6*, 361–375.

Dennis, P.P., and Shimmin, L.C. (1997). Evolutionary divergence and salinity-mediated selection in halophilic archaea. Microbiol. Mol. Biol. Rev. *61*, 90–104.

Di Giulio, M. (2006). The non-monophyletic origin of the tRNA molecule and the origin of genes only after the evolutionary stage of the last universal common ancestor (LUCA). J. Theor. Biol. *240*, 343–352.

Fedorov, A., and Fedorova, L. (2004). Introns: mighty elements from the RNA world. J. Mol. Evol. *59*, 718–721.

Fedorov, A., Cao, X., Saxonov, S., de Souza, S.J., Roy, S.W., and Gilbert, W. (2001). Intron distribution difference for 276 ancient and 131 modern genes suggests the existence of ancient introns. Proc. Natl. Acad. Sci. U.S.A. *98*, 13177–13182.

Fedorov, A., Roy, S., Cao, X., and Gilbert, W. (2003). Phylogenetically older introns strongly correlate with module boundaries in ancient proteins. Genome Res. *13*, 1155–1157.

Fitch, W.M., and Upper, K. (1987). The phylogeny of tRNA sequences provides evidence for ambiguity reduction in the origin of the genetic code. Cold Spring Harbour Symp. Quant. Biol. *52*, 759–767.

Forterre, P. (1995). Thermoreduction, a hypothesis for the origin of prokaryotes. C. R. Acad. Sci. III *318*, 415–422.

Fournier, G.P., and Gogarten, J.P. (2007). Signature of a primitive genetic code in ancient protein lineages. J. Mol. Evol. *65*, 425–436.

Fournier, G.P., and Gogarten, J.P. (2010). Rooting the Ribosomal Tree of Life. Mol. Biol. Evol. epub ahead of print.

Fournier, G.P., Huang, J., and Gogarten, J.P. (2009). Horizontal gene transfer from extinct and extant lineages: biological innovation and the coral of life. Phil. Trans. R. Soc. Lond. B *364*, 2229–2239.

Fujishima, K., Sugahara, J., Kikuta, K., Hirano, R., Sato, A., Tomita, M., and Kanai, A. (2009). Tri-split tRNA is a transfer RNA made from 3 transcripts that provides insight into the evolution of fragmented tRNAs in Archaea. Proc. Natl. Acad. Sci. U.S.A. *106*, 2683–2687.

Fukuchi, S., and Nishikawa, K. (2001). Protein surface amino acid compositions distinctively differ between thermophilic and mesophilic bacteria. J. Mol. Biol. *309*, 835–843.

Galtier, N., and Daubin, V. (2008). Dealing with incongruence in phylogenomic analyses. Phil. Trans. R. Soc. Lond B *363*, 4023–4029.

Gao, B., and Gupta, R.S. (2005). Conserved indels in protein sequences that are characteristic of the phylum Actinobacteria. Int. J. Syst. Evol. Microbiol. *55*, 2401–2412.

Gentle, I., Gabriel, K., Beech, P., Waller, R., and Lithgow, T. (2004). The Omp85 family of proteins is essential for outer membrane biogenesis in mitochondria and bacteria. J. Cell Biol. *164*, 19–24.

Glansdorff, N. (2000). About the last common ancestor, the universal life-tree and lateral gene transfer: a reappraisal. Mol. Microbiol. *38*, 177–185.

Glansdorff, N., Xu, Y., and Labedan, B. (2008). The last universal common ancestor: emergence, constitution and genetic legacy of an elusive forerunner. Biol. Direct *3*, 29.

Glibert, W. (1986). Origin of life, the RNA world. Nature *319*, 618.

Gogarten, J.P. (1995). The early evolution of cellular life. Trends Ecol. Evol. *10*, 147–151.

Gogarten, J.P., and Taiz, L. (1992). Evolution of proton pumping ATPases: Rooting the tree of life. Photosynth. Res. *33*, 137–146.

Gogarten, J.P., Kibak, H., Dittrich, P., Taiz, L., Bowman, E.J., Bowman, B.J., Manolson, M.F., Poole, R.J., Date, T., Oshima, T., *et al.* (1989b). Evolution of the vacuolar $H^+$-ATPase: implications for the origin of eukaryotes. Proc. Natl. Acad. Sci. U.S.A. *86*, 6661–6665.

Gogarten, J.P., Doolittle, W.F., and Lawrence, J.G. (2002). Prokaryotic evolution in light of gene transfer. Mol. Biol. Evol. *19*, 2226–2238.

Gogarten, J.P., Fournier, G., and Zhaxybayeva, O. (2007). Gene transfer and the reconstruction of life's early history from the molecular record. Space Sci. Rev. *135*, 115–131.

Gogarten-Boekels, M., Hilario, E., and Gogarten, J. (1995). The effects of heavy meteorite bombardment on the early evolution – the emergence of the three domains of life. Orig. Life Evol. Biosph. *25*, 251–264.

Gouy, M., and Li, W.H. (1989). Phylogenetic analysis based on rRNA sequences supports the archaebacterial rather than the eocyte tree. Nature *339*, 145–147.

Gribaldo, S., and Cammarano, P. (1998a). The root of the universal tree of life inferred from anciently duplicated genes encoding components of the protein-targeting machinery. J. Mol. Evol. *47*, 508–516.

Gribaldo, S., and Cammarano, P. (1998b). The root of the universal tree of life inferred from anciently duplicated genes encoding components of the protein-targeting machinery. J. Mol. Evol. *47*, 508–516.

Gribaldo, S., Lumia, V., Creti, R., de Macario, E.C., Sanangelantoni, A., and Cammarano, P. (1999). Discontinuous occurrence of the hsp70 (dnaK) gene among Archaea and sequence features of HSP70 suggest a novel outlook on phylogenies inferred from this protein. J. Bacteriol. *181*, 434–443.

Gupta, R.S. (1998). Protein phylogenies and signature sequences: A reappraisal of evolutionary relationships among archaebacteria, eubacteria, and eukaryotes. Microbiol. Mol. Biol. Rev. *62*, 1435–1491.

Hartman, H., and Fedorov, A. (2002). The origin of the eukaryotic cell: a genomic investigation. Proc. Natl. Acad. Sci. U.S.A. *99*, 1420–1425.

Higgs, P.G. (2009). A four-column theory for the origin of the genetic code: tracing the evolutionary pathways that gave rise to an optimized code. Biol. Direct *4*, 16.

Horiike, T., Hamada, K., Kanaya, S., and Shinozawa, T. (2001). Origin of eukaryotic cell nuclei by symbiosis of Archaea in Bacteria is revealed by homology-hit analysis. Nature Cell Biol. *3*, 210–214.

Islas, S., Hernández-Morales, R., and Lazcano, A. (2007). Question 7: Comparative genomics and early cell evolution: a cautionary methodological note. Orig. Life Evol. Biosph, *37*, 415–418.

Iwabe, N., Kuma, K., Hasegawa, M., Osawa, S., and Miyata, T. (1989). Evolutionary relationship of archaebacteria, eubacteria, and eukaryotes inferred from phylogenetic trees of duplicated genes. Proc. Natl. Acad. Sci. U.S.A. *86*, 9355–9359.

Javaux, E.J., Knoll, A.H., and Walter, M.R. (2001). Morphological and ecological complexity in early eukaryotic ecosystems. Nature *412*, 66–69.

Knight, R.D., Freeland, S.J., and Landweber, L.F. (2001). Rewiring the keyboard: evolvability of the genetic code. Nature Rev. Genet. *2*, 49–58.

Koonin, E.V., and Martin, W. (2005). On the origin of genomes and cells within inorganic compartments. Trends Genet. *21*, 647–654.

Koonin, E.V., Wolf, Y.I., and Puigbo, P. (2009). The phylogenetic forest and the quest for the elusive tree of life. Cold Spring Harbour Symp Quant Biol., in press.

Kurland, C.G., Canback, B., and Berg, O.G. (2003). Horizontal gene transfer: a critical view. Proc. Natl. Acad. Sci. U.S.A. *100*, 9658–9662.

Kurland, C.G., Canback, B., and Berg, O.G. (2007). The origins of modern proteomes. Biochimie *89*, 1454–1463.

Lake, J.A. (1987). Prokaryotes and archaebacteria are not monophyletic: rate invariant analysis of rRNA genes indicates that eukaryotes and eocytes form a monophyletic taxon. Cold Spring Harbor Symp. Quant. Biol. *52*, 839–846.

Lake, J.A. (2009). Evidence for an early prokaryotic endosymbiosis. Nature *460*, 967–971.

Lake, J.A., Henderson, E., Oakes, M., and Clark, M.W. (1984). Eocytes: a new ribosome structure indicates a kingdom with a close relationship to eukaryotes. Proc. Natl. Acad. Sci. U.S.A. *81*, 3786–3790.

Lake, J.A., Herbold, C.W., Rivera, M.C., Servin, J.A., and Skophammer, R.G. (2007). Rooting the tree of life using nonubiquitous genes. Mol. Biol. Evol. *24*, 130–136.

Lake, J.A., Skophammer, R.G., Herbold, C.W., and Servin, J.A. (2009). Genome beginnings: rooting the tree of life. Phil. Trans. R. Soc. Lond. B *364*, 2177–2185.

Lawson, F., Charlebois, R., and Dillon, J. (1996). Phylogenetic analysis of carbamoylphosphate synthetase genes: complex evolutionary history includes an internal duplication within a gene which can root the tree of life. Mol. Biol. Evol. *13*, 970–977.

Logsdon, J.M., Jr., and Palmer, J.D. (1994). Origin of introns – early or late? Nature *369*, 526; author reply 527–528.

Logsdon, J.M., Jr., Tyshenko, M.G., Dixon, C., J, D.J., Walker, V.K., and Palmer, J.D. (1995). Seven newly discovered intron positions in the triose-phosphate isomerase gene: evidence for the introns-late theory. Proc. Natl. Acad. Sci. U.S.A. *92*, 8507–8511.

Margulis, L. (1996). Archaeal-eubacterial mergers in the origin of Eukarya: phylogenetic classification of life. Proc. Natl. Acad. Sci. U.S.A. *93*, 1071–1076.

Martin, W., and Müller, M. (1998). The hydrogen hypothesis for the first eukaryote. Nature *392*, 37–41.

Miranda, I., Silva, R., and Santos, M.A. (2006). Evolution of the genetic code in yeasts. Yeast *23*, 203–213.

Nather, D.J., and Rachel, R. (2004). The outer membrane of the hyperthermophilic archaeon *Ignicoccus*: dynamics, ultrastructure and composition. Biochem. Soc. Trans. *32*, 199–203.

Olendzenski, L., and Gogarten, J.P. (1998). Deciphering the molecular record for the early evolution of life: Gene duplication and horizontal gene transfer. In Thermophiles: The keys to molecular evolution and the origin of life?, J. Wiegel, and M.W.W. Adams, eds. (Philadelphia, Taylor & Francis), pp. 165–176.

Olsen, G.J., and Woese, C.R. (1989). A brief note concerning archaebacterial phylogeny. Can. J. Microbiol. *35*, 119–123.

Ouzounis, C.A., Kunin, V., Darzentas, N., and Goldovsky, L. (2006). A minimal estimate for the gene content of the last universal common ancestor – exobiology from a terrestrial perspective. Res. Microbiol. *157*, 57–68.

Paila, U., Kondam, R., and Ranjan, A. (2008). Genome bias influences amino acid choices: analysis of amino acid substitution and re-compilation of substitution matrices exclusive to an AT-biased genome. Nucl. Acids Res. *36*, 6664–6675.

Pallen, M.J., Beatson, S.A., and Bailey, C.M. (2005). Bioinformatics, genomics and evolution of non-flagellar type-III secretion systems: a Darwinian perspective. FEMS Microbiol. Rev. *29*, 201–229.

Penny, D., Hoeppner, M.P., Poole, A.M., and Jeffares, D.C. (2009). An overview of the introns-first theory. J. Mol. Evol., in press.

Penny, D., and Poole, A. (1999). The nature of the last universal common ancestor. Curr. Opin. Genet. Dev. *9*, 672–677.

Pereto, J., Lopez-Garcia, P., and Moreira, D. (2004). Ancestral lipid biosynthesis and early membrane evolution. Trends Biochem. Sci. *29*, 469–477.

Puigbo, P., Wolf, Y.I., and Koonin, E.V. (2009). Search for a 'Tree of Life' in the thicket of the phylogenetic forest. J. Biol. *8*, 59.

Ragan, M.A. (2001). On surrogate methods for detecting lateral gene transfer. FEMS Microbiol. Lett. *201*, 187–191.

Randau, L., Munch, R., Hohn, M.J., Jahn, D., and Söll, D. (2005a). *Nanoarchaeum equitans* creates functional tRNAs from separate genes for their 5'- and 3'-halves. Nature *433*, 537–541.

Randau, L., Pearson, M., and Söll, D. (2005b). The complete set of tRNA species in *Nanoarchaeum equitans*. FEBS Lett. *579*, 2945–2947.

Rivera, M.C., and Lake, J.A. (1992). Evidence that eukaryotes and eocyte prokaryotes are immediate relatives. Science *257*, 74–76.

Rivera, M.C., and Lake, J.A. (2004). The ring of life provides evidence for a genome fusion origin of eukaryotes. Nature *431*, 152–155.

Roy, S.W., and Gilbert, W. (2006). The evolution of spliceosomal introns: patterns, puzzles and progress. Nature Rev. Genet. *7*, 211–221.

Schofield, J.P. (1993). Molecular studies on an ancient gene encoding for carbamoyl-phosphate synthetase. Clin Sci (Lond) *84*, 119–128.

Schopf, J.W. (1999). Cradle of Life: The Discovery of Earth's Earliest Fossils (Princeton: Princeton University Press).

Servin, J.A., Herbold, C.W., Skophammer, R.G., and Lake, J.A. (2008). Evidence excluding the root of the tree of life from the actinobacteria. Mol. Biol. Evol. *25*, 1–4.

Singer, G.A., and Hickey, D.A. (2003). Thermophilic prokaryotes have characteristic patterns of codon usage, amino acid composition and nucleotide content. Gene *317*, 39–47.

Skophammer, R.G., Herbold, C.W., Rivera, M.C., Servin, J.A., and Lake, J.A. (2006). Evidence that the root of the tree of life is not within the Archaea. Mol. Biol. Evol. *23*, 1648–1651.

Skophammer, R.G., Servin, J.A., Herbold, C.W., and Lake, J.A. (2007). Evidence for a Gram-positive, eubacterial root of the tree of life. Mol. Biol. Evol. *24*, 1761–1768.

Sorek, R., Zhu, Y., Creevey, C.J., Francino, M.P., Bork, P., and Rubin, E.M. (2007). Genome-wide experimental determination of barriers to horizontal gene transfer. Science *318*, 1449–1452.

Stoltzfus, A., Spencer, D.F., Zuker, M., Logsdon, J.M., Jr., and Doolittle, W.F. (1994). Testing the exon theory of genes: the evidence from protein structure. Science *265*, 202–207.

Stoltzfus, A., Spencer, D.F., and Doolittle, W.F. (1995). Methods for evaluating exon-protein correspondences. Comput. Appl. Biosci. *11*, 509–515.

Sun, F.J., and Caetano-Anolles, G. (2008a). Evolutionary patterns in the sequence and structure of transfer RNA: early origins of archaea and viruses. PLoS Comput. Biol. *4*, e1000018.

Sun, F.J., and Caetano-Anolles, G. (2008b). The origin and evolution of tRNA inferred from phylogenetic analysis of structure. J. Mol. Evol. *66*, 21–35.

Tekaia, F., Yeramian, E., and Dujon, B. (2002). Amino acid composition of genomes, lifestyles of organisms, and evolutionary trends: a global picture with correspondence analysis. Gene *297*, 51–60.

Trifonov, E.N. (2000). Consensus temporal order of amino acids and evolution of the triplet code. Gene *261*, 139–151.

Valas, R.E., and Bourne, P.E. (2008). Rethinking proteasome evolution: two novel bacterial proteasomes. J. Mol. Evol. *66*, 494–504.

Valas, R.E., and Bourne, P.E. (2009). Structural analysis of polarizing indels: an emerging consensus on the root of the tree of life. Biol. Direct *4*, 30.

Vellai, T., and Vida, G. (1999). The origin of eukaryotes: the difference between prokaryotic and eukaryotic cells. Proc. Biol. Sci. *266*, 1571–1577.

Wächtershäuser, G. (1990). Evolution of the first metabolic cycles. Proc. Natl. Acad. Sci. U.S.A. *87*, 200–204.

Wächtershäuser, G. (1992). Groundworks for an evolutionary biochemistry: the iron-sulfur world. Prog. Biophys. Mol. Biol. *58*, 85–201.

Wächtershäuser, G. (2000). Origin of life. Life as we don't know it. Science *289*, 1307–1308.

Watanabe, K., Chishiro, K., Kitamura, K., and Suzuki, Y. (1991). Proline residues responsible for thermostability occur with high frequency in the loop regions of an extremely thermostable oligo-1,6-glucosidase from *Bacillus thermoglucosidasius* KP1006. J. Biol. Chem. *266*, 24287–24294.

Wilkinson, M., Cotton, J.A., Lapointe, F.J., and Pisani, D. (2007). Properties of supertree methods in the consensus setting. Syst. Biol. *56*, 330–337.

Woese, C.R. (2002). On the evolution of cells. Proc. Natl. Acad. Sci. U.S.A. *99*, 8742–8747.

Woese, C.R., and Fox, G.E. (1977). Phylogenetic structure of the prokaryotic domain: the primary kingdoms. Proc. Natl. Acad. Sci. U.S.A. *74*, 5088–5090.

Woese, C.R., and Olsen, G.J. (1986). Archaebacterial phylogeny: perspectives on the urkingdoms. Syst. Appl. Microbiol. *7*, 161–177.

Wu, M., and Eisen, J.A. (2008). A simple, fast, and accurate method of phylogenomic inference. Genome Biol. *9*, R151.

Xue, H., Tong, K.L., Marck, C., Grosjean, H., and Wong, J.T. (2003). Transfer RNA paralogs: evidence for genetic code-amino acid biosynthesis coevolution and an archaeal root of life. Gene *310*, 59–66.

Zhaxybayeva, O., and Gogarten, J.P. (2004). Cladogenesis, coalescence and the evolution of the three domains of life. Trends Genet. *20*, 182–187.

Zhaxybayeva, O., Lapierre, P., and Gogarten, J.P. (2005). Ancient gene duplications and the root(s) of the tree of life. Protoplasma *227*, 53–64.

Zhaxybayeva, O., Swithers, K.S., Lapierre, P., Fournier, G.P., Bickhart, D.M., DeBoy, R.T., Nelson, K.E., Nesbø, C.L., Doolittle, W.F., Gogarten, J.P., and Noll, K.M. (2009). On the chimeric nature, thermophilic origin, and phylogenetic placement of the Thermotogales. Proc. Natl. Acad. Sci. U.S.A. *106*, 5865–5870.

Zhou, X.X., Wang, Y.B., Pan, Y.J., and Li, W.F. (2008). Differences in amino acids composition and coupling patterns between mesophilic and thermophilic proteins. Amino Acids *34*, 25–33.

# Applications of Conserved Indels for Understanding Microbial Phylogeny

7

Radhey S. Gupta

Abstract

Comparative analysis of genome sequences is leading to discovery of large numbers of novel molecular markers that are proving very helpful in understanding many important aspects of microbial phylogeny. Of these molecular markers, the conserved inserts or deletions (indels) in protein sequences provide particularly useful means for identifying different groups of microbes in clear molecular terms and for understanding how they have branched off from a common ancestor. This chapter provides an overview of the utility of conserved indels and other novel molecular markers (viz. lineage-specific proteins) for understanding microbial phylogeny at different phylogenetic depths with specific examples. Genetic and biochemical studies of these markers should also lead to identification of novel properties that are unique to different groups of microbes.

## Microbial phylogeny: overview and key unresolved issues

Our current understanding of the evolutionary relationships among prokaryotes is largely based on 16S rRNA sequences (Woese, 1987; Olsen *et al.*, 1994; Maidak *et al.*, 2001; Ludwig and Klenk, 2005; Garrity *et al.*, 2005). The analyses of 16S rRNA and other gene/protein sequences have led to division of prokaryotic organisms into two main domains: *Bacteria* and *Archaea* (Woese *et al.*, 1990). Of these, *Bacteria* constitute >98% of the known prokaryotic organisms and most of the discussion in this chapter is limited to them (Maidak *et al.*, 2001; Garrity *et al.*, 2005). The branching pattern of *Bacteria* in the 16S rRNA trees has led to their division into many main groups or phyla. The second edition of *Bergey's Manual* of Systematic Bacteriology, which represents our currently accepted view of microbial taxonomy, divides all cultured bacteria into 24 main phyla: *Aquificae, Thermotogae, Thermodesulfobacteria, Deinococcus-Thermus, Chrysiogenetes, Chloroflexi, Thermomicrobia, Nitrospirae, Deferribacteres, Cyanobacteria, Chlorobi, Proteobacteria, Firmicutes, Actinobacteria, Planctomycetes, Chlamydiae, Spirochaetes, Fibrobacteres, Acidobacteria, Bacteroidetes, Fusobacteria, Verrucomicrobia, Dictyoglomi* and *Gemmatimonadetes* (Garrity *et al.*, 2005). Some of these phyla viz. *Thermodesulfobacteria, Thermomicrobia, Chrysiogenetes, Dictyoglomi, Deferribacters*, consist of only a few species, whereas others such as *Proteobacteria, Cyanobacteria, Firmicutes, Actinobacteria* and *Bacteroidetes* are made up of thousands of species accounting for more than 95% of all cultured bacteria (Maidak *et al.*, 2001; Garrity *et al.*, 2005).

Most of these bacterial phyla were first described when sequence information was available for only a limited number of species and these groups could be clearly distinguished from each other by long naked branches in the 16S rRNA trees (Woese *et al.*, 1985; Woese, 1987; Olsen *et al.*, 1994). However, with the enormous increase in the number of sequences in the past 15–20 years, most of the naked branches have been filled, making it increasingly difficult to clearly demarcate these groups in phylogenetic trees (Ludwig and Schleifer, 1999; Ludwig and Klenk, 2005). However, except for their branching pattern in phylogenetic trees, which is affected by a large number of variables (Gupta, 1998b; Felsenstein, 2004; Delsuc *et al.*, 2005), for most bacterial phyla, no molecular, biochemical or physiological properties are known that are unique to them (Ludwig and Klenk, 2005; Garrity *et al.*, 2005). Hence, it is of much importance to develop other reliable means to circumscribe different main groups of bacteria in more definitive terms. In this context, an important question that remains to be understood is: '*In what respects do different main groups of bacteria differ from each other? Are species from these groups represent merely phylogenetically distinct entities or do they commonly share certain unique molecular, biochemical, structural or physiological properties that are specific for them?*'

It should be pointed out that the division of *Bacteria* into these 24 or so main phyla is arbitrary and it is not based on any objective evolutionary or taxonomic considerations (Ludwig and Klenk, 2005; Stackebrandt, 2006). Currently, there are no clearly stated criteria as to what constitute a phylum or other higher taxonomic groups (viz. Class, Order or Family) within *Bacteria* or prokaryotes (Ludwig and Klenk, 2005; Stackebrandt, 2006). The arbitrariness of this classification is illustrated by the following example. The bacterial groups known as the Alpha ($\alpha$)-, Beta ($\beta$)- and Gamma ($\gamma$)- Proteobacteria are amongst the largest groups within prokaryotes accounting, respectively, for about 12%, 8% and 26% of all cultured bacteria (Maidak *et al.*, 2001). Species from these groups can be clearly distinguished from all other bacteria, both in phylogenetic trees and based on many other discrete molecular characteristics (Gupta, 2000b, 2005b, 2006; Kersters *et al.*, 2006; Gupta and Mok, 2007; Cutino-Jimenez *et al.*, 2009; Gao *et al.*, 2009a). Additionally, based upon different studies, it is now established that these groups have branched off from a common ancestor in the following order: $\delta/\varepsilon \rightarrow \alpha \rightarrow \beta \rightarrow \gamma$ (Gupta, 2000b; Kersters *et al.*, 2006; Gupta and Sneath, 2007). However, despite the enormity, distinctness and importance of these groups, they are presently recognized as subdivisions (taxonomic ranks Class) of the Proteobacteria phylum (Murray *et al.*, 1990; Garrity *et al.*, 2005). In contrast, many other poorly characterized groups *Thermodesulfobacteria*, *Thermomicrobia*, *Chrysiogenetes*, *Dictyoglomi*, *Deferribacteres*, *Verrucomicrobia* and *Nitrospirae*, consisting of only small numbers of species are presently regarded as the main phyla of Bacteria (Garrity *et al.*, 2005). Hence, in order to develop a better understanding of microbial phylogeny it is necessary to develop objective criteria based on sound evolutionary principles, for assignment of higher taxonomic ranks within microbes.

Another important aspect of microbial evolution that needs to be clarified is the branching order of different main groups of bacteria from a common ancestor. In view of the fact that prokaryotic organisms were the sole inhabitants of this planet for a large part (~ 1.5 to 2.0 Gy) of the history of life (Schopf, 1978; Gupta, 1998b; Knoll, 1999; Rasmussen *et al.*, 2008), they hold the key for understanding the beginning of life and how different biochemical processes, which are essential for life have evolved. An understanding of the evolutionary relationships among prokaryotes is also of critical importance in understanding how more complex organisms, such as eukaryotes, have evolved (Gupta and Golding, 1996; Martin and Müller, 1998; Rasmussen *et al.*, 2008). The phylogenetic trees based on 16S rRNA or other gene/protein sequences have not been able to resolve the branching order or interrelationships among different main groups of bacteria. This has led to the notion that this important question perhaps will never be understood (Doolittle, 1999; Boucher *et al.*, 2003) and all of the main phyla have branched off from a common ancestor at about the same time (Olsen *et al.*, 1994; Ludwig and Schleifer, 1999; Ludwig and Klenk, 2005; Ciccarelli *et al.*, 2006). Hence, it is of critical importance to develop additional means to clarify the branching order and interrelationships among different bacterial phyla (Gupta and Griffiths, 2002; Oren and Stackebrandt, 2002; Shah *et al.*, 2009).

## Conserved indels as novel tools for systematic and evolutionary studies

The availability of genomic sequences from large number of microbes has provided an unprecedented opportunity for discovering novel molecular characteristics that could be used to understand various unresolved aspects of microbial phylogeny (Nelson *et al.*, 2001; Gupta and Griffiths, 2002; Korbel *et al.*, 2005). The markers that are ideally suited for such studies are homologous apomorphic characters that evolved only once (synapomorphy) during the course of evolution (Gupta, 1998b; Stackebrandt, 2006). Further, these markers should be such that their presence or absence in orthologous sequences could be readily ascertained and it should be minimally affected by factors such as long-branch attraction effect, differences in evolutionary rates, lateral gene transfers etc., which confound inferences from phylogenetic trees (Felsenstein, 2004; Delsuc *et al.*, 2005). The conserved inserts and deletions (indels) in widely distributed proteins that are limited to well-defined groups of organisms provide an important class of molecular markers that generally satisfy these criteria (Rivera and Lake, 1992; Gupta, 1998b; Rokas and Holland, 2000; Gupta and Griffiths, 2002). Although homologues of the proteins where these indels are found are present in various organisms, the indels of interest are limited to particular groups of microbes.

The indels that provide useful phylogenetic markers generally are of defined size and they are flanked on both sides by conserved regions to ensure that they constitute reliable characteristics, which are not resulting from sequence alignment artifacts (Gupta, 1998b, 2000b; Gupta and Griffiths, 2002). Because, the genetic changes that give rise to a given conserved indel are highly specific and extremely rare in occurrence, such changes are less likely to arise independently in different groups by convergent evolution (Gupta, 1998b; Rokas and Holland, 2000). Hence, when a conserved signature indel (CSI) of defined size is uniquely found in a phylogenetically defined group(s) of species, the simplest explanation for this is that the genetic change responsible for this CSI occurred only once in a common ancestor of this group of species and then passed on to various descendents. Further, based upon the presence or absence of a particular CSI in various outgroup species, it is also possible to infer whether the CSI under consideration is an insert or a deletion in a given group, i.e. which of its two character states, insert containing or insert-lacking, is ancestral and which is derived (Rivera and Lake, 1992; Gupta, 1998b; Gupta and Griffiths, 2002). Because genetic changes leading to CSIs can occur at various stages in evolution, it is possible to identify CSIs in gene/protein sequences at different phylogenetic depths corresponding to different taxonomic groupings (e.g. phylum, order, family or genus). Thus, making use of different CSIs, which have been introduced at various stages in evolution, a rooted evolutionary relationship among different taxa under consideration can be derived (Gupta, 1998b, 2001; Lake *et al.*, 2007). Indeed, many CSIs have proven useful in clarifying the branching order and interrelationships among the bacterial phyla (Gupta, 1998b, 2003; Gupta and Griffiths, 2002; Griffiths and Gupta, 2004b; Lake *et al.*, 2007). Some examples of how these CSIs are proving useful in clarifying different aspects of microbial evolution are provided below.

## Conserved indels that are specific for different taxonomic groups of Bacteria

One of the major applications of conserved indels has been in identifying different groups of bacteria in clear molecular terms. In the past 6–7 years large numbers of conserved indels that are specific for different bacterial groups have been identified. Based upon these CSIs, most of the major groups within *Bacteria* including *Alphaproteobacteria* (Kainth and Gupta, 2005; Gupta and Mok, 2007), *Gammaproteobacteria* (Gupta, 2000b; Cutino-Jimenez *et al.*, 2009; Gao *et al.*, 2009a), *Epsilonproteobacteria* (Gupta, 2006), *Cyanobacteria* (Gupta *et al.*, 2003; Gupta, 2009), *Deinococcus-Thermus* (Griffiths and Gupta, 2004a, 2007a), *Chlamydiae–Verrucomicrobia* (Griffiths *et al.*, 2005, 2006; Gupta and Griffiths, 2006; Griffiths and Gupta, 2007b), *Fibrobacter–Chlorobi–Bacteroidetes* (Gupta, 2004; Gupta and Lorenzini, 2007), *Actinobacteria* (Gao and Gupta, 2005; Gao *et al.*, 2006), *Aquificales* (Griffiths and Gupta, 2004b, 2006b) and Firmicutes (Gupta, 1998b; Gupta and Gao, 2009), as well as many of their subgroups, can now be identified and circumscribed in more definitive molecular terms. The usefulness of these signatures for understanding microbial phylogeny is illustrated by some signatures indels for the phylum *Cyanobacteria*.

Cyanobacteria, which comprise the sole prokaryotic phylum capable of carrying out oxygenic photosynthesis, are highly diverse in terms of their morphology, physiology and other characteristics (Rippka *et al.*, 1979; Wilmotte and Golubic, 1991; Kondratieva *et al.*, 1992; Castenholz, 2001; Sanchez-Baracaldo *et al.*, 2005). In the 16S rRNA trees, the cyanobacterial species/strains form 14 unresolved clusters; however, their taxonomy and evolutionary relationship is poorly understood (Wilmotte and Herdman, 2001; Hoffmann *et al.*, 2005; Oren and Tindall, 2005). Recent studies have led to identification of >40 CSIs that are specific for either all cyanobacteria or their different main clades (Gupta, 2009; Gupta and Mathews, 2009). Fifteen of these CSIs are uniquely found in all sequenced cyanobacteria and the genetic changes responsible for them were likely introduced in a common ancestor of this phylum. One example of a conserved indel that is specific for cyanobacteria is shown in Fig. 7.1. In this case, a two amino acids insert in a highly conserved region of the SecA protein is present in all sequenced cyanobacteria, but not in any other bacteria (Gupta *et al.*, 2003). The absence of this indel in all other bacterial phyla provides evidence that this indel represents an insert in *Cyanobacteria,* rather than a deletion in other phyla. Interestingly this insert, which is specific for cyanobacteria, is also present in SecA homologues from various plastids including *Rhodophyta, Chlorophyta, Chromophyta,* and also *Cryptophyta* (Gupta *et al.*, 2003), thereby providing evidence that the primary or secondary plastids found in these organisms have originated from cyanobacterial ancestors (Barbrook *et al.*, 1998; Palmer and Delwiche, 1998).

Fig. 7.2 shows another large conserved indel that is specific for cyanobacteria. In this case, an 18 amino acids indel in DNA polymerase I (Pol I) is present in all other cyanobacteria with the exception of *Gloeobacter violaceus* and *Synechococcus* spp. (JA-3-3Ab and JA2-3-B′a). Phylogenetic analyses of sequenced cyanobacteria based on different sets of protein sequences provide evidence that the cyanobacteria lacking this indel, which are referred to as Clade A, form the deepest branching lineage within cyanobacteria (Sanchez-Baracaldo *et al.*, 2005; Zhaxybayeva *et al.* 2006; Swingley *et al.*, 2008; Gupta, 2009; Gupta and Mathews, 2009). The species distribution pattern of the Pol I insert strongly suggests that the genetic change leading to this CSI occurred in a common ancestor of various other cyanobacteria after the branching of Clade A. Likewise, many other conserved indels that are specific for various other subgroups or clades of cyanobacteria have been identified. Based upon their species distribution, the evolutionary stages where genetic changes responsible for these CSIs likely occurred are indicated in Fig. 7.3.

Based upon these CSIs, all of the main clades of cyanobacteria, which are observed in phylogenetic trees (Sanchez-Baracaldo *et al.*, 2005; Swingley *et al.*, 2008; Gupta, 2009; Gupta and Mathews, 2009), can now be identified in molecular terms. Importantly, the identified CSIs are also instrumental in identifying some clades of cyanobacteria, which are not clearly resolved in phylogenetic trees. For example, three conserved indels (19 amino acids insert in DnaE, 13 amino acids deletion in GDP-mannose pyrophosphorylase and 22–27 amino acids insert in NAD(P)H-quinone oxidoreductase subunit D; see Fig. 7.3) are uniquely shared by various sequenced cyanobacteria belonging to the orders *Nostocales, Oscillatoriales* and *Chroococcales.* These three orders of cyanobacteria also uniquely share a number of conserved proteins that are not present in any other cyanobacteria or other phyla of bacteria (Gupta and Mathews, 2009). Although these cyanobacterial orders branch close to each other in phylogenetic trees, a specific relationship among them has not been previously demonstrated (Sanchez-Baracaldo *et al.*, 2005; Hoffmann *et al.*, 2005; Swingley *et al.*, 2008; Gupta, 2009; Gupta and Mathews, 2009). The uniquely shared presence of these CSIs and other conserved cyanobacteria-specific proteins (Gupta and Mathews, 2009) strongly suggest that species/strains from these three orders shared a common ancestor exclusive of other cyanobacteria and that a clade consisting of them represents a deeper branching grouping within cyanobacteria (Fig. 7.3). Similar detailed analyses based on conserved indels have been carried out for many other groups/phyla within *Bacteria* enabling identification and circumscription of species from these groups in molecular terms (Gao *et al.*, 2006, 2009a; Gupta and Griffiths, 2006; Gupta and Lorenzini, 2007; Gupta and Mok, 2007).

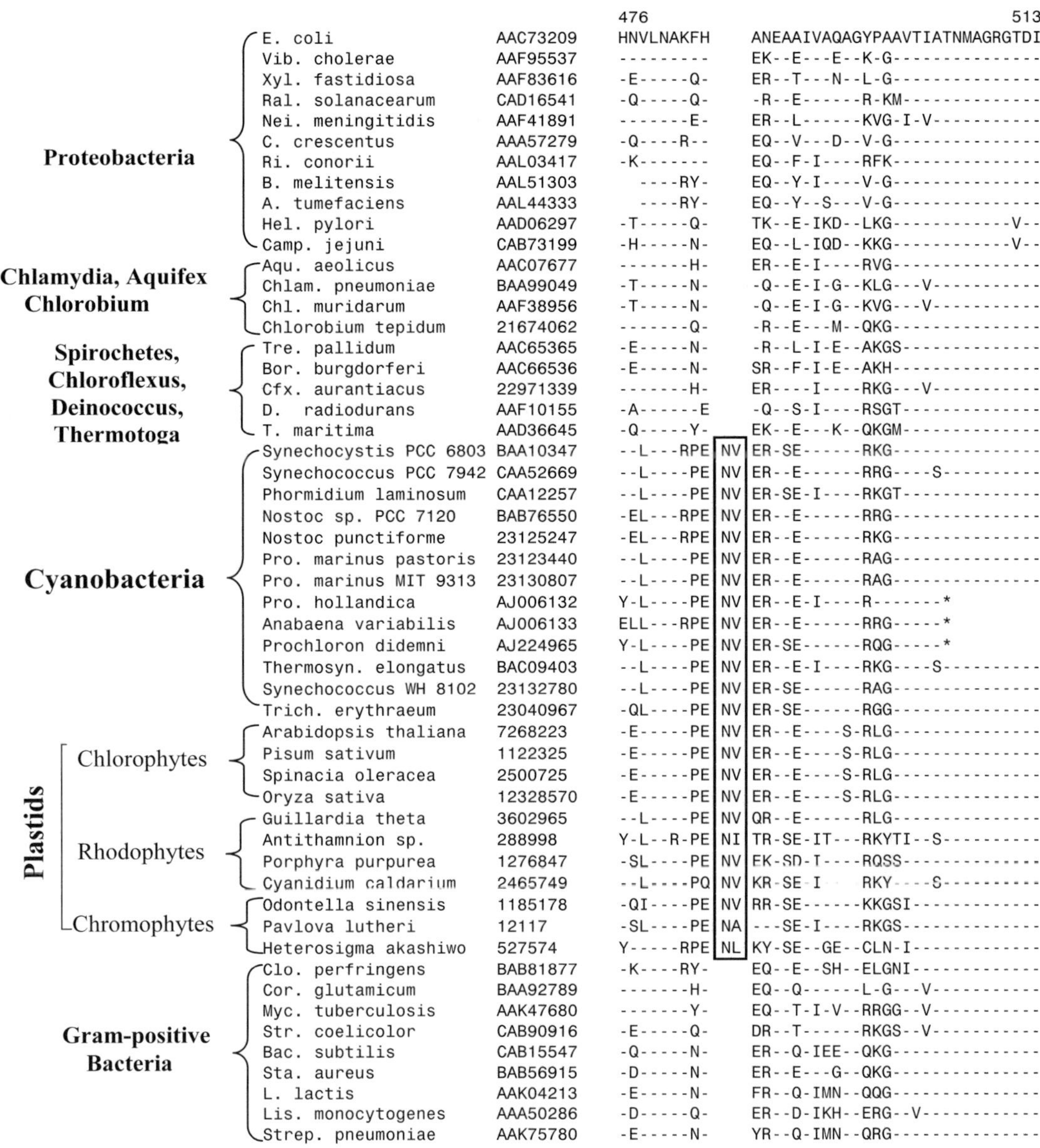


**Figure 7.1** Partial alignment of SecA protein sequences showing a 2-aa insert (boxed) that is uniquely found in various Cyanobacteria and the plastid homologues. Dashes in this and all other sequence alignments indicate identity with the amino acid on the top line. Accession numbers for the sequences are given in the second column. The numbers on the top indicate the position of this sequence. Sequence information for only representative species is presented here. Adapted from Gupta *et al.* (2003).

## Application of conserved indels to determine the branching order of the main phyla

It has proven difficult to resolve the relative branching order of the main bacterial phyla based on phylogenetic trees (Doolittle, 1999; Ludwig and Klenk, 2005). The conserved indels in protein sequences that have been introduced at various stages in evolution, and which are commonly shared by species from different phyla, have again provided a powerful mean to deduce their relative branching orders (Gupta *et al.*, 1999; Gupta, 2001, 2003; Griffiths and Gupta, 2004b). Fig. 7.4 shows example of a 21–23 amino

| Group | Organism | Accession | 845 | (boxed insert) | 910 |
|---|---|---|---|---|---|
| All other Cyano-bacteria | Nostoc PCC 7120 | NP_485297 | GYVETILGRRRYFDFTNNSLRKL | KGSKPEDIDLSKLKNLGPY | DAGLLRSAANAPIQGSSADIIKIA |
| | Nostoc PCC 73102 | ZP_00107254 | ---------------------R- | ---N--------------- | ----------------N------- |
| | Anabaena variabilis | YP_321100 | -------------N-DST--LN- | ---N-------------AK | ------------------------ |
| | Lyngbya PCC 8106 | ZP_01624605 | ---T-L-------H-ESP--KA- | -E-PL---P-DE--KI-QN | --QF--A----------------- |
| | Nodularia spumigena | ZP_01631596 | -----------V--K-D------ | -N-----------------E | ------------------------ |
| | Synechocystis PCC6803 | NP_442677 | ---T--V------N-VTEA--Q- | R-KTVTEL--VDV-MNYN | --Q--------------------- |
| | Crocosphaera watsonii | ZP_00516744 | ---T---------N-FDDN-YRF | R-HD-QN---N-ININYN | -SQ---A----------------- |
| | Cyanothece CCY0110 | ZP_01729743 | ---T---------N-FDDN-YRF | R-HD-QT---NQ-NINYN | -SQ---A----------------- |
| | Microcystis aeruginosa | YP_001656825 | -------------N-D-PL-KR- | R-LSLH----DS--LNNE | E-Q------------------V- |
| | Acaryochloris marina | YP_001516479 | -------------E-DSRG-K-Y | R-KAL-ELAEVD-SDIKM* | -R----A-----------L---- |
| | Thermosyn. elongatus | NP_682129 | -------------A-ESRE-QS- | R-KPLDVLADVDPSK-KM* | ER----A--------------C- |
| | Trichodes. erythraeum | YP_720469 | -------------ELESKTI-D- | -DKE--E-N-QE--FSQN | --QI-----S------------ |
| | Syn. elongatus PCC7942 | YP_399213 | -----V---------EDTG-Q-- | R--D--S---D-IRPSRF | E-Q---A--------------V- |
| | Syn. elongatus PCC6301 | YP_172027 | -----V---------EDTG-Q-- | R--D--S---D-IRPSRF | E-Q---A--------------V- |
| | Synechococcus WH8102 | NP_896795 | -----------P-H-DR-G-GR- | L-KE-LE---DVARRG-M | E-QQ--A--------------V- |
| | Synechococcus RS9916 | ZP_01470390 | -----------P-H-DR-G-GR- | L-KD-LE---DVARRG-M | E-QQ--A--------------L- |
| | Synechococcus WH5701 | ZP_01085666 | -----------P-A-DPGG-GR- | R-KP--E-E-EVARRA-M | E-QQ--A--------------L- |
| | Synechococcus RS9917 | ZP_01080692 | -----------P-H-DR-G-GR- | L-KD-LE---EVARRG-M | E-QQ--A--------------L- |
| | Synechococcus WH7803 | YP_001225340 | ------M----P-H-DR-G-GR- | L-KD-FE---DVARRG-M | E-QQ--A--------------L- |
| | Synechococcus RCC307 | YP_001227925 | -----LM----P-H-DPQG-GR- | --TD-AE---DVARRG-M | E-QQ--A--------------K- |
| | Pro. marinus CCMP1375 | NP_875626 | ---Q-L-------N-DK-G-GR- | L--D-MN---KRARRA-M | E-QQ--A---------------- |
| | Pro. marinus NATL1A | YP_001015410 | -F---L-------H-NK-G-GR- | L-TP-NE---TTARRA-M | E-QQ--A--------------L- |
| | Pro. marinus MIT9211 | YP_001551099 | -F---------H-H-DK-G-GR- | L-KN-LE---KLARRA-M | E-QQ--A--------------V- |
| | Pro. marinus MIT9301 | YP_001091553 | ---K--F--K-E-K-DK-G-GR- | I-KD-YE---QSARRA-M | E-QS--A---------------- |
| | Pro. marinus AS9601 | YP_001009706 | ---K--F--K-E-K-DK-G-GR- | L-KD-YE---QAARRA-M | E-QS--A---------------- |
| | Pro. marinus MIT9312 | YP_397731 | ---K--F--K-E-K-DK-G-GR- | I-KD-YE---QSARRA-M | E-QS--A---------------- |
| | Pro. marinus CCMP1986 | NP_893257 | ------F--K-E-K-DK-G-GR- | I-KD-YE---QTARKA-M | E-QS--A---------------- |
| | Pro. marinus MIT9313 | NP_894987 | -----L-----P-H-DR-G-GR- | L-KD-M----DVARRG-M | E-QQ--A--------------V- |
| Clade A Cyano-bacteria | Synechococcus JA-3-3A | YP_474011 | -----L-------PNLSQLSGHR | | KQAE--A-V-------AS----V- |
| | Synechococcus JA-2-3B | YP_476390 | -------------PNLSRLTGHR | | KQAE--A-V-------AS----V- |
| | Gloeobacter violaceus | NP_923582 | -----L-------RGLGQLNQRD | | RE-A--A-F------TA------- |
| All other Bacteria | Escherichia coli | 16131704 | -----LD---L-LPDIKS-NGAR | | R-AAE-A-I---M--TA-----R- |
| | Xylella fastidiosa | 15837705 | -----VF---L-LNSIASGNQTQ | | R--AE-A-I---M--TA-----R- |
| | Nitrosomonas europaea | 22955357 | -----V----LQLSDIRSNQ-NR | | QM-AE-A-I---M--TA-----L- |
| | Rhodospirillum rubrum | 22965813 | -W-A-PF---IPIPGIQDKNPAA | | R-FAE-Q-I------GA-----R- |
| | Chlorobium tepidum | 21674485 | ---T-LM-----VPDL-SANSNI | | RKAAE-VTM-T----TA-----F- |
| | Borrelia burgdorferi | 15594893 | --S----K----IKEI-SNNYLE | | RSAAE-I-I-SI----A---M--- |
| | Deinococcus radiodurans | 15806710 | -----LY-----VPGLSSRN-VQ | | REAEE-L-Y-M----TA---M-L- |
| | Thermotoga maritima | 15644367 | ---R-LF--K-DIPQLMARD-NT | | Q-EGE-I-I-T----TA-----L- |
| | Clostridium perfringens | 18310976 | ---L-M-N----IPEA-A-NKIV | | K-LGE-L-M------TA-----L- |
| | Streptomyces coelicolor | 21220485 | --TA-LF-----LPDL-SDN-QR | | REAAE-M-L------TA---V--- |
| | Staphylococcus aureus | 21283362 | -----L-H----IPDITSRNFN- | | RGFAE-T-M-T-----A-----L- |
| | Bacillus subtilis | 16079961 | ---T-LMH----IPELTSRNFNI | | RSFAE-T-M-T-----A-----K- |

**Figure 7.2.** Partial sequence alignment of DNA polymerase I (Pol I) showing a large conserved insert (boxed) that is uniquely present in various cyanobacteria except those from the Clade A species/strains. Sequence information for only some bacteria from other phyla is presented, however, this insert is not found in any of them. Due to space considerations, sequence information for some closely related cyanobacterial species/strains is also not shown. At the position marked by * three extra amino acids are present in *Acaryochloris marina* and *Thermosynechococcus elongatus*. Adapted from Gupta (2009).

acids conserved indel in the Hsp70 protein that has been extensively studied. In this case, the indicated indel (boxed) is commonly shared by all orthologous sequences from different phyla of Gram-negative bacteria (viz. *Proteobacteria, Aquificae, Bacteroidetes, Chlorobi, Fibrobacteres, Chlamydia, Verrucomicrobia, Spirochaetes, Cyanobacteria, Nitrospirae, Chloroflexi* and *Deinococcus-Thermus*, but it is absent in all available sequences from *Firmicutes, Thermotogae* and *Fusobacteria* as well as most *Actinobacteria* (Gupta, 1998b; Singh and Gupta, 2009). However, some *Actinobacteria* have two copies of this gene, one of which contains this indel; this gene copy has likely been acquired by means of lateral gene transfer (Singh and Gupta, 2009). The absence of this insert in various *Archaea* provides evidence that the absence of this insert is the ancestral character state of this protein (Gupta, 1998b, 1998c). It is also of much importance that in contrast to the bacterial groups lacking this indel, which are surrounded by a single lipid bilayer membrane (Monoderm prokaryotes), all bacterial phyla containing this indel contain both an inner as well as an outer lipid bilayer membrane (Diderm bacteria). We have suggested that the bacterial cells bounded by a single membrane are ancestral to those containing two different membranes (Gupta, 1998b, 1998c, 2000a, 2005a). Hence, it is very likely that the rare genetic change leading to this conserved indel likely occurred in a common ancestor of various Gram-negative (Diderm) phyla as indicated in Fig. 7.5 (Gupta, 1998a,b, 2000a). It should be noted that some authors have suggested that Diderm bacteria evolved first (Cavalier-Smith, 2002; Valas and Bourne, 2009). However, it is difficult to envisage any scenario where a cell with two membrane can evolve

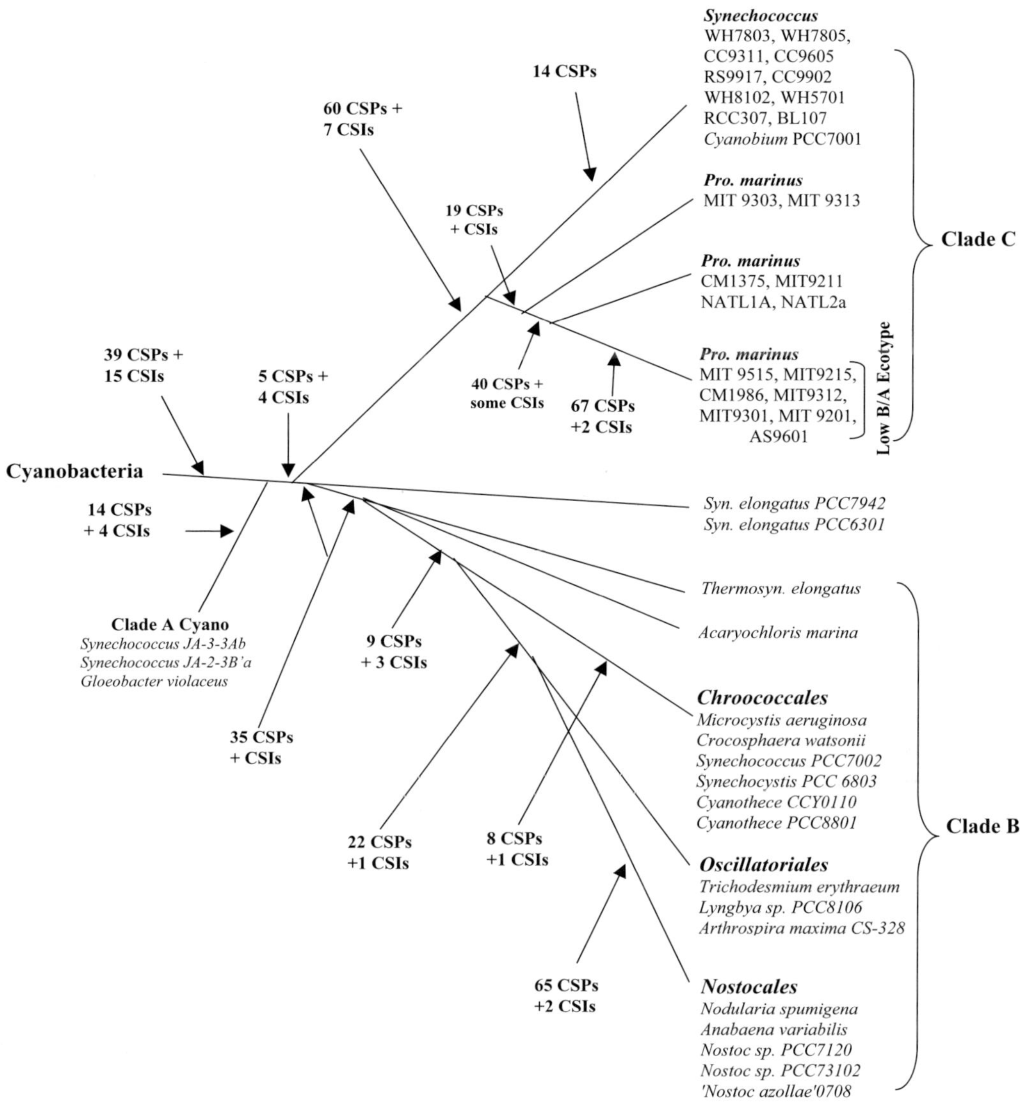


**Figure 7.3** A summary diagram showing the evolutionary stages where different conserved indels (CSIs) as well as conserved signature proteins (CSPs) that are specific for the indicated cyanobacterial were likely introduced (Gupta *et al.*, 2003; Gupta, 2009; Gupta and Mathews, 2009). All of the clades identified by these signatures are also supported by phylogenetic analyses (Swingley *et al.*, 2008; Gupta, 2009; Gupta and Mathews, 2009). Pro = *Prochlorococcus*.

without the initial development of a cell with one membrane (Gupta, 1998b, 2000a).

Likewise, the conserved insert in the Hsp60 protein is absent in virtually all *Firmicutes, Actinobacteria, Thermotogae, Fusobacteria, Deinococcus-Thermus* and *Chloroflexi,* but it is present in various other bacterial phyla, without any exception (Gupta, 2000b; Singh and Gupta, 2009). Based upon its species distribution, this insert was introduced at a later stage in bacterial evolution in comparison to the insert in the Hsp70 protein as indicated in Fig. 7.5. Fig. 7.4 also shows

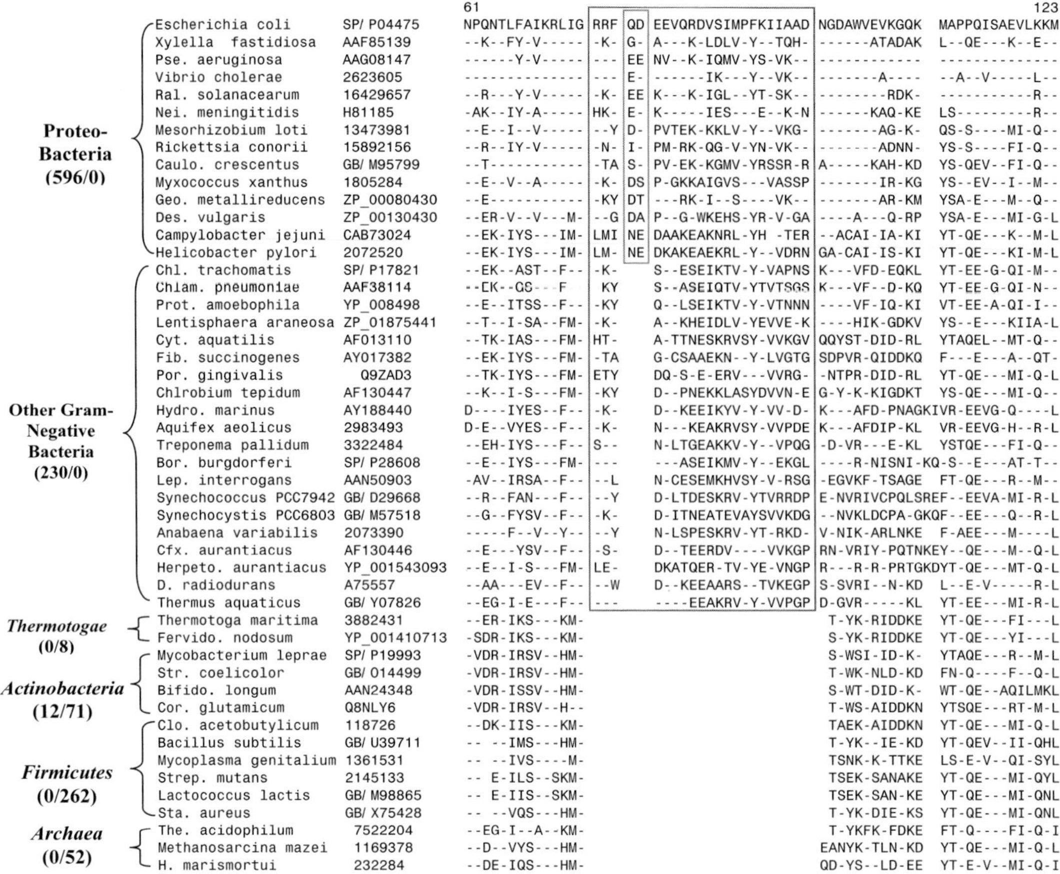


**Figure 7.4** Partial sequence alignment of the Hsp70/DnaK protein showing a large insert (boxed) that is specific for various Gram-negative bacteria (i.e. diderm bacteria bounded by two membranes) and the eukaryotic homologues, but not found in other bacteria. The numbers with the group's name indicate the presence or absence of this insert in various blast hits from each group. For example, for *Proteobacteria* a total of 596 hits were observed and all of them contained this insert. The distribution (presence/absence) of this insert in various subgroups of Gram-negative bacteria was as follows: *Gammaproteobacteria* 213/0; *Betaproteobacteria*, 120/0; *Alphaproteobacteria*, 147/0; *Delta-Epsilonproteobacteria*, 116/0; *Chlamydiae-Verrucomicrobiae*, 10/0; *Bacteroidetes–Chlorobi*, 54/0; *Cyanobacteria*, 121/0; *Aquificae*, *Spirochaetes*, *Chloroflexi*, *Deinococcus–Thermus*, 45/0; All of the sequences shown here are for the DnaK homologues; however, this insert is also present in other Hsp70 homologues (viz. HscA or HscC) that are present in some bacteria (mainly proteobacteria). The smaller box inside the larger box indicates a 2 aa insert that is specific for *Proteobacteria* and not found in other Gram-negative bacteria (Gupta, 1998b, 2000b).

another insert (two amino acids) in the Hsp70 protein, which is present within the larger insert. This smaller insert (boxed) is uniquely found in all *Proteobacteria* (different Classes) and provides one of the few markers that is specific for this vast group of bacteria (Gupta, 1998b, 2000b; Singh and Gupta, 2009). In a similar manner, the various other CSIs shown in Fig. 7.5 are hypothesized to be introduced at the indicated stages of evolution. Various conserved indels in this diagram are present in different bacterial phyla that lie above the indicated insertion points, but they are generally not found in the groups that are noted below. Thus, based upon the presence or absence of these CSIs in different bacterial groups, their branching order from a common ancestor can be deduced (Gupta, 1998b, 2001, 2003; Griffiths and Gupta, 2004b). All of the main branches in this tree, as well as their sub-branches, can also be clearly distinguished from each other based

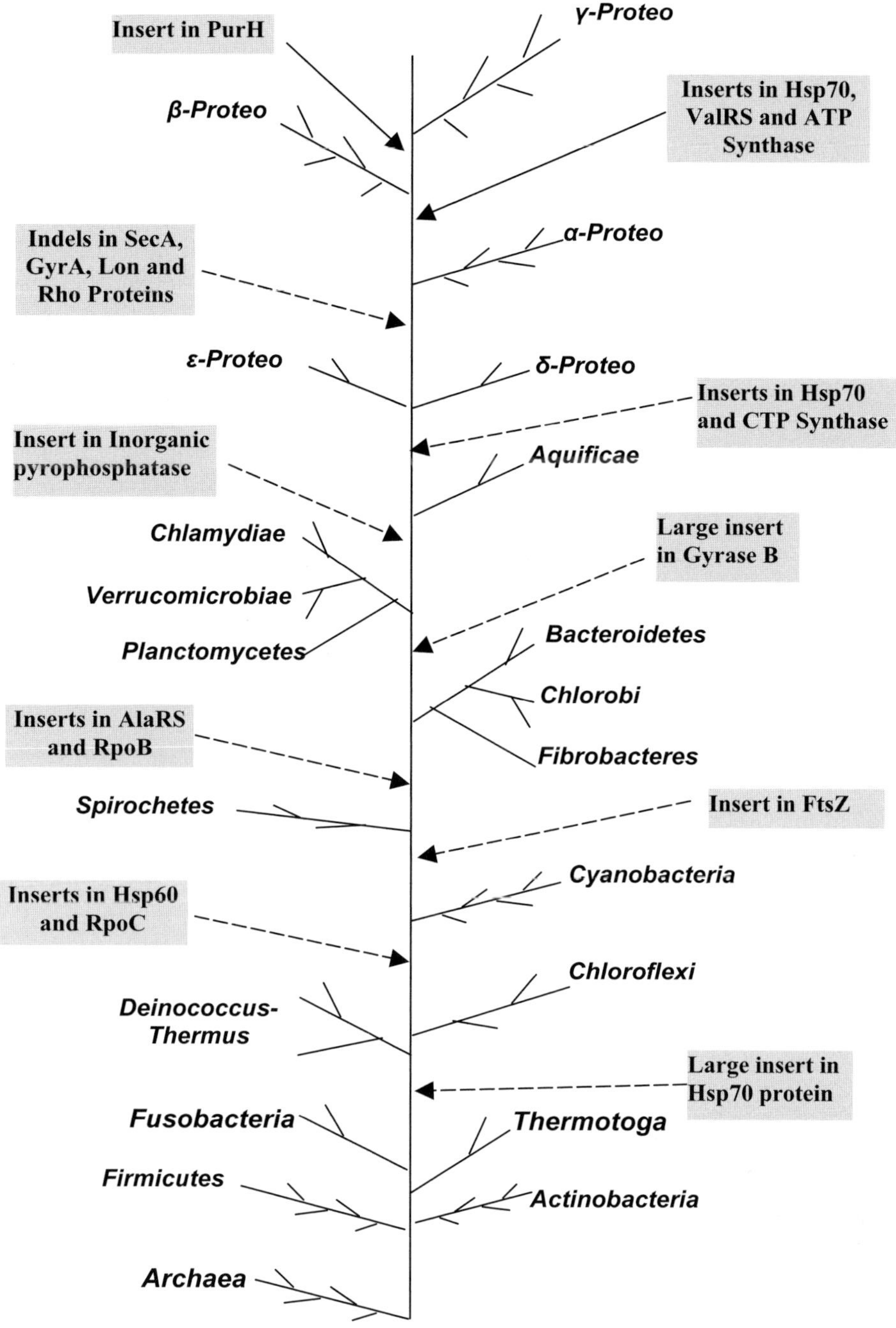


**Figure 7.5** A summary diagram showing the branching pattern of various bacterial phyla based upon the species distribution patterns of conserved indels in various proteins. The arrows mark the evolutionary branch points where the indicated CSIs in some of these widely distributed proteins were introduced. These CSIs are present in all of the species from various bacterial phyla (or groups) that lie above the indicated insertion points (i.e. those diverging later), but they are absent in virtually all of the species from various phyla noted below, which are inferred to have diverged earlier. The inserts in Hsp70, AlaRS, RpoB and RpoC are also not found in any archaeal homologues indicating that the groups at the base of this tree are ancestral. The sequence information for most of these signatures has been presented in our earlier work (Gupta, 1998b, 2003, 2005a; Griffiths and Gupta, 2004b, 2007b; Singh and Gupta, 2009). Numerous signatures (not shown) those are specific for different bacterial phyla or their subgroups have also been identified (see Figure 7.3 for cyanobacteria).

upon large numbers of conserved indels that are specific for them (not shown; see publications noted above). This branching order deduced from these indels is internally highly consistent and it was first deduced >10 years ago, when sequence information for only a handful of genomes was available. However, this model makes very specific predictions regarding the presence or absence of these indels in various genomes. In the past 10 years, genome sequences for >1000 prokaryotic organisms have become available and the presence or absence of these CSIs in various genomes shows excellent agreement (>95% accuracy) with that predicted by this model (Gupta, 2001, 2003, 2005a; Gupta and Griffiths, 2002; Griffiths and Gupta, 2004b) (and unpublished results). These results strongly indicate that this branching pattern is reliable.

It should be mentioned that the branching order of bacterial phyla deduced based on conserved indels differs significantly from that based on 16S rRNA (Olsen *et al.*, 1994; Griffiths and Gupta, 2004). In particular, these results do not support the very deep branching of *Aquificae* species as seen in the rRNA trees (Olsen *et al.*, 1994; Ludwig and Klenk, 2005). In contrast to the rRNA tree, in a phylogenetic tree based on concatenated sequences for large numbers of universally distributed proteins the *Aquificae* species branch in a similar position as indicated by the indel data (Ciccarelli *et al.*, 2006). The deep branching of the *Aquificae* in the 16S rRNA and some other phylogenetic trees could result from a variety of factors, including the long-branch length effect and LGTs (Gogarten *et al.*, 2002; Felsenstein, 2004; Beiko *et al.*, 2005; Delsuc *et al.*, 2005; Griffiths and Gupta, 2004b). The *Aquifex* genome appears to be rapidly evolving (Deckert *et al.*, 1998), and thus many *Aquifex* genes are subjects of long branch-length effects leading to their abnormal branching in phylogenetic trees. About 10% of *A. aeolicus* genes exhibit extensive sequence identity to homologues from various Archaea, implicating massive LGT between these groups (Deckert *et al.*, 1998). If this is the case, then in phylogenetic trees constructed from homologues of the transferred genes, *Aquificae* species will branch near the root of the tree, as their sequences would closely resemble those of the Archaea. It is also important in this context that in contrast to the *A. aeolicus* genome, which has a G+C content of 43.4%, the G+C content of 16S–23S–5S operon in this species is 65% (Deckert *et al.*, 1998), which suggests that either the rRNA genes in this species have selectively evolved at a very rapid rate, or that they have been acquired from a high G + C species by means of LGT. Meyer and Bansal (2005) have shown that the high G+C content of Archaea and hyperthermophilic bacteria is necessary to prevent thermal denaturation of their rRNAs at elevated temperatures. It should be pointed out that although the genomic G + C content of most *Actinobacteria* is quite high, the G + C contents of their rRNA are much lower than those observed for *Aquificae* and other hyperthermophiles (e.g. *Thermotogae*). Thus, the abnormally high G + C content of *Aquificae* rRNA is in large part responsible for their artefactual grouping with the deep branching Archaea in the rRNA trees.

The deduced branching pattern indicates that the bacterial phyla comprising of *Fibrobacteres*, *Chlorobi* and *Bacteroidetes* branch in the same position and they shared a common ancestor exclusive of all other bacteria (Griffiths and Gupta, 2001; Gupta, 2004). This inference is strongly supported by two conserved indels, a 5–7 amino acids insert in the RNA polymerase β′ subunit and a 14–16 aa insert in serine hydroxymethyltransferase, that are uniquely present in all available sequences from these three groups of bacteria (Gupta, 2004; Shah *et al.*, 2009). Three additional conserved indels in the proteins FtsK, UvrB and ATP synthase alpha-subunit provide evidence that of these bacterial groups, *Bacteroidetes* and *Chlorobi* are more closely related and they shared a common ancestor exclusive of *Fibrobacteres* (Gupta, 2004). In a similar manner, *Chlamydiae*, *Verrucomicrobiae* and *Planctomycetes* branch in the same position and their grouping together is supported by the shared presence of conserved indels (Griffiths and Gupta, 2007b). We have suggested that the bacterial groups that branch in the same position (i.e. form a distinct branch of the tree) and whose grouping together is supported by both phylogenetic analysis as well as by the shared presence of conserved indels should be recognized as part of a single phylum (or superphylum) (Gupta, 2000b, 2004, 2005a; Griffiths and Gupta, 2007b). Using the same criteria, Alpha (α)-, Beta (β)-, Gamma (γ)- and δ/ε-Proteobacteria, which are currently placed into to a single phylum *Proteobacteria*, should be recognized as distinct phyla of bacteria. Based upon conserved indels in several proteins and detailed phylogenetic analyses, the species from these groups, which can be clearly distinguished from all other bacteria, have branched off from a common ancestor at different times and thus they correspond to distinct branches of the microbial tree (Gupta, 2000b, 2005b; 2006; Kersters *et al.*, 2006; Gupta and Sneath, 2007; Gupta and Mok, 2007; Cutino-Jimenez *et al.*, 2009; Gao *et al.*, 2009a). Thus, by judicious application of genealogical criteria to the results from conserved indels and other forms of analyses, it should now be possible to define the higher taxonomic groups in a more objective manner reflecting the evolution of microbes.

### Identification of lateral gene transfers using conserved indels

Lateral gene transfer (LGT) is indicated to be an important force in shaping the evolution of prokaryotes (Gogarten *et al.*, 2002; Raymond *et al.*, 2002; Beiko *et al.*, 2005). The lateral gene transfer, particularly among closely related species, is indicated to be very common; however, numerous cases of LGT among distantly related phyla or even within different domains have been identified (Griffiths and Gupta, 2002; Gogarten *et al.*, 2002; Raymond *et al.*, 2002; Beiko *et al.*, 2005). The occurrence of LGT between two taxa is inferred by a variety of observations including abnormal branching in phylogenetic trees, results of BLAST searches, or atypical base composition of the gene (Ragan, 2001). The vast majority of conserved indels that have thus far been identified are specific for different monophyletic clades indicating that in these cases no LGTs of the genes under considerations have occurred. However, we have also come across a limited number of cases where a given conserved indel is commonly shared by two otherwise very distinct and distantly related groups of bacteria. Fig. 7.6 shows a three amino acids conserved insert in the enzyme serine hydroxymethyltransferase (GlyA protein), which is uniquely shared by various chlamydiae species as well as in a subset of *Actinobacteria* and *Treponema* species (Griffiths and Gupta, 2006a). The *Chlamydiae* and *Actinobacteria* are very distantly related phyla and large numbers of conserved indels and other molecular signatures that are specific for each of these groups have been identified (Griffiths *et al.*, 2005; Gao and Gupta, 2005; Gao *et al.*, 2006, 2009a; Gupta and Griffiths, 2006). In phylogenetic trees based on GlyA, the branching of *Chlamydiae* species with the insert containing subclade of *Actinobacteria* provides evidence that the shared presence of this insert in these two groups is due to LGT (Griffiths and Gupta, 2006a). The lateral transfers of *murA* gene between *Chlamydiae* and *Actinobacteria*, and of *glmU* gene between *Chlamydiae* and *Archaea* was also first detected due to shared presence of conserved indels (Griffiths and Gupta, 2002). In another instance, a large 51 aa insert in the SecA protein is commonly shared by *Aquificae* and *Thermotogae* (Griffiths and Gupta, 2006b). These observations indicate that conserved indels are also useful in the detection of LGT events.

## Lineage-specific proteins as markers for evolutionary and taxonomic studies

In addition to the conserved indels, comparative genomic analyses have also been instrumental in identifying yet another important class of molecular markers that are proving very useful for systematic and evolutionary studies. These markers consist of whole proteins that are uniquely found in particular groups or subgroups of bacteria (Gao *et al.*, 2006; Gupta and Lorenzini, 2007; Gupta and Mok, 2007; Dutilh *et al.*, 2008). The analyses of genome sequences have revealed that a large fraction of genes (open reading frames) in different organisms encodes for proteins that are of unknown functions (Danchin, 1999; Siew and Fischer, 2003; Daubin and Ochman, 2004; Doerks *et al.*, 2004; Galperin and Koonin, 2004). While most of these proteins are unique to a given species or closely related strains (Siew and Fischer, 2003; Daubin and Ochman, 2004; Lerat *et al.*, 2005; Dutilh *et al.*, 2008), a substantial fraction of these proteins are present in various species from a given phylum at different phylogenetic depths (Gao *et al.*, 2006; Gao and Gupta, 2007; Gupta and Lorenzini, 2007; Gupta and Mok, 2007; Dutilh *et al.*, 2008). These clade- or lineage-specific proteins, which we refer to as conserved

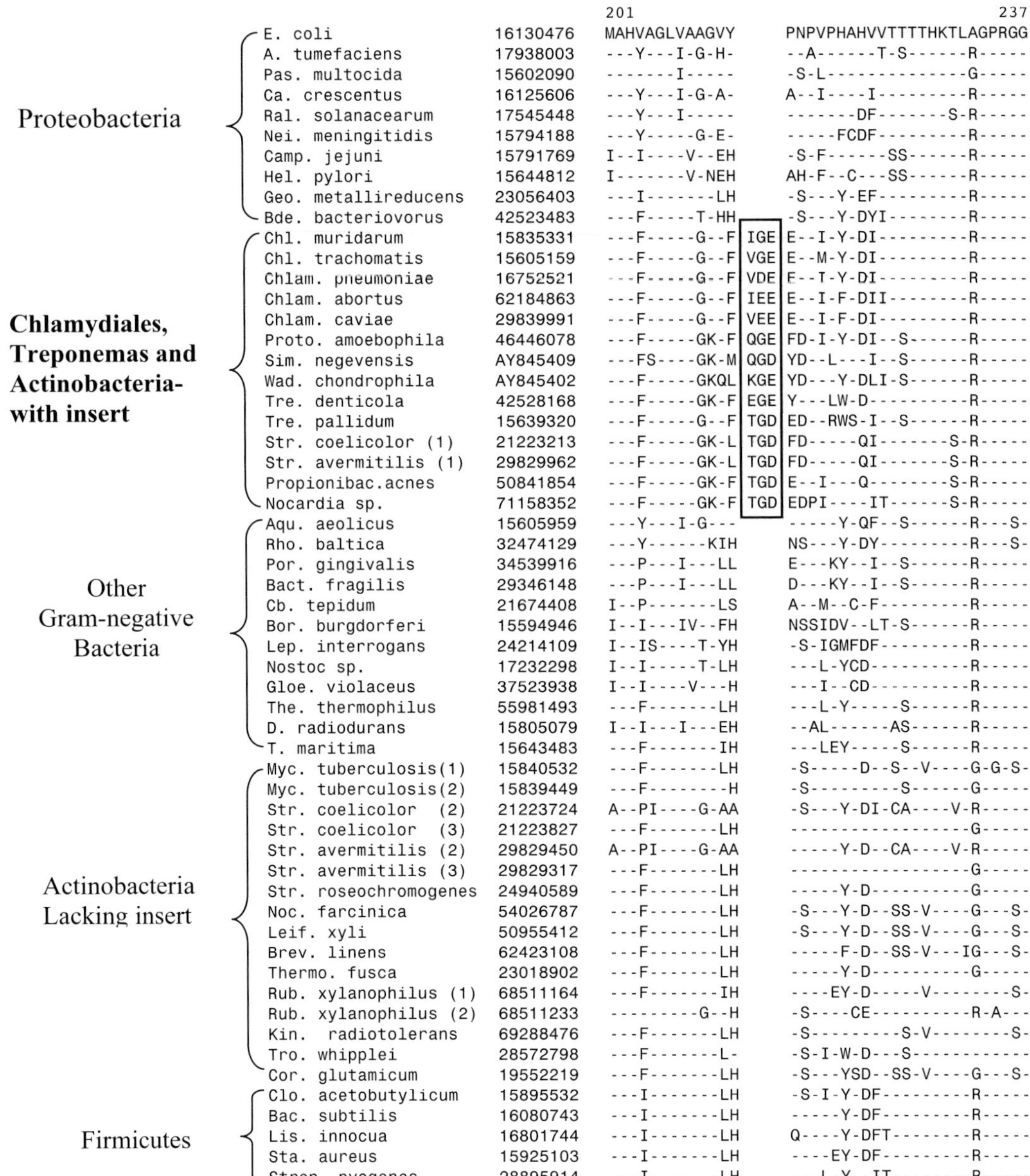


**Figure 7.6** Excerpts from the GlyA sequence alignment showing a three amino acid conserved insert which is uniquely found in various *Chlamydiales*, some actinobacteria and the *Treponema*. A number of actinobacterial species contain more than one homologue of this protein. The phylogenetic analysis of GlyA sequences show that the insert-containing homologues from different bacteria branch together as a clade within the *Actinobacteria*, indicating that the shared presence of this insert is due to lateral gene transfer (Griffiths and Gupta, 2006a). Adapted from Griffiths and Gupta (2006a).

signature proteins (CSPs), again provide useful molecular markers for taxonomic/evolutionary studies. The detailed analyses of these CSPs in conjunction with phylogenetic and other forms of analyses (viz. indel-based studies), have revealed that a large number of these proteins are distinctive characteristics of different monophyletic clades of bacteria (Griffiths *et al.*, 2006; Gao *et al.*, 2006, 2009a; Gupta, 2006; Mulkidjanian *et al.*, 2006; Gupta and Lorenzini, 2007; Gupta and Mok, 2007; Griffiths and Gupta, 2007a; Gupta and Mathews, 2009). Because of their unique presence in various species from these monophyletic clades, it is likely that the genes for these proteins first evolved in a common ancestor of these clades and then retained by all descendents. Similar to the work described above on identification of conserved indels, extensive work has been carried out on identifying CSPs that are specific for different monophyletic clades of bacteria at various phylogenetic depths. The evolutionary relationships suggested by species distribution patterns of these CSPs are in excellent agreement with the inferences derived from CSIs and phylogenetic analyses and together they are leading to a more holistic picture of microbial phylogeny (Gao *et al.*, 2006, 2009a; Gupta and Lorenzini, 2007; Gupta and Mok, 2007; Gupta and Mathews, 2009). These newly identified molecular markers are also providing valuable means for identification of different groups of bacteria in clear and more definitive terms based upon multiple molecular characteristics.

## Genetic and biochemical significance of CSIs and CSPs

In addition to their utility for evolutionary and systematic studies, the identified CSIs and CSPs also provide novel and valuable tools for genetic and biochemical studies. For the vast majority of taxonomic groups within bacteria (above genus level) very few or no biochemical or physiological characteristics are known that are unique to them. In this context, functional studies on these CSIs and CSPs, which are specific for different bacterial groups, hold the promise for discovering novel biochemical and physiological characteristics that are unique to them. Most of the discovered CSIs are present in proteins that carry out essential cellular functions (viz. RpoB, RpoC, Pol I, Hsp70, SecA, EF-Tu, Hsp60, Gyrase A, Gyrase B, etc; see Fig. 7.6) (Singh and Gupta, 2009). The primary functions of these proteins (viz. RNA synthesis, DNA synthesis, protein folding, etc.) are vital for cell survival and they are expected to remain the same in all organisms. Hence, the question arises: what is the functional significance of these lineage-specific CSIs? Our recent work on a number of conserved indels in the Hsp60 and Hsp70 proteins has shown that identified CSIs in these proteins are essential for the growth of these bacteria (Singh and Gupta, 2009). Since most of the conserved indels in protein sequences are present in surface loops (Akiva *et al.*, 2008; Singh and Gupta, 2009), it is likely they play important role in the interactions of these proteins with other cellular proteins that are essential for these bacteria (Singh and Gupta, 2009). In contrast to the CSIs which are found in widely distributed proteins, nearly all of the lineage-specific CSPs are of unknown functions. The discovery of these proteins points to our lack of knowledge regarding many fundamental aspects of cells, particularly functions that are specific for different bacterial groups. Hence, an important challenge for the future is to understand the cellular functions of these lineage-specific genes/proteins (Danchin, 1999; Galperin and Koonin, 2004; Roberts, 2004; Gupta and Griffiths, 2006; Gao *et al.*, 2009b), which should provide valuable insights into the biochemical and physiological characteristics that are unique to different bacterial groups.

## Acknowledgements

The research work from the author's lab has been supported by research grants from the Canadian Institute of Health Research and Natural Sciences and Engineering Research Council of Canada.

## References

Akiva, E, Itzhaki Z., and Margalit, H. (2008). Built-in loops allow versatility in domain-domain interactions: lessons from self-interacting domains. Proc. Natl. Acad. Sci. U.S.A. *105*, 13292–13297.

Barbrook, A.C., Lockhart, P.J., and Howe, C.J. (1998). Phylogenetic analysis of plastid origins based on secA sequences. Curr. Genet. *34*, 336–341.

Beiko, R.G., Harlow, T.J., and Ragan, M.A. (2005). Highways of gene sharing in prokaryotes. Proc. Natl. Acad. Sci. U.S.A. *102*, 14332–14337.

Boucher, Y., Douady, C.J., Papke, R.T., Walsh, D.A., Boudreau, M.E., Nesbø, C.L., Case, R.J., and Doolittle, W.F. (2003), Lateral gene transfer and the origins of prokaryotic groups. Annu. Rev. Genet. *37*, 283–328.

Castenholz, R.W. (2001). Phylum BX. Cyanobacteria: oxygenic photosynthetic bacteria. In *Bergey's Manual* of Systematic Bacteriology, 2nd ed., Vol. 1, D.R. Boone, R.W. Castenholz, and G.M. Garrity, eds. (New York: Springer), pp. 474–487.

Cavalier-Smith, T. (2002). The neomuran origin of Archaebacteria, the negibacterial root of the universal tree and bacterial megaclassification. Int. J. Syst. Evol. Microbiol. *52*, 7–76.

Ciccarelli, F.D., Doerks T., von Mering, C., Creevey, C.J., Snel, B., and Bork, P. (2006). Toward automatic reconstruction of a highly resolved tree of life. Science *311*, 1283–1287.

Cutino-Jimenez, A.M., Martins-Pinheiro, M., Lima, W.C., Martin-Tornet, A., Morales, O.G., and Menck, C.F. (2009). Evolutionary placement of Xanthomonadales based on conserved protein signature sequences. Mol. Phylog. Evol., in press.

Danchin, A. (1999). From protein sequence to function. Curr. Opin. Struct. Biol. *9*, 363–367.

Daubin, V., and Ochman, H. (2004). Bacterial genomes as new gene homes: the genealogy of ORFans in *E. coli*. Genome Res. *14*, 1036–1042.

Deckert, G., Warren, P.V., Gaasterland, T., Young, W.G., Lenox, A.L., Graham, D.E., Overbeek, R., Snead, M.A., Keller, M., Aujay, M., Huber, R., Feldman, R.A., Short, J.M., Olsen, G.J., and Swanson, R.V. (1998). The complete genome of the hyperthermophilic bacterium *Aquifex aeolicus*. Nature *392*, 353–358.

Delsuc, F., Brinkmann, H., and Philippe, H. (2005). Phylogenomics and the reconstruction of the tree of life. Nature Rev. Genet. *6*, 361–375.

Doerks, T., von Mering, C., and Bork, P. (2004). Functional clues for hypothetical proteins based on genomic context analysis in prokaryotes. Nucl. Acids Res. *32*, 6321–6326.

Doolittle, W.F. (1999). Phylogenetic classification and the universal tree. Science *284*, 2124–2128.

Dutilh, B.E., Snel, B., Ettema, T.J., and Huynen, M.A. (2008). Signature genes as a phylogenomic tool. Mol. Biol. Evol. *25*, 1659–1667.

Felsenstein, J. (2004). Inferring Phylogenies (Sunderland, MA: Sinauer Associates).

Galperin, M.Y., and Koonin, E.V. (2004). 'Conserved hypothetical' proteins: prioritization of targets for experimental study. Nucl. Acids Res. *32*, 5452–5463.

Gao, B., and Gupta, R.S. (2005). Conserved indels in protein sequences that are characteristic of the phylum *Actinobacteria*. Int. J. Syst. Evol. Microbiol. *55*, 2401–2412.

Gao B., and Gupta, R.S. (2007). Phylogenomic analysis of proteins that are distinctive of *Archaea* and its main subgroups and the origin of methanogenesis. BMC Genomics *8*, 86.

Gao, B., Parmanathan, R., and Gupta, R.S. (2006). Signature proteins that are distinctive characteristics of Actinobacteria and their subgroups. Antonie van Leeuwenhoek *90*, 69–91.

Gao, B., Mohan, R., and Gupta, R.S. (2009a). Phylogenomics and protein signatures elucidating the evolutionary relationships among the *Gammaproteobacteria*. Int. J. Syst. Evol. Microbiol. *59*, 234–247.

Gao, B., Sugiman-Marangos, S., Junop, M.S., and Gupta, R.S. (2009b). Structural and phylogenetic analysis of a conserved actinobacteria-specific protein (ASP1; SCO1997) from *Streptomyces coelicolor*. BMC Struct. Biol. *9*, 40.

Garrity, G.M., Bell, J.A., and Lilburn T.G. (2005). The revised road map to the manual. In Bergey's Manual of Systematic Bacteriology, 2nd ed., Vol. 2, Part A, D.J. Brenner, N.R. Krieg, and J.T. Staley, eds. (New York: Springer), pp. 159–220.

Gogarten, J.P., Doolittle, W.F., and Lawrence J.G. (2002). Prokaryotic evolution in light of gene transfer. Mol. Biol. Evol. *19*, 2226–2238.

Griffiths, E., and Gupta, R.S. (2001). The use of signature sequences in different proteins to determine the relative branching order of bacterial divisions: evidence that *Fibrobacter* diverged at a similar time to *Chlamydia* and the *Cytophaga–Flavobacterium–Bacteroides* division. Microbiology *147*, 2611–2622.

Griffiths, E., and Gupta, R.S. (2002). Protein signatures distinctive of chlamydial species: Horizontal transfer of cell wall biosynthesis genes *glmU* from Archaebacteria to Chlamydiae, and *murA* between Chlamydiae and *Streptomyces*. Microbiology *148*, 2541–2549.

Griffiths, E., and Gupta, R.S. (2004a). Distinctive protein signatures provide molecular markers and evidence for the monophyletic nature of the *Deinococcus-Thermus* phylum. J. Bacteriol. *186*, 3097–3107.

Griffiths, E., and Gupta R.S. (2004b). Signature sequences in diverse proteins provide evidence for the late divergence of the order *Aquificales*. Int. Microbiol. *7*, 41–52.

Griffiths, E., and Gupta, R.S. (2006a). Lateral transfers of serine hydroxymethyl transferase (*glyA*) and UDP-N-acetylglucosamine enolpyruvyl transferase (*murA*) genes from free-living Actinobacteria to the parasitic chlamydiae. J. Mol. Evol. *63*, 283–296.

Griffiths, E., and Gupta, R.S. (2006b). Molecular signatures in protein sequences that are characteristics of the phylum *Aquificae*. Int. J. Syst. Evol. Microbiol. *56*, 99–107.

Griffiths, E., and Gupta, R.S. (2007a). Identification of signature proteins that are distinctive of the Deinococcus-Thermus phylum. Int. Microbiol. *10*, 201–208.

Griffiths, E., and Gupta, R.S. (2007b). Phylogeny and shared conserved inserts in proteins provide evidence that *Verrucomicrobia* are the closest known free-living relatives of chlamydiae. Microbiology *153*, 2648–2654.

Griffiths, E., Petrich, A., and Gupta, R.S. (2005). Conserved indels in essential proteins that are distinctive characteristics of Chlamydiales and provide novel means for their identification. Microbiology *151*, 2647–2657.

Griffiths, E., Ventresca, M.S., and Gupta, R.S. (2006). BLAST screening of chlamydial genomes to identify signature proteins that are unique for the *Chlamydiales, Chlamydiaceae, Chlamydophila* and *Chlamydia* groups of species. BMC Genomics *7*, 14.

Gupta, R.S. (1998a). Life's third domain (*Archaea*): An established fact or an endangered paradigm? A new proposal for classification of organisms based on protein sequences and cell structure. Theor. Popul. Biol. *54*, 91–104.

Gupta, R.S. (1998b). Protein phylogenies and signature sequences: a reappraisal of evolutionary relationships among Archaebacteria, Eubacteria, and Eukaryotes. Microbiol. Mol. Biol. Rev. *62*, 1435–1491.

Gupta, R.S. (1998c). What are Archaebacteria: Life's third domain or monoderm prokaryotes related to Gram-positive bacteria? A new proposal for the classification of prokaryotic organisms. Mol. Microbiol. *29*, 695–708.

Gupta, R.S. (2000a). The *natural* evolutionary relationships among Prokaryotes. Crit. Rev. Microbiol. *26*, 111–131.

Gupta, R.S. (2000b). The phylogeny of proteobacteria: relationships to other eubacterial phyla and eukaryotes. FEMS Microbiol. Rev. *24*, 367–402.

Gupta, R.S. (2001). The branching order and phylogenetic placement of species from completed bacterial genomes, based on conserved indels found in various proteins. Int. Microbiol. *4*, 187–202.

Gupta, R.S. (2003). Evolutionary relationships among photosynthetic bacteria. Photosynth. Res. *76*, 173–183.

Gupta, R.S. (2004). The phylogeny and signature sequences characteristics of *Fibrobacteres, Chlorobi* and *Bacteroidetes*. Crit. Rev. Microbiol. *30*, 123–143.

Gupta, R.S. (2005a). Molecular sequences and the early history of life. In Microbial Phylogeny and Evolution: Concepts and Controversies, J. Sapp, ed. (New York: Oxford University Press), pp. 160–183.

Gupta, R.S. (2005b). Protein signatures distinctive of Alpha proteobacteria and its subgroups and a model for α-proteobacterial evolution. Crit. Rev. Microbiol. *31*, 101–135.

Gupta, R.S. (2006). Molecular signatures (unique proteins and conserved indels) that are specific for the epsilon proteobacteria (*Campylobacterales*). BMC Genomics 7, 167.

Gupta, R.S. (2009). Protein signatures (molecular synapomorphies) that are distinctive characteristics of the major cyanobacterial clades. Int. J. Syst. Evol. Microbiol. *59*, 2510–2526.

Gupta, R.S., and Gao, B. (2009). Phylogenomic analyses of clostridia and identification of novel protein signatures that are specific to the genus *Clostridium* sensu stricto (cluster I). Int. J. Syst. Evol. Microbiol. *59*, 285–294.

Gupta, R.S., and Golding, G.B. (1993). Evolution of HSP70 gene and its implications regarding relationships between archaebacteria, eubacteria, and eukaryotes. J. Mol. Evol. *37*, 573–582.

Gupta, R.S., and Golding, G.B. (1996). The origin of the eukaryotic cell. Trends Biochem. Sci. *21*, 166–171.

Gupta, R.S., and Griffiths, E. (2002). Critical issues in bacterial phylogenies. Theor. Popul. Biol. *61*, 423–434.

Gupta, R.S., and Griffiths, E. (2006). Chlamydiae-specific proteins and indels: novel tools for studies. Trends Microbiol. *14*, 527–535.

Gupta, R.S., and Lorenzini, E. (2007). Phylogeny and molecular signatures (conserved proteins and indels) that are specific for the Bacteroidetes and Chlorobi species. BMC Evol. Biol. *7*, 71.

Gupta, R.S., and Mathews, D.W. (2009). Signature proteins for the major clades of Cyanobacteria. BMC Evol. Biol., in press.

Gupta, R.S., and Mok, A. (2007). Phylogenomics and signature proteins for the alpha Proteobacteria and its main groups. BMC Microbiol. *7*, 106.

Gupta, R.S., and Sneath, P.H.A. (2007). Application of the character compatibility approach to generalized molecular sequence data: branching order of the proteobacterial subdivisions. J. Mol. Evol. *64*, 90–100.

Gupta, R.S., Mukhtar, T., and Singh, B. (1999). Evolutionary relationships among photosynthetic prokaryotes (*Heliobacterium chlorum, Chloroflexus aurantiacus*, cyanobacteria, *Chlorobium tepidum* and proteobacteria): implications regarding the origin of photosynthesis. Mol. Microbiol. *32*, 893–906.

Gupta, R.S., Pereira, M., Chandrasekera, C., and Johari, V. (2003). Molecular signatures in protein sequences that are characteristic of Cyanobacteria and plastid homologues. Int. J. Syst. Evol. Microbiol. *53*, 1833–1842.

Hoffmann, L., Komárek, J., and Kaštovsky, J. (2005). System of cyanoprokaryotes (cyanobacteria) – State in 2004. Algol. Stud. *117*, 95–115.

Kainth, P., and Gupta, R.S. (2005). Signature proteins that are distinctive of alpha proteobacteria. BMC Genomics *6*, 94.

Kersters, K., Devos, P., Gillis, M., Swings, J., Vandamme, P., and Stackebrandt, E. (2006) Introduction to the Proteobacteria. In The Prokaryotes: A Handbook on the Biology of Bacteria, M. Dworkin, S. Falkow, E. Rosenberg, K.H. Schleifer, and E. Stackebrandt, eds. (New York: Springer), pp. 3–37.

Knoll, A.H. (1999). A new molecular window on early life. Science *285*, 1025–1026.

Kondratieva, E.N., Pfennig, N., and Trüper, H.G. (1992). The phototrophic prokaryotes. In The Prokaryotes A Handbook on the Biology of Bacteria: Ecophysiology and Biochemistry, A. Balows, H.G. Trüper, M. Dworkin, W. Harder, and K.H. Schleifer, eds. (New York: Springer-Verlag), pp. 312–330.

Korbel, J.O., Doerks, T., Jensen, L.J., Perez-Iratxeta, C., Kaczanowski, S., Hooper S.D., Andrade, M.A., and Bork, P. (2005). Systematic association of genes to phenotypes by genome and literature mining. PLoS Biol. 3, e134.

Lake, J.A., Herbold, C.W., Rivera, M.C., Servin, J.A., and Skophammer, R.G. (2007). Rooting the tree of life using nonubiquitous genes. Mol. Biol. Evol. *24*, 130–136.

Lerat, E., Daubin, V., Ochman, H., and Moran, N.A. (2005). Evolutionary origins of genomic repertoires in bacteria. PLoS Biol. *3*, e130.

Ludwig, W., and Klenk, H.-P. (2005). Overview: A phylogenetic backbone and taxonomic framework for prokaryotic systematics. In *Bergey's Manual* of

Systematic Bacteriology, D.J. Brenner, N.R. Krieg, J.T. Staley, and G.M. Garrity, eds. (Berlin: Springer-Verlag), pp. 49–65.

Ludwig, W., and Schleifer, K.H. (1999). Phylogeny of *Bacteria* beyond the 16S rRNA standard. ASM News *65*, 752–757.

Maidak, B.L., Cole, J.R., Lilburn, T.G., Parker, C.T., Jr., Saxman, P.R., Farris, R.J., Garrity, G.M., Olsen, G.J., Schmidt, T.M., and Tiedje, J.M. (2001). The RDP-II (Ribosomal Database Project). Nucl. Acids Res. *29*, 173–174.

Martin, W., and Müller, M. (1998). The hydrogenosome hypothesis for the first eukaryote. Nature *392*, 37–41.

Meyer, T.E., and Bansal, A.K. (2005). Stabilization against hyperthermal denaturation through increased CG content can explain the discrepancy between whole genome and 16S rRNA analyses. Biochemistry *44*, 11458–11465.

Mulkidjanian, A.Y., Koonin, E.V., Makarova, K.S., Mekhedov, S.L., Sorokin, A., Wolf, Y.I., Dufresne, A., Partensky, F., Burd, H., Kaznadzey, D., Haselkorn, R., and Galperin, M.Y. (2006). The cyanobacterial genome core and the origin of photosynthesis. Proc. Natl. Acad. Sci. U.S.A. *103*, 13126–13131.

Murray, R.G.E., Brenner, D.J., Colwell, R.R., De Vos, P., Goodfellow, M., Grimont, P.A.D., Pfennig, N., Stackebrandt, E., and Zavarzin, G.A. (1990). Report of the ad hoc committee on approaches to taxonomy within the Proteobacteria. Int. J. Syst. Bacteriol. *40*, 213–215.

Nelson, K.E., Paulsen, I.T., and Fraser, C.M. (2001). Microbial genome sequencing: a window into evolution and physiology. ASM News *67*, 310–317.

Olsen, G.J., Woese, C.R., and Overbeek, R. (1994). The winds of (evolutionary) change: breathing new life into microbiology. J. Bacteriol. *176*, 1–6.

Oren, A., and Stackebrandt, E. (2002). Prokaryote taxonomy online: challenges ahead. Nature *419*, 15.

Oren, A., and Tindall, B.J. (2005). Nomenclature of the cyanophyta/cyanobacteria/cyanoprokaryotes under the International Code of Nomenclature of Prokaryotes. Algol. Stud. *117*, 39–52.

Palmer, J.D., and Delwiche C.F. (1998). The origin and evolution of plastids and their genomes. In Molecular Systematics of Plants. II. DNA Sequencing, D.E. Sotis, P.E. Soltis, and J.J. Doyle, eds. (Norwell, MA: Kluwer Academic Publishers), pp. 375–409.

Ragan, M.A. (2001). Detection of lateral gene transfer among microbial genomes. Curr. Opin. Genet. Dev. *11*, 620–626.

Rasmussen, B., Fletcher, I.R., Brocks, J.J., and Kilburn, M.R. (2008). Reassessing the first appearance of eukaryotes and cyanobacteria. Nature *455*, 1101–1104.

Raymond, J., Zhaxybayeva, O., Gogarten, J.P., Gerdes, S.Y., and Blankenship, R.E. (2002). Whole-genome analysis of photosynthetic prokaryotes. Science *298*, 1616–1620.

Rippka, R., Deruelles, J., Waterbury, J.B., Herdman, M., and Stanier, R.Y. (1979). Generic assignments, strain histories and properties of pure cultures of cyanobacteria. J. Gen. Microbiol. *111*, 1–61.

Rivera, M.C., and Lake, J.A. (1992). Evidence that eukaryotes and eocyte prokaryotes are immediate relatives. Science *257*, 74–76.

Roberts, R.J. (2004). Identifying protein function – a call for community action. PLoS Biol. *2*, E42.

Rokas, A., and Holland, P.W. (2000). Rare genomic changes as a tool for phylogenetics. Trends Ecol. Evol. *15*, 454–459.

Sanchez-Baracaldo, P., Hayes, P.K., and Blank, C.E. (2005). Morphological and habitat evolution in the Cyanobacteria using a compartmentalization approach. Geobiology *3*, 145–165.

Schopf, J.W. (1978). The evolution of the earliest cells. Sci. Am. *239*, 110–120.

Shah, H.N., Olsen, I., Bernard, K., Finegold, S.M., Gharbia, S.E., and Gupta, R.S. (2009). Approaches to the study of the systematics of anaerobic, Gram-negative, non-spore-forming rods: current status and perspectives. Anaerobe *15*, 179–194.

Siew, N., and Fischer, D. (2003). Analysis of singleton ORFans in fully sequenced microbial genomes. Proteins *53*, 241–251.

Singh, B., and Gupta, R.S. (2009). Conserved inserts in the Hsp60 (GroEL) and Hsp70 (DnaK) proteins are essential for cellular growth. Mol. Genet. Genomics *281*, 363–371.

Stackebrandt, E. (2006). Defining taxonomic ranks. In: The Prokaryotes, A Handbook on the Biology of Bacteria: Ecophysiology and Biochemistry, M. Dworkin, S. Falkow, E. Rosenberg, K.-H. Schleifer, and E. Stackebrandt, eds. (New York: Springer), pp. 29–57.

Swingley, W.D., Blankenship, R.E., and Raymond, J. (2008). Integrating Markov clustering and molecular phylogenetics to reconstruct the cyanobacterial species tree from conserved protein families. Mol. Biol. Evol. *25*, 643–654.

Valas, R.E., and Bourne, P.E. (2009). Structural analysis of polarizing indels: an emerging consensus on the root of the tree of life. Biol. Direct *4*, 30.

Wilmotte, A., and Golubic, S. (1991). Morphological and genetic criteria in the taxonomy of Cyanophyta/Cyanobacteria. Arch. Hydrobiol. *64*, 1–24.

Wilmotte, A., and Herdman, M. (2001). Phylogenetic relationships among the Cyanobacteria based on 16S rRNA sequences. In *Bergey's Manual* of Systematic Bacteriology, D.R. Boone, R.W. Castenholz, and G.M. Garrity, eds. (New York: Springer), pp. 487–493.

Woese, C.R. (1987). Bacterial evolution. Microbiol. Rev. *51*, 221–271.

Woese, C.R., Stackebrandt, E., Macke, R.J., and Fox, G.E. (1985). A phylogenetic definition of the major eubacterial taxa. Syst. Appl. Microbiol. *6*, 143–151.

Woese, C.R., Kandler, O., and Wheelis, M.L. (1990). Towards a natural system of organisms: proposal for the domains Archaea, Bacteria, and Eucarya. Proc. Natl. Acad. Sci. U.S.A. *87*, 4576–4579.

Zhaxybayeva, O., Gogarten, J.P., Charlebois, R.L., Doolittle, W.F., and Papke, R.T. (2006) Phylogenetic analyses of cyanobacterial genomes: quantification of horizontal gene transfer events. Genome Res. *16*, 1099–1108.

# Construction and Deconstruction: The Influence of Lateral Gene Transfer on the Evolution of the Tree of Life

Maureen A. O'Malley

Abstract

Efforts to construct the tree of life take their conceptual motivation from Charles Darwin's theory of evolution. Until the advent of molecular biology, however, a universal tree of life was well beyond the scope of the data and methods of traditional organismal phylogeny. The rapid development of these methods and bodies of genetic sequence from the 1970s onwards resulted in major reclassifications of life and revived ambitions to represent all organismal lineages by one true tree of life. Subsequent realization of the significance of lateral gene transfer and other non-vertical processes has subtly reconceptualized and reoriented attempts to construct this universal phylogeny. This chapter sets out these shifts of construction, deconstruction and reconstruction, with an eye towards understanding the future of the tree of life.

## Introduction

> *Old prejudices tend to inhibit, distort, or otherwise shape new ideas, and historical analysis helps to eliminate much of the negative impact of the status quo (Woese, 1987).*

The Tree of Life is a powerful symbol of the unity of evolutionary process and pattern. From branches of vertical descent emerge species bifurcations, which go on to further bifurcate or end in extinction. Proposed by Charles Darwin as both the phenomenon to be explained by evolutionary theory, as well as proof of evolution by natural selection (Doolittle and Bapteste, 2007; Doolittle, 2009a), the Tree of Life for today's evolutionary biologists is both a fact and a logical necessity (e.g. Cracraft and Donoghue, 2004; www.tolweb.org; Eldredge, 2005). Although its history has in fact much deeper roots than Darwin (Ragan *et al.*, 2009; Archibald, 2009; Pallen, 2009), there is little in-depth examination of what such a Tree has meant to the communities that have employed it. One scientific area in which the metaphor of the Tree of Life has re-emerged and been closely examined is the triumphant molecular microbial phylogeny of the last few decades. This chapter will outline this recent history, examine why a Tree was so central to the three-domain proposal of life, and why challenges to such a Tree structure are so vigorously contested right up to the present day. The analysis will conclude with an outline of the future prospects of gaining knowledge of evolutionary history through the Tree metaphor.

## Darwin as a basis for the Tree of Life

Anyone who has thought at all about the Tree of Life, and everyone who has examined Darwin's texts, has taken note of Darwin's sole diagram in *On the Origin of Species* (1859) and its accompanying interpretation.

> *The affinities of all the beings of the same class have sometimes been represented by a great tree. I believe this simile largely speaks the truth. The green and budding twigs may represent existing species; and those produced during each former*

*year may represent the long succession of extinct species.*

Darwin expanded on this metaphor in the 6th edition of the *Origin* (1872):

> *As limbs give rise by growth to fresh buds, and these, if vigorous, branch out and overtop on all sides many a feebler branch, so by generation I believe it has been with the great Tree of Life, which fills with its dead and broken branches the crust of the earth, and covers the surface with its ever branching and beautiful ramifications.*

For biologists, the Tree did not provide a central organizing metaphor immediately after Darwin, despite or perhaps because of Ernst Haeckel's attempts to draw detailed trees of life shortly after Darwin (Haeckel, 1866; Dayrat, 2003). Haeckel, known for his unorthodox appropriation of Darwinian thinking, may have tainted the legitimacy of endeavours to understand the history of all evolution in one general mapping process. Moreover, the terminology of 'tree of life' had strong religious connotations, both from Biblical references and a variety of other cultural sources (Hacking, 2007). Classification of the time was still largely taxonomical as opposed to evolutionary, and branching patterns, if they were suggested, were derived straightforwardly from existing taxonomical schema (Stevens, 1984; de Queiroz, 1988; Mayr, 1942). 'Our phylogenies are invented to account for our taxonomic facts or theories', complained botanist Harry Allan, as he discussed the 'new systematics' of the 1940s (Allan, 1940).

All of this would change with the advent of contemporary phylogenetic methods, especially as formalized by cladism in the 1960s and 70s. But these transformations applied primarily to animals, plants and occasionally fungi (Hennig, 1966; Mayr, 1982; Futuyma, 2004). Unicellular organisms known as bacteria, and even unicellular eukaryotes, were still difficult to classify, let alone to make them divulge their evolutionary histories. Nevertheless, influential microbiologists of the 1940s and 1950s saw the development of a 'natural' evolution-based classification system as an imperative for the scientific advancement of microbiology (Stanier and van Niel, 1941; Sapp, 2009). But even a sketch of the Tree of Life, necessarily rooted in the microbial world, was beyond the grasp of the methods and means of data collection until well into the second half of the twentieth century (Fernholm *et al.*, 1989). And for most botanists and zoologists, incorporating the evolution of microbes into a universal representation of speciating lineages was not a pressing or relevant task: trees of angiosperms or arthropods were demanding enough. More general talk of 'The Tree of Life' thus fell outside the disciplinary commitments of most evolutionary biologists and phylogeneticists.

## Constructing the Tree of Life

The ambition to represent, at least schematically, the evolutionary relationships of all organismal lineages found its succour in the 1970s. Despite the limited integration of microbiology and evolutionary biology, and the lower institutional status of evolutionary understandings of microbes than of pandas and orchids, the unifying urge spread *from* microbiology into the rest of biology. In some respects this should not be surprising, because it is undeniable that deep phylogeny will always be concerned with microbes. But most zoologists and botanists had little compulsion or ability to do such deep phylogeny, even if they made vague speculations about basal eukaryotes or prokaryotes or their properties. One such example comes from the work of ornithologist and co-architect of the modern synthesis of evolution, Ernst Mayr. In his arguments about evolutionary phylogeny and the nature of species, he frequently claimed that the original organisms on the earth must have been sexual reproducers, and that asexuality was therefore a derived and not a primitive condition (Mayr, 1963). He could only make this argument theoretically, however. What the Tree of Life shift required was a method for substantiating such speculations, and that method was found in molecular approaches to phylogeny.

Emile Zuckerkandl and Linus Pauling provided the rationale for such methods by arguing and demonstrating the efficacy of using molecular sequences as repositories of evolutionary records. Building on several earlier efforts to construct animal phylogenies from protein and antiserum data, Zuckerkandl and Pauling declared that amino acid changes could serve as a molecular clock, and that comparison of such changes across lineages would allow accurate and objective measurements of evolutionary distance (Pauling and Zuckerkandl, 1963; Zuckerkandl and Pauling, 1965; Zuckerkandl, 1987). They thus amplified the suggestions of earlier molecular biologists, such as Frederick Sanger, Francis Crick, and Emanuel Margoliash, who had anticipated the creation of sequence-based taxonomies and the evolutionary interpretations such taxonomies would enable (Harris *et al.*, 1956; Crick, 1958; Margoliash, 1963; Sibley, 1962). In the same paper that outlined his early views on the sequence hypothesis and central dogma, Crick (1958) foretold that

> *before long we shall have a subject that might be called 'protein taxonomy' – the study of the amino acid sequences of the proteins of an organism and the comparison of them between species. It can be argued that these sequences are the most delicate expression possible of the phenotype of an organism, and that vast amounts of evolutionary information may be hidden away within them'.*

But in even more fundamentally transformative ways, Margaret Dayhoff and colleagues opened up the possibility of constructing 'a biologically comprehensive phylogenetic tree' (Dayhoff and Schwartz, 1981; Dayhoff *et al.*, 1974). They provided a nucleic acid and protein sequence database, which they advertised with exhortations to combine multiple gene analyses through computational methods. The 'phylogenetic tree of all life' (Dayhoff *et al.*, 1974) must be a 'composite tree', argued Robert Schwartz and Dayhoff (1978), because single gene trees could not be expected to depict in full major evolutionary events and relationships between many different lineages. A composite tree could sketch out an evolutionary framework that could be constantly expanded with new sequence data, they argued, and this could be collectively compiled by the sequencing community.

Carl Woese, who was developing his own database of oligonucleotide sequences, took this advice to heart but came ultimately to rely on particular macromolecules, small subunit ribosomal RNAs and their genes, as the primary determinants of evolutionary history. Everybody reading this textbook will recognize the revelatory impact Woese had on microbiology, microbial phylogenetics, and organismal classification in general. His most obvious achievement was to challenge fundamentally a previously popular scheme depicting a five-kingdom division of lifeforms that was focused on 'modes of nutrition' for multicellular organisms, and unicellularity for everything left over (Whittaker, 1959, 1969; Woese, 1987). Woese used signature sequences followed by biochemical investigation to show that a previously undistinguished form of unicellular life was apparently different enough to constitute its own domain (a new Woesian hierarchical level above that of kingdoms): that of archaebacteria, later Archaea (Woese and Fox, 1977; Woese, 2005). While earlier it had been taken for granted that prokaryotes and eukaryotes represented the most fundamental division in cellular organization and evolutionary history (e.g. Stanier and van Niel, 1962; Mayr, 1982; Doolittle and Brown, 1994), Woese's three-fold division of the living world immediately required a more sophisticated historical narrative and representation than could be obtained by a 'simple to complex' story of evolution. This challenge was, of course, resisted (e.g. Mayr, 1990; Margulis and Guerrero, 1991: Margulis and Schwartz, 1998), but even these attempts to hold on to the five-kingdom view of the tree of life accommodated Archaea as an important group of evolutionarily and biologically distinct organisms.

With the incorporation of Archaea as one of the primary domains of life, a new universal model of evolutionary pattern became necessary for synthesizing and reinforcing these basic divisions. Woese attempted to establish a universal phylogeny that showed the fundamental domains of life as given by Nature and her history, rather than devised by human aims and interests. He argued that Bacteria, Archaea and Eukarya constituted the 'primary tripartite division of the living world' (Woese *et al.*, 1990: Woese, 1987). All life was included and the totality of evolutionary relationships could be more effectively unravelled once this basic division was understood, claimed Woese. Basic questions about the nature of the first organisms and major evolutionary transitions could be treated scientifically once a universal phylogeny had been constructed (Fox *et al.*, 1980; Olsen *et al.*, 1994). In other words, the acceptance of the three domains and domain-level classification opened up a broad outlook on how evolutionary history and life could be represented, and this was the molecular realization of 'Darwin's dream, a phylogenetic map covering all life' (Woese, 1996; Wheelis *et al.*, 1992).

It is at this point that two transformations come together fully: the use of molecules and molecular databases to understand all evolutionary history, and a global representation of that history. But despite the comprehensive view of evolutionary relationships that had now become possible and desirable, any search of PubMed for the term 'tree of life' turns up very few instances of its use until the early 1990s (Fig. 8.1). Woese may have been drawing on the Tree of Life as a guiding narrative in which to understand his findings, but he persistently called his own representation and goal 'the universal phylogenetic tree' or 'the universal tree' that 'encompasses all extant life' (e.g. Woese, 1987; Woese *et al.*, 1990, 2000). In one of his few uses of 'tree of life', Woese contrasted his universal phylogeny with Darwin's more timid 'genealogies' of life. 'Perhaps even Darwin dared not dream the synthesis of the great kingdoms into the universal tree of life', suggested Woese (1994a). Dayhoff and colleagues likewise emphasized the comprehensiveness of such a tree, but they too shied away from using the more colloquial term of 'tree of life'.

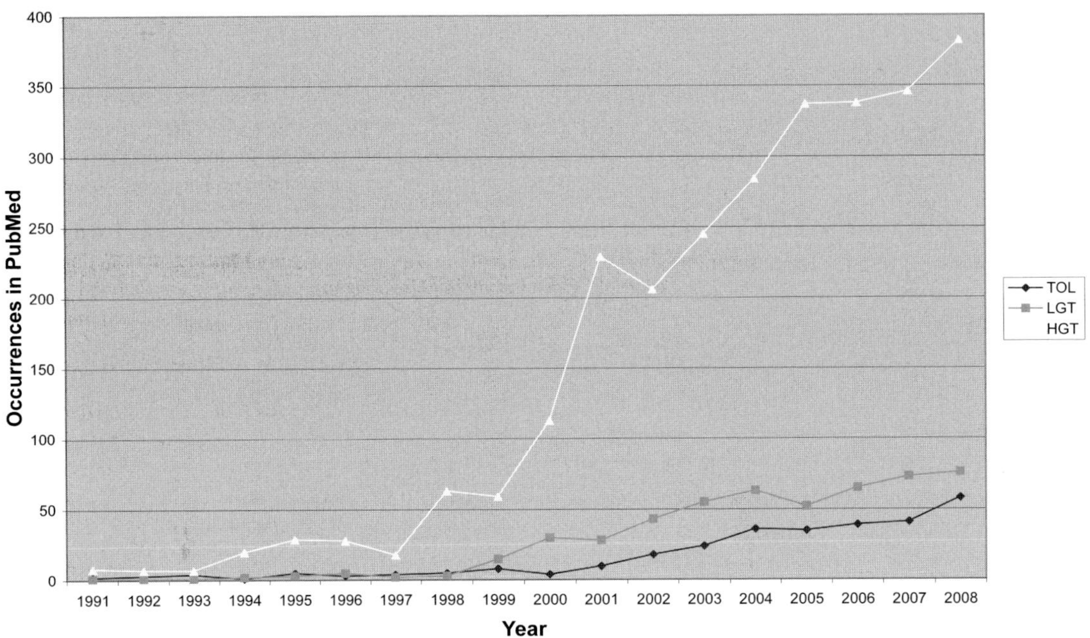


**Figure 8.1** Occurrences of ToL, LGT and HGT in PubMed titles and abstracts, 1991–2008. Uses of terms begin in 1982 (HGT), 1989 (ToL) and 1991 (LGT) and continue thereafter at very low rates, which are not represented on this graph until 1991. 2009 data has been excluded due to incompleteness, but indications from data gathered in mid-August, 2009, are for 2008 rates of occurrence to increase. Prior to the terminology of HGT and LGT, common expressions included 'genetic transfer' 'genetic exchange', 'chromosome transfer' and 'gene transfer' (the last the most prolific, due to being employed for animal genetics in laboratory situations). These broader terms begin in the 1950s and increase in frequency in the 1960s and 70s (and onwards), but are only represented here when prefaced by 'horizontal' or 'lateral' in relation to microorganismal genetics after 1991. Graph constructed by Cheryl Sutton, Egenis, University of Exeter.

One of the earliest references to a universal 'tree of life' can be found in the work of Allan Wilson, a preeminent pioneer of molecular approaches to the interpretation of evolutionary histories.[1] Although most well known for his once controversial interpretations of molecular evidence of human and other primate evolution (e.g. Wilson and Sarich, 1969), Wilson more broadly conceived of a universal phylogeny that could accurately relate all organisms to one another, no matter the extent of their evolutionary distance or the depth of those relationships in time (Sidow and Wilson, 1990; Wilson *et al.*, 1977). In 1990, he (along with Arend Sidow) began a tradition of using the term 'Tree of Life' to cover all three domains as he attempted to resolve deep branching order in the microbial origins of all life (Sidow and Wilson, 1990).

Although Wilson may have been an early advocate of the 'tree of life', it was Peter Gogarten and colleagues who popularized the term in the early 1990s, as they reflected on the implications of horizontal gene transfer (HGT) for a universal tree (Linkkila and Gogarten, 1991; Hilario and Gogarten, 1993; the latter paper marks the first intersection of ToL and HGT as terms used in the same paper). Their use of 'tree of life' occurred very shortly after they had rooted it in accordance with Dayhoff's suggestions for using primeval gene duplications to trace back the earliest divergences between lineages (Gogarten *et al.*, 1989; Iwabe *et al.*, 1989; Dayhoff and Schwartz, 1981). Gogarten comments that his choice of 'tree of life' terminology was not a particularly conscious or determined decision at the time. He simply thought that Woese's term, 'rooting the universal phylogeny of life', had much less rhetorical and metaphorical appeal than 'rooting the Tree of Life' (2009, personal communication). As well as sounding more attractive, Gogarten believes that his affinity to the tree image was likely to have been influenced by his training as a botanist, during which he worked on water transport in trees.

Whatever Gogarten's personal reasons, his imagery resonated with the existing community of microbial and other phylogeneticists. The metaphor caught on rapidly. The Darwinian affinities of constructing such trees was taken for granted, but Darwin was seldom if ever interrogated for his ideas about this metaphor until very recently (Doolittle and Bapteste, 2007). By the late-1990s, the reintroduction of 'tree of life' terminology to modern biology had been successfully achieved (see Fig. 8.1), as it was extensively propagated by researchers trying to piece together a global understanding of evolutionary relationships especially in regard to rooting this universal tree of life or detecting the origin of eukaryotes (e.g. Sogin, 1991; Sogin *et al.*, 1993; Margulis and Guerrero, 1991; Forterre *et al.*, 1993; Brown and Doolittle, 1995; Saccone *et al.*, 1995; Baldauf *et al.*, 1996; Gribaldo and Cammarano, 1998). Converts to the project of generating a universal understanding of life were by no means exclusively microbial geneticists or microbial phylogeneticists, although the most fundamental efforts were driven by those communities (e.g. Embley *et al.*, 1994; Forterre *et al.*, 1992; Benachehou-Lahfa *et al.*, 1993). Many of the later uses of 'tree of life', especially in the 2000s, are partial, applying to particular groups of organisms (e.g. 'avian tree of life', 'angiosperm tree of life').

But accompanying these very earliest uses of ToL were increasingly sophisticated discussions of HGT and lateral gene transfer (LGT). HGT and LGT are synonymous terms, although the former is now more commonly used (see Fig. 8.1). Although there are occasional remarks made that LGT is favoured by more radical evolutionary microbiologists (O'Malley and Boucher, 2005), some of the strongest challenges to a single Tree of Life come from researchers who use HGT as their preferred term (e.g. Gogarten, 1995). Even the very first papers putting such terms together (e.g. Linkkila and Gogarten, 1991; Hilario and Gogarten, 1993) pointed out the implications of LGT and how it could undermine the very idea

1 Paleobiologist Michael Benton, making a commentary on the meeting 'Major Evolutionary Radiations', used the metaphor of 'pruning' the tree of life to express how diversity has shaped and pruned by adaptation, extinction and opportunism (Benton, 1989). The meeting was concerned primarily with animals and other organisms that had left paleobiological traces.

of a single universal tree. Once thought of as a process that occurred due to laboratory manipulations, microbial phylogeneticists and other biologists quickly became convinced that LGT existed and was important in natural environments, such as the human body (Anderson, 1968; Jones and Sneath, 1970; Reanney, 1977; Coughter and Stewart, 1989). Cases of cross-lineage genetic exchange rapidly gained a great deal of attention, suggested one commentator, because 'they challenge common perceptions about inheritance and the sanctity of species ... [they have the] tantalizing ... aura of heresy' (Sprague, 1991). One of the very first proposals that a tree-of-life representation of evolutionary history was not only challenged but deeply problematized by HGT was the Gogarten paper suggesting a 'web of life' structure to evolutionary history rather than a tree (Hilario and Gogarten, 1993). But for most microorganismal tree builders, a discovery of gene transfer was an anomaly that might need to be reported but would not fundamentally endanger the basic tree structure (e.g. Wilson *et al.*, 1977; Schwartz and Dayhoff, 1978; Dayhoff and Schwartz, 1981; Wheelis *et al.*, 1992; Woese, 1987, 2000; Woese *et al.*, 1980).

By the late 1990s a steady stream of papers on HGT and LGT had become a flood, with the frequency of the term HGT quickly overwhelming that of LGT (Fig. 8.1). But the ToL concept grew along with them, even if not at the same pace. Two publications that contributed to cementing the term 'Tree of Life' into general use were W. Ford Doolittle's influential papers that drew attention to the implications of LGT and predicted the demise of the Tree. One of them was in a popular publication, *American Scientist* (Doolittle, 2000); the other a highly cited paper in *Science* (Doolittle, 1999; also notable for the first intersection of ToL and LGT in the same paper). This key publication summed up diagrammatically the issues facing microbial phylogeny and suggested that the new Woesian paradigm was besieged. Its diagnosis as well as aim was the 'uprooting' of the Tree of Life. That message was reinforced by a further powerful visualization in another article, by Bill Martin, who also aimed to convince a broader readership of the deeply perturbing consequences of LGT (Martin, 1999). These papers and a slew of associated publications not only deconstructed in important ways the project of microbial phylogeny but also began to reconstruct it in ways that are still now setting the agenda, whichever 'side' participants are on.

## The deconstruction of the Tree of Life

Throughout the 1980s and 1990s, and into the 2000s, the understanding of what molecules could do for phylogeny made a huge epistemological shift: from being tools that could test existing theories about organismal relationships to becoming the very source of novel hypotheses about the evolutionary history of organismal groups (Doolittle, 1996). Numerous papers, both original research and reviews, set out scenarios of rampant and promiscuous LGT (e.g. Syvanen, 1987; Doolittle and Brown, 1994; Lan and Reeves, 1996; Martin, 1999; Gogarten *et al.*, 2002). The more sequence data that was accumulated, the less resolution to messy branching patterns they seemed to offer – quite contrary to the earlier expectations of Woese and colleagues (e.g. Woese *et al.*, 1980). There were increasing reports of conflict between phylogenies from different genes or sets of genetic data, and between genetic data and accepted organism-based classifications. These incongruent findings indicated that different evolutionary processes, including LGT, had been at work at different levels of biology, and that rRNA trees were disappointingly inadequate for the purpose of constructing a representative tree of life.

An additional complication to LGT between species (and genera, families, phyla and even domains) came from deepening recognition of the intra-species recombination of genetic material in prokaryotes (mostly bacteria). In a number of taxa, a single prokaryote species label was found to cover a multitude of genomically differentiated strains due to the homologous recombination of acquired DNA fragments. Although these acquisitions within species are more similar than those gained from evolutionarily distant groups, the acquired and recombined DNA nevertheless further reticulates the evolutionary history of the organisms involved and refuses the straightforward mapping of the histories of genes, genomes and organisms onto one another (Spratt *et al.*, 2001; Feil *et al.*, 2001; Feil and Spratt, 2001;

Lawrence, 2002; Lawrence and Retchless, 2009; Maynard Smith *et al.*, 2000). As troublesome as this reticulation (inter- and intra-species) seemed with limited genetic datasets, it became even worse with the advent of genomics in the 1990s.

Genomics, which has transformed microbiology in numerous ways (Koonin, 2009; Ward and Fraser, 2005; Gogarten *et al.*, 2009), has made extensive contributions to comparative evolutionary analyses of microbes – possibly the technology's greatest success story so far. Analyses of single microbial genomes have revealed many to have mosaic or patchwork genomes. Acquisitions of DNA from other lineages (often other domains) constitute at least 17% of the *Escherichia coli* K12 genome, 24% of the genome of the thermophilic bacterium, *Thermotoga maritima*, and 34% of the genome of the mesophilic archeon, *Methanosarcina mazei* (Lawrence and Ochman, 1998; Nelson *et al.*, 1999 Deppenmeier *et al.*, 2002). Recently, a detailed analysis of four Thermotogales genomes in addition to *T. maritima* has found that only a tiny proportion of the genes of these organisms has *not* been transferred at some point in their evolutionary histories (Zhaxybayeva *et al.*, 2009a). Not only do such transfers cross large phylogenetic distances, but many make functionally crucial contributions to the lifestyle of the recipient. Furthermore, strains in some prokaryote taxa have been found to vary hugely in gene content and to have wide-ranging phenotypic differences, with more variation expected in every genome sequenced from the same taxon (Welch *et al.*, 2002; Medini *et al.*, 2005; Tettelin *et al.*, 2008; Lapierre and Gogarten, 2008).

Many prokaryote to eukaryote transfers have also been detected (Andersson *et al.*, 2006; Alsmark *et al.*, 2009; Loftus *et al.*, 2005; Hotopp *et al.*, 2007) and a few eukaryote to prokaryote donations (Keeling and Palmer, 2008). There also appear to be some eukaryote to eukaryote exchanges that have crossed small and large evolutionary distances (Busslinger *et al.*, 1982; Lang, 1984; Syvanen, 1984; Andersson, 2005, 2009; Richards *et al.*, 2009; not all of the earlier cases have withstood subsequent scrutiny). Ambitious LGT searches have even ventured into the human genome analyses and found – quite erroneously – bacterial transfers into the vertebrate lineage (Stanhope *et al.*, 2001; Salzberg *et al.*, 2001; Andersson *et al.*, 2001). But despite some excesses and questionable interpretations (Martin, 2005), LGT and HGT findings and credibility have flourished (Fig. 8.1), accompanied by growing doubts about the very project of constructing a single universal tree of life.

Central to the Tree doubters' assault is the claim that there is no universal tree, whether this is called a universal phylogeny or more dramatically The Tree of Life. Trees exist for doubters, to be sure, but as partial representations of a much more complex evolutionary process, especially in the prokaryotic world. But these LGT-based challenges to the very possibility of a tree of life came up against an entrenched opposition. A certain amount of the interest in LGT, as depicted in Fig. 8.1, was aimed at showing such findings to have been put to the service of tenuous and overinterpreted theoretical conclusions. These more conservative analyses of genetic and phylogenetic discordance consisted of a variety of attempts to save the concept of a universal tree and to develop methods that would deal effectively with incongruences. For at least one group of microbial phylogeneticists, species and the universal tree are crucial for the development of a truly Darwinian microbiology, and too many theoretical concessions to the existence of LGT are merely destructive (e.g.: Kurland *et al.*, 2003). From the point of view of these commentators, LGT may occasionally blur vertical patterns (the tree-like structure), but its effects should not be exaggerated to the extent that the tree disappears (Snel *et al.*, 2005). And the fact that LGT is usually detected against the background of a reference tree, particularly the 16S rRNA tree (Gogarten, 1995), seemed contradictory to tree supporters, who saw no way forward without the framework a universal tree provided.

In whatever way LGT is acknowledged, the pursuit of a global phylogeny – or at least a phylogeny that imposes evolutionary order on early Archaea, Bacteria and Eukarya – is a major undertaking. Problematic in any endeavour to construct this tree is finding sufficient signal of the process of bifurcation so that it can be recovered in a biologically meaningful representation (Beiko and Ragan, 2009). All methods that attempt to tease out this pattern from LGT, gene duplication and differential loss, poor signal and phylogenetic artefact have to rationalize numerous exclusions and layers of interpretations to arrive at a tree, which may in the end be neither the history of any single gene nor the history of the organism itself (Swithers *et al.*, 2009; Haggerty *et al.*, 2009; Lawrence and Retchless, 2009). Nevertheless, it is undeniable that traces of vertical signal are often found through a variety of analyses, and the question then becomes one of whether such signal should be taken as the central truth of the evolutionary past or just one amongst several measures and representations of genetic relatedness. 'Highways' of gene exchange may in fact have more to say than vertical descent about major events in the evolutionary history of many groups of organisms (Beiko *et al.*, 2005; Huang and Gogarten, 2006). LGT may be in many cases the creative force that structures organismal relationships and the patterns detected by phylogenetic analyses (Zhaxybayeva *et al.*, 2009b). Presuming coherence and congruence to be produced only by vertical descent will result in an inadequate understanding of evolutionary history and organismal relationships.

Different approaches to preserving the tree, with different degrees of recognition of the evolutionary importance of LGT, have developed over the last decade (Ragan and Beiko, 2009; Brown, 2003; House, 2009). For conceptual convenience we can label some of the more recent efforts 'core genome' approaches (concerned with a biologically existent stable core of genes that can be taken to represent the organismal lineage) and 'central trend' approaches (focused on finding methods for weaving together vertical signals that may not agree in all of their histories). The minimal phylogenetic core approach seeks to identify genes that have a wide representation and also, that produce congruent phylogenetic signal (e.g. Lerat *et al.*, 2003; Daubin *et al.*, 2003). Selection of such genes is crucial. There is little doubt that groups of genes can represent particular and recent evolutionary histories – usually at the genus or family level, sometimes class – but much more scepticism that sufficient genes can be found to constitute the entire tree of life. One well-known core analysis examined 191 species genomes from all three domains of life and was able to identify 31 universal genes (Ciccarelli *et al.*, 2006). But because the total number of genes constituting each prokaryote numbers around 3000 (conservatively), the tree constructed by 31 genes is a very limited tree – 'a tree of 1%' at the most, and less if much larger eukaryote genomes are considered (Dagan and Martin, 2006).

Although a flippant response to this problem might be that phylogeneticists are very lucky to have even a small core (Gribaldo and Brochier, 2009), in general the objection that all existing genes in prokaryote genomes have undergone at least one LGT in the 3.5 billion year history of cellular genomes means that no pure untransferred core exists, and no tree uncontaminated by LGT is constructable (McInerney *et al.*, 2008). A very practical reinterpretation of the core approach is to relax the ubiquity requirement for the categorization of genes as 'core', and aim instead for very common and broadly distributed genes (Charlebois and Doolittle, 2004). This strategy has some conceptual overlap with what is sometimes called the central trend approach, in which the aim is to piece together whatever signal there is in a large body of data and see how much of it constitutes a universal tree.

A conceptually intriguing example of a central trend approach by Eugene Koonin and colleagues formulates its method as an effort to highlight vertical 'tree' patterns against the 'forest' of life (Koonin *et al.*, 2009; Puigbò *et al.*, 2009). This study thoroughly acknowledges the pervasiveness of LGT but nevertheless finds a central trend of vertical descent from the consensus trees of highly conserved genes. Each tree is assessed against all other trees (the forest) for inconsistency, and the most consistent trees put in a category of 'nearly universal trees' or NUTs. The central trend, composed of NUTs, is very faint at deep phylogenetic levels, except for the signal of bifurcation between archaea and bacteria. The overall conclusions of this analysis are that recovery of a universal tree-like structure is possible from some parts of genomes and for some part of life's history (Puigbò *et al.*, 2009; Koonin *et al.*, 2009). The authors urge their study to be seen not as a failure to recover a complete Tree, but as a success. Not only has it given further reason to pursue such methodologies, but the nature of limitations of the Tree itself are now understood more clearly. Is such a tree a total representation of evolutionary history? No. Does this analysis imply that LGT events should be removed from the picture? Again, no: this is an example of an integrated approach that opens up a constructive route of ongoing inquiry and shows its current limits.

> *Whether or not this central trend is denoted a tree of life could be a matter of convention and convenience, but the nature of this trend as well as the other trends that can be discerned in the forest merit investigation* (Puigbò *et al.*, 2009).

In other words, the authors leave open the ontological status of the tree (the nature of its existence), in favour of advancing its usefulness as an epistemological tool – something that generates knowledge.

## What are the current state and future prospects of the Tree of Life?

If we think further about conceptions of the Tree of Life as a core or central trend, it is clear that these notions are already only conceptual cousins to early Tree of Life rationales, in which vertical descent by modification was supposedly revealed straightforwardly by most of the data. This basic understanding of evolution has been profoundly enriched by insights into dynamic genome-based evolutionary processes, in which a range of entities play major adaptive roles (Norman *et al.*, 2009; Jain *et al.*, 2003; Gogarten and Townsend, 2005; Ragan and Beiko, 2009; Brüssow, 2009; Bapteste and Boucher, 2009). But such conceptual transitions imply, according to some analyses, that the microbial phylogeny community should see the Tree of Life as a ladder, and that it should now be kicked away because it has taken the community to the top of its usefulness and can go no further (Doolittle and Bapteste, 2007). Simply clinging to it for security is not a good scientific option, runs this argument, and a population genetics approach to understanding the evolutionary history of groups of organisms is the effective substitute (Doolittle, 2009a).

For many tree analysts, however, challenges to the Tree of Life from LGT have led not to its abandonment but to a different ontologically based shift in conceptualizing the tree. Rather than seeing it as a tree of *species*, based on a tree of genes or genomes, the Tree has been reinvented as a tree of *organisms*, or equivalently, a tree of cells (e.g. Woese, 2002). One reason for taking the latter option is that the history of life is in fact a history of bifurcating cell divisions and genome replication: The 'Tree of Cells' (Puigbò *et al.*, 2009). For some modellers, a tree structure simply represents the history of cells, and it is the organismal backbone on which the web-like structure of genome evolution can be draped (Gribaldo and Brochier, 2009). 'The species tree could still [therefore] be a useful concept even if incongruent with *every* gene tree' (Galtier and Daubin, 2008 – emphasis added). What phylogeny is doing from this perspective, it is claimed, is taking the true history of organisms and not being deceived by the wayward history of some (or all) parts of the constitutive genomes (Gribaldo and Brochier, 2009). The rationale seems to be that the tree of organisms is (somehow) available as a reference tree, even though any tree of cell divisions has to be interpreted as a tree of species in order to make any evolutionary sense (a tree of all individual organisms would not be useful), and for most unicellular organisms the evidence is still going to be gathered almost exclusively from genomes. None of the problems of the traditional bifurcating tree of species is avoided in this conceptual reconstruction, but it constitutes an interesting appeal to the existence of evolutionarily necessary processes.

A major philosophical question that arises here is the distinction between the epistemology (or methodology) of trees and their ontology or the nature of their existence in the world. For many commentators the two are conflated: the postulated existence of the Tree means that it must be knowable. Richard Dawkins (2003) sums up this sort of conflation very aptly:

> *For there is, after all, one true tree of life, the unique pattern of evolutionary branchings that actually happened. It exists. It is in principle knowable. We don't know it all yet. By 2050 we should – or if we do not, we shall have been defeated only at the terminal twigs, by the sheer number of species.*

For anyone who has thought at all about prokaryote evolution is it clear that Dawkins may be right about a tree of animals (Dagan and Martin, 2009), but he is unlikely to be right about a tree of all life (Eldredge, 2005). And even though many microbiologists might be willing to acknowledge that some evolutionary processes form a fundamentally tree-like pattern, they are just as likely to accompany this acknowledgement with questions about its *knowability* in relation to prokaryote evolution.

But Dawkins does express quite clearly the pervading background assumptions to the idea of a universal tree of life. These are echoed by more abstract statements by philosophers of evolution that Dobzhansky's famous dictum should be interpreted to read, 'Nothing in biology makes any sense except in the context of its place in phylogeny, its context in the tree of life ... reconstructing that tree is critical to understanding the living world' (Sterelny and Griffiths, 1999). What they are articulating is a deeply held intuitive conviction that of course all evolution involves groups arising out of groups, and that every organism should belong to one of those groups. But the claim does not address how much of this process can be known, and whether in fact it is the primary thing to be known about evolutionary processes. As Ragan and fellow authors put it,

> *LGT is a central modality of genome evolution, and treating it purely as a distraction from vertical (parent-to-offspring) transmission hinders us from appreciating the plurality of mechanism and pattern beyond a unitary tree of life* (Ragan et al., 2009).

Not only does the conventional schema of a tree need to be accurately supplemented by LGT processes – another methodological complication (Zhaxybaeva, 2009; Poptsova, 2009) – but it also needs to accommodate other patterns from major evolutionary events, such as endosymbiosis, hybridization, co-evolving symbioses and other such instances of lineage fusion and innovation (Gogarten and Townsend, 2005; Dagan and Martin, 2006; Martin and Müller, 2007; Lake, 2009; Fournier *et al.*, 2009; Foster *et al.*, 2009; McInerney *et al.*, 2008). All these processes are of major evolutionary importance, and their exclusion by strict tree conceptions would seem to be far more problematic than recognizing the 'universal' tree as a representation of some but not all evolutionary history. An additional issue is whether to separate representations and theories of eukaryote and prokaryote evolution, due to the different tempos, modes and outcomes involved (Dagan and Martin, 2009; Bapteste *et al.*, 2009).

For many purposes, a tree of life is still a valuable ambition. It serves as a general metaphor of evolutionary relatedness, even if those relationships cannot be captured by a strictly bifurcating pattern. It accepts fundamental distinctions in cell type and physiology that are central to the evolution of life on earth. It provides a way in which to order biological knowledge for both scientific and broader social purposes (e.g. www.tolweb.org). Even some pro-Tree commentators are able to agree that any construction of a universal tree is in fact a human-made conceptual tool that is useful for *some* relationships and for heuristically imposing order on the world.

> *In my view, a tree is just a human-made conceptual tool that we might decide to adopt if it means something to us, like any other graphical representation, irrespective of its 'existence' in the real world* (Galtier, 2009).

In reaching this conventionalist viewpoint, a conceptual rapprochement becomes possible between those who would persist with the idea of a universal tree, and those who have long been arguing against the unilateral deployment of trees to represent prokaryote evolution:

> *I have no objection to the continued use of an rRNA tree (or of any other agreed upon averaging or gene core-based TOCD&S [tree of cell division and speciation]) as a conventional framework for classification, provided everyone knows that that is all that it might be, a conventional taxonomic framework, not the TOL with all its baggage. Other ways of classifying microbes (for instance by gene content or ecological role or indeed by relative position in a multidimensional network) might well have more predictive value, but still this relatively stable hierarchical scheme would serve a very useful organizing function. In fact, I think this*

*is the posture that many microbiologists have already accepted* (Doolittle, 2009b).

This pragmatic stance, adopted by microbiologists in a variety of forms, may explain why the Tree of Life is thriving as never before despite all the challenges to its realness and epistemological legitimacy. Hundreds of biologists participate in an online attempt to understand the totality of genealogical relationships (www.tolweb.org), and as Fig. 8.1 shows, the trend towards increasing citation of the term 'tree of life' continues into the end of the first decade of this millennium. Part of this popularity is inspired by the 'Darwin year' of 2009, of course (the 200th anniversary of his birth, and the 150th of the publication of the *Origin*), but much is probably due to the unifying capacity of this metaphor. We live in times of both increasingly fine-grained and high-volume data on biodiversity, which at a genetic level indicate far more complex evolutionary relationships than there are clear visualizations for. And, as science becomes more specialized and technical, the existence of a unifying metaphor for at least some biological knowledge is of deep appeal – both inside and outside science. As Woese noted, the tree of life might also unite disciplines, notably microbiology 'provided at last with a phylogenetic articulating framework, microbiology can now grow to become a complete biological discipline' (Woese, 1994b).

## Conclusions

As theories and disciplines mature, the idea that a single framework is vital to success becomes increasingly questioned. From some points of view in microbial phylogeny, contemporary understandings of LGT mean that it is time, finally, to give up on any 'unifying metanarrative' such as the tree (Doolittle and Bapteste, 2007), or that it may be appropriate to find another one, such as a Web of Life (Doolittle, 2009a; Dagan and Martin, 2009) or even a Ring of Life (Lake *et al.*, 2009). The more general question that might be raised at the end of such an overview is whether the Tree itself still relevant. Does it continue to provoke challenges, defences and an ongoing parade of valuable findings, both for and against its existence? From a practical perspective, the Tree of Life can be conceived as a central illustration of the process of scientific inquiry. As a heuristic, it has advanced understanding, despite its many (now obvious) inadequacies. And as a metaphor and theory, it functions not only to guide further investigation but to orient and integrate a multiplicity of communities and technologies. As in all enduring and powerful scientific avenues of inquiry, the pertinent question is not the abstract one of 'Is there a tree?', but the practical ones of 'What is achieved by thinking about trees? What is not achieved by thinking about trees?' As an emblem of how science works, the universal tree is not and has never been purely about wrongness and rightness, but about its practical knowledge-producing value. As long as the notion of a Tree stimulates inquiry it is valuable; the moment it is taken for granted it loses its scientific power and becomes just an everyday assumption or an unchallengeable metaphysical assertion. As a deeply contested concept, informing a battery of different approaches and a variety of biological goals, the Tree's demise is still far off in the future. But it too has a history that has branched into different conceptualizations, methods, and representations, so the future is inevitably one of trees and not a Tree.

## References

Allan, H.H. (1940). Natural hybridization in relation to taxonomy. In The New Systematics, J. Huxley, ed. (Oxford: Clarendon), pp. 515–528.

Alsmark, U.C., Sicheritz-Ponten, T., Foster, P.G., Hirt, R.P., and Embley, T.M. (2009). Horizontal gene transfer in eukaryotic parasites: A case study of *Entamoeba histolytica* and *Trichomonas vaginalis*. In Horizontal Gene Transfer: Genomes in Flux, M.B. Gogarten, J.P. Gogarten, and L.C. Olendzenski, eds. (NY: Humana Press), pp. 489–500.

Anderson, E.S. (1968). The ecology of transferable drug resistance in the enterobacteria. Annu. Rev. Microbiol. *22*, 131–180.

Andersson, J.O. (2005). Lateral gene transfer in eukaryotes. Cell. Mol. Life Sci. *62*, 1182–1197.

Andersson, J.O. (2009). Horizontal gene transfer between microbial eukaryotes. In Horizontal Gene Transfer: Genomes in Flux, M.B. Gogarten, J.P. Gogarten, and L.C. Olendzenski, eds. (NY: Humana Press), pp. 473–487.

Andersson, J.O., Doolittle, W.F., and Nesbø, C.L. (2001). Are there bugs in our genome? Science *292*, 1848–1850.

Andersson, J.O., Hirt, R.P., Foster, P.G., and Roger, A.J. (2006). Phylogenetic analyses of diplomonad genes reveal frequent lateral gene transfers affecting eukaryotes. Curr. Biol. *13*, 94–104.

Archibald, J.D. (2009). Edward Hitchcock's pre-Darwin (1840) "Tree of life." J. Hist. Biol. *42*, 561–592.

Baldauf, S.L., Palmer, J.D., and Doolittle, W.F. (1996). The root of the universal tree and the origin of eukaryotes based on elongation factor phylogeny. Proc. Natl. Acad. Sci. U.S.A. *93*, 7749–7754.

Bapteste, E., and Boucher, Y. (2009). Epistemological aspects of horizontal gene transfer on classification in microbiology. In Horizontal Gene Transfer: Genomes in Flux, M.B. Gogarten, J.P. Gogarten, and L.C. Olendzenski, eds. (NY: Humana Press), pp. 55–72.

Bapteste, E., O'Malley, M.A., Beiko, R.M., Ereshefsky, M., Gogarten, J.P., Franklin-Hall, L., Lapointe, F.J., Dupré, J., Dagan, T., Boucher, Y., and Martin, W. (2009). Prokaryote evolution and the tree of life are two different things. Biol. Direct *4*, 34.

Beiko, R.G., and Ragan, M.A. (2009). Untangling hybrid phylogenetic signals: Horizontal gene transfer and artefacts of phylogenetic reconstruction. In Horizontal Gene Transfer: Genomes in Flux, M.B. Gogarten, J.P. Gogarten, and L.C. Olendzenski, eds. (NY: Humana), pp. 241–256.

Beiko, R.G., Harlow, T.J., and Ragan, M.A. (2005). Highways of gene sharing in prokaryotes. Proc. Natl. Acad. Sci. USA. *102*, 14332–14337.

Benachenhou-Lahfa, N., Forterre, P., and Labedan, B. (1993). Evolution of glutamate dehydrogenase genes: Evidence for two paralagous protein families and unusual branching patterns of the archaebacteria in the universal tree of life. J. Mol. Evol. *36*, 335–346.

Benton, M.J. (1989). Pruning the tree of life. Nature *342*, 129–130.

Brown, J.R. (2003). Ancient horizontal gene transfer. Nature Rev. Genet. *4*, 121–132.

Brown, J.R., and Doolittle, W.F. (1995). Root of the universal tree of life based on an ancient aminoacyl-tRNA synthase gene duplications. Proc. Natl. Acad. Sci. U.S.A. *92*, 2441–2445.

Brüssow, H. (2009). The not so universal tree of life *or* the place of viruses in the living world. Phil. Trans. R. Soc. B *364*, 2263–2274.

Busslinger, M., Rusconi, S., and Birnstiel, M.L. (1982). An unusual evolutionary behaviour of a sea urchin histone cluster. EMBO J. *1*, 27–33.

Charlebois, R.L., and Doolittle, W.F. (2004). Computing prokaryotic gene ubiquity: Rescuing the core from extinction. Genome Res. *14*, 2469–2477.

Ciccarelli, F.D., Doerks, T., von Mering, C., Creevey, C.J., Snel, B., and Bork, P. (2006). Toward automatic reconstruction of a highly resolved tree of life. Science *311*, 1283–1287.

Coughter, J.P., and Stewart, G.J. (1989). Genetic exchange in the environment. Antonie van Leeuwenhoek *55*, 15–22.

Cracraft, J., and Donoghue, M.J., eds. 2004. Assembling the Tree of Life (New York: Oxford University Press).

Crick, F.H.C. (1958). On protein synthesis. Symp. Soc. Exp. Biol. *12*, 138–163.

Dagan, T., and Martin, W. (2006). The tree of one percent. Genome Biol. *7*, 118.

Dagan, T., and Martin, W. (2009). Getting a better picture of microbial evolultion en route to a network of genomes. Phil. Trans. R. Soc. B *364*, 2187–2196.

Darwin, C. (1859). On the Origin of Species by Means of Natural Selection, or the Preservation of Favoured Races in the Struggle for Life, 1st ed. (London: John Murray).

Darwin, C. (1872). On the Origin of Species by Means of Natural Selection, or the Preservation of Favoured Races in the Struggle for Life, 6th ed. (London: John Murray).

Daubin, V., Moran, N.A., and Ochman, H. (2003). Phylogenetics and the cohesion of bacterial genomes. Science *301*, 829–832.

Dawkins, R. (2003). A Devil's Chaplain (New York: Mariner).

Dayhoff, M.O., Barker, W.C., and McLaughlin, P.J. (1974). Inferences from protein and nucleic acid sequences: Early molecular evolution, divergence of kingdoms and rates of change. Orig. Life *5*, 311–330.

Dayhoff, M.O., and Schwartz, R.M. (1981). Evidence on the origin of eukaryotic mitochondria from protein and nucleic acid sequences. Ann. N. Y. Acad. Sci. *361*, 92–104.

Dayrat, B. (2003). The roots of phylogeny: How did Haeckel build his trees? Syst. Biol. *52*, 515–527.

Deppenmeier, U., Johann, A., Hartsch, T., Merkl, R., Schmitz, R.A., Martinez-Arias, R., Henne, A., Wiezer, A., Bäumer, S., Jacobi, C., Brüggemann, H., Lienard, T., Christmann, A., Bömeke, M., Steckel, S., Bhattacharyya, A., Lykidis, A., Overbeek, R., Klenk, H.P., Gunsalus, R.P., Fritz, H.J., and Gottschalk, G. (2002). The genome of *Methanosarcina mazei*: evidence for lateral gene transfer between bacteria and archaea. J. Mol. Microbiol. Biotechnol. *4*, 453–461.

de Queiroz, K. (1988). Systematics and the Darwinian revolution. Phil. Sci. *55*, 238–259.

Doolittle, W.F. (1996). At the core of the Archaea. Proc. Natl. Acad. Sci. U.S.A. *93*, 8797–8799.

Doolittle, W.F. (1999). Phylogenetic classification and the universal tree. Science *284*, 2124–2128.

Doolittle, W.F. (2000). Uprooting the tree of life. Sci. Am. *282*, 90–95.

Doolittle, W.F. (2009a.) The practice of classification and the theory of evolution, and what the demise of Charles Darwin's tree of life hypothesis means for both of them. Phil. Trans. R. Soc. B *364*, 2221–2228.

Doolittle, W.F. (2009b). Response to Eric Bapteste *et al.* Biol. Direct *4*, 34.

Doolittle, W. F., and Bapteste, E. (2007). Pattern pluralism and the Tree of Life hypothesis. Proc. Natl. Acad. Sci. U.S.A. *104*, 2043–2049.

Doolittle, W.F., and Brown, J.R. (1994). Tempo, mode, the progenote, and the universal root. Proc. Natl. Acad. Sci. U.S.A. *91*, 6721–6728.

Eldredge, N. 2005. Darwin: Discovering the Tree of Life (New York: Norton).

Embley, T.M., Hirt, R.P., and Williams, D.M. (1994). Biodiversity at the molecular level: The domains, kingdoms and phyla of life. Phil. Trans. R. Soc. B *345*, 21–33.

Feil, E.J., and Spratt, B.G. (2001). Recombination and the population structures of bacterial pathogens. Annu. Rev. Microbiol. *55*, 561–90.

Feil, E.J., Holmes, E.C., Bessen, D.E., Chan, M.S., Day, N.P., Enright, M.C., Goldstein, R., Hood, D.W., Kalia, A., Moore, C.E., Zhou. J., and Spratt, B.G. (2001). Recombination within natural populations of pathogenic bacteria: Short-term empirical estimates and long-term phylogenetic consequences. Proc. Natl. Acad. Sci. U.S.A. *98*, 182–187.

Fernholm, B., Bremer, K., and Jörnvall, H., eds. (1989). The Hierarchy of Life: Molecules and Morphologies in Phylogenetic Analysis (Amsterdam: Elsevier).

Forterre, P., Benachenhou-Lahfa, N., and Confalonieri, F. (1992). The nature of the last universal ancestor and the root of the tree of life, still open questions. Biosystems *28*, 15–32.

Forterre, P., Benachenhou-Lahfa, N., and Labedan, B. (1993). Universal tree of life. Nature *362*, 795.

Foster, P.G., Cox, C.J., and Embley, T.M. (2009). The primary divisions of life: A phylogenomic approach employing composition-heterogeneous methods. Phil. Trans. R. Soc. B *364*, 2197–2207.

Fournier, G.P., Huang, J., and Gogarten, J.P. (2009). Horizontal gene transfer from extinct and extant lineages: Biological innovation and the tree of life. Phil. Trans. R. Soc. B *364*, 2229–2239.

Fox, G.E., Stackebrandt, E., Hespell, R.B., Gibson, J., Maniloff, J., Dyer, T.A., Wolfe, R.S., Balch, W.E., Tanner, R.S., Magrum, L.J., Zablen, L.B., Blakemore, R., Gupta, R., Bonen, L., Lewis, B.J., Stahl, D.A., Luehrsen, K.R., Chen, K.N., and Woese, C.R. (1980). The phylogeny of prokaryotes. Science *209*, 457–463.

Futuyma, D. (2004). The fruit of the tree: Insights into evolution and ecology. In J. Cracraft, and M.J. Donoghue, eds. Assembling the Tree of Life (New York: Oxford University Press), pp. 25–39.

Galtier, N. (2009). Comment on Bapteste *et al.*, Biol. Direct *4*, 34.

Galtier, N., and Daubin, V. (2008). Dealing with incongruence in phylogenetic analysis. Phil. Trans. R. Soc. B *363*, 4023–4029.

Gogarten, J.P. (1995). The early evolution of cellular life. Trends Ecol. Evol. *10*, 147–151.

Gogarten, J.P., and Townsend, J.P. (2005). Horizontal gene transfer, genome innovation and evolution. Nature Rev. Microbiol. *3*, 679–687.

Gogarten, J.P., Kibak, H., Dittrich, P., Taiz, L., Bowman, E.J., Bowman, B.J., Manolson, M.F., Poole, R.J., Date, T., and Oshima, T. (1989). Evolution of the vacuolar $H^+$-ATPase: Implications for the origins of the eukaryotes. Proc. Natl. Acad. Sci. U.S.A. *86*, 6661–6665.

Gogarten, J.P., Doolittle, W.F., and Lawrence, J.G. (2002). Prokaryotic evolution in light of gene transfer. Mol. Biol. Evol. *19*, 2226–2238.

Gogarten, M.B., Gogarten, J.P., and Olendzenski, L.C. eds. (2009). Horizontal Gene Transfer: Genomes in Flux (New York: Humana Press).

Gribaldo, S., and Brochier, C. (2009). Phylogeny of prokaryotes: Does it exist and why should we care? Res. Microbiol. epub ahead of print, doi:10.1016/j.resmic.2009.07.006

Gribaldo, S., and Cammarano, P. (1998). The root of the universal tree of life inferred from anciently duplicated genes encoding components of the protein-targeting machinery. J. Mol. Evol. *47*, 508–516.

Hacking, I. (2007). Root and branch. The Nation, October 8. http://www.thenation.com/doc/20071008/hacking

Haeckel, E. (1866). Generelle Morphologie der Organismen (Berlin: Reimer).

Haggerty, L.S., Martin, F.J., Fitzpatrick, D.A., and McInerney, J.O. (2009). Gene and genome trees conflict at many levels. Phil. Trans. R. Soc. B *364*, 2209–2219.

Harris, J.I., Sanger, F., and Naughton, M.A. (1956). Species differences in insulin. Arch. Biochem. Biophys. *65*, 427–438.

Hennig, W. (1966). Phylogenetic Systematics (Urbana: University of Illinois Press).

Hilario, E., and Gogarten, J.P. (1993). Horizontal transfer of ATPase genes – the tree of life becomes a net of life. BioSystems *31*, 111–119.

Hotopp, J.C., Clark, M.E., Oliveira, D.C., Foster, J.M., Fischer, P., Torres, M.C., Giebel, J.D., Kumar, N., Ishmael, N., Wang, S., Ingram, J., Nene, R.V., Shepard, J., Tomkins, J., Richards, S., Spiro, D.J., Ghedin, E., Slatko, B.E., Tettelin, H., and Werren, J.H. (2007). Widespread lateral gene transfer from intracellular bacteria to multicellular eukaryotes. Science *317*, 1753–1756.

House, C.H. (2009). The tree of life viewed through the contents of genomes. In Horizontal Gene Transfer: Genomes in Flux, M.B. Gogarten, J.P. Gogarten, and L.C. Olendzenski, eds. (New York: Humana Press), pp. 141–161.

Huang, J., and Gogarten, J.P. (2006). Ancient horizontal gene transfer can benefit phylogenetic reconstruction. Trends Genet. *22*, 361–366.

Iwabe, N., Kuma, K., Hasegawa, M., Osawa, S., and Miyata, T. (1989). Evolutionary relationship of archaebacteria, eubacteria and eukaryotes inferred from phylogenetic trees of duplicated genes. Proc. Natl. Acad. Sci. U.S.A. *86*, 9335–9359.

Jain, R., Rivera, M.C., Moore, J.E., and Lake, J.A. (2003). Horizontal gene transfer accelerates genome innovation and evolution. Mol. Biol. Evol. *20*, 1598–1602.

Jones, D., and Sneath, P.H.A. (1970). Genetic transfer and bacterial taxonomy. Bacteriol. Rev. *34*, 40–81.

Koonin, E.V. (2009). Darwinian evolution in light of genomics. Nucl. Acids Res. *37*, 1011–1034.

Koonin, E.V., Wolf, Y.I., and Puigbò, P. (2009). The phylogenetic forest and the quest for the elusive tree of life. Cold Spring Harb. Symp. Quant. Biol. *74*, doi: 10.1101/sqb.2009.74.006

Keeling, P.J., and Palmer, J.D. (2008). Horizontal gene transfer in eukaryotic evolution. Nat. Rev. Genet. *9*, 605–618.

Kurland, C.G., Canback, B., and Berg, O.G. (2003). Horizontal gene transfer: A critical view. Proc. Natl. Acad. Sci. U.S.A. *100*, 9658–9662.

Lake, J.A. (2009). Evidence for an early prokaryotic endosymbiosis. Nature *460*, 967–971.

Lake, J.A., Skophammer, R.G., Herbold, C.W., and Servin, J.A. (2009). Genome beginnings: Rooting the tree of life. Phil. Trans. R. Soc. B *364*, 2177–2185.

Lan, R., and Reeves, P.R. (1996). Gene transfer is a major factor in bacterial evolution. Mol. Biol. Evol. *13*, 47–55.

Lang, B.F. (1984). The mitochondrial genome of the fission yeast *Schizosaccharomyces pombe*: Highly homologous introns are inserted at the same position of the otherwise less conserved cox1 genes in *Schizosaccharomyces pombe* and *Aspergillus nidulans*. EMBO J. 3, 2120–2136.

Lapierre, P., and Gogarten, J.P. (2008). Estimating the size of the bacterial pan-genome. Trends Genet. *25*, 107–110.

Lawrence, J.G. (2002). Gene transfer in bacteria: speciation without species. Theor. Popul. Biol. *61*, 449–460.

Lawrence, J.G., and Ochman, H. (1998). Molecular archaeology of the *Escherichia coli* genome. Proc. Natl. Acad. Sci. U.S.A. *95*, 9413–9417.

Lawrence, J.G., and Retchless, A.C. (2009). The interplay of homologous recombination and horizontal gene transfer in bacterial speciation. In Horizontal Gene Transfer: Genomes in Flux, M.B. Gogarten, J.P. Gogarten, and L.C. Olendzenski, eds. (New York: Humana Press), pp. 29–53.

Lerat, E., Daubin, V., and Moran, N.A. (2003). From gene trees to organismal phylogeny in prokaryotes: The case of the γ-proteobacteria. PLoS Biol. *1*, 101–109.

Linkkila, T.P., and Gogarten, J.P. (1991). Tracing origins with molecular sequences: Rooting the tree of life. Trends Biochem. Sci. *16*, 287–288.

Loftus, B., Anderson, I., Davies, R., Alsmark, U.C., Samuelson, J., Amedeo, P., Roncaglia, P., Berriman, M., Hirt, R.P., Mann, B.J., Nozaki, T., Suh, B., Pop, M., Duchene, M., Ackers, J., Tannich, E., Leippe, M., Hofer, M., Bruchhaus, I., Willhoeft, U., Bhattacharya, A., Chillingworth, T., Churcher, C., Hance, Z., Harris, B., Harris, D., Jagels, K., Moule, S., Mungall, K., Ormond, D., Squares, R., Whitehead, S., Quail, M.A., Rabbinowitsch, E., Norbertczak, H., Price, C., Wang, Z., Guillén, N., Gilchrist, C., Stroup, S.E., Bhattacharya, S., Lohia, A., Foster, P.G., Sicheritz-Ponten, T., Weber, C., Singh, U., Mukherjee, C., El-Sayed, N.M., Petri, W.A. Jr, Clark, C.G., Embley, T.M., Barrell, B., Fraser, C.M., and Hall, N. (2005). The genome of the protist parasite *Entamoeba histolytica*. Nature *433*, 865–868.

Margoliash, E. (1963). Primary structure and evolution of cytochrome c. Proc. Natl. Acad. Sci. U.S.A. *50*, 672–679.

Margulis, L., and Guerrero, R. (1991). Kingdoms in turmoil. New Sci. *1761*, 46–50.

Margulis, L., and Schwartz, K.V. (1998). Five Kingdoms: An Illustrated Guide to the Phyla of Life on Earth, 3rd ed. (New York: W.H. Freeman).

Martin, W. (1999). Mosaic bacterial chromosomes: a challenge en route to a tree of genomes. BioEssays *21*, 99–104.

Martin, W. (2005). Molecular evolution: lateral gene transfer and other possibilities. Heredity *94*, 565–566.

Martin, W.F., and Müller, M. (2007). Origin of Mitochondria and Hydrogenosomes (New York: Springer).

Maynard Smith, J., Feil, E.J., and Smith, N.H. (2000). Population structure and evolutionary dynamics of pathogenic bacteria. BioEssays *22*, 1115–1122.

Mayr, E. (1942). Systematics and the Origin of Species: From the Viewpoint of a Zoologist (New York: Columbia University Press).

Mayr, E. (1963). Animal Species and Evolution (Cambridge, MA: Harvard University Press).

Mayr, E. (1982). The Growth of Biological Thought: Diversity, Evolution, and Inheritance (Cambridge, MA: Harvard University Press).

Mayr, E. (1990). A natural system of organisms. Nature *348*, 491.

McInerney, J.O., Cotton, J.A., and Pisani, D. (2008). The prokaryotic tree of life: Past, present ... and future? Trends Ecol. Evol. *23*, 276–281.

Medini, D., Donati, Tettelin, H., Masignani, V., and Rappuoli, R. (2005). The microbial pan-genome. Curr. Opin. Genet. Dev. *15*, 589–594.

Nelson, K.E., Clayton, R.A., Gill, S.R., Gwinn, M.L., Dodson, R,J,, Haft, D.H., Hickey, E.K., Peterson, J.D., Nelson, W.C., Ketchum, K,A,, McDonald. L,, Utterback, T.R., Malek, J.A., Linher, K.D., Garrett, M.M., Stewart, A.M., Cotton, M.D., Pratt, M.S., Phillips, C.A., Richardson, D., Heidelberg, J., Sutton, G.G., Fleischmann, R.D., Eisen, J.A., White, O., Salzberg, S.L., Smith, H.O., Venter, J.C., and Fraser, C.M. (1999). Evidence for lateral gene transfer between Archaea and Bacteria from genome sequence of *Thermotoga maritima*. Nature *399*, 323–329.

Norman, A., Hansen, L.H., and Sørensen, S.J. (2009). Conjugative plasmids: Vessels of the communal gene pool. Phil. Trans. R. Soc. B *364*, 2275–2289.

Olsen, G.J., Woese, C.R., and Overbeek, R. (1994). The winds of (evolutionary) change: Breathing new life into microbiology. J. Bacteriol. *176*, 1–6.

O'Malley, M.A., and Boucher, Y. (2005). Paradigm change in evolutionary microbiology. Stud. Hist. Philos. Biol. Biomed. Sci. *36*, 183–208.

Pallen, M. (2009). Lamarck, Darwin and the Tree of Life. http://roughguidetoevolution.blogspot.com/2009/08/lamarck-darwin-and-tree-of-life.html (accessed 18.09.09).

Pauling, L., and Zuckerkandl, E. (1963). Chemical paleogenetics: Molecular 'restoration studies' of extinct forms of life. Acta Chem. Scand. *17*, S9-S16.

Poptsova, M. (2009). Testing phylogenetic methods to identify horizontal gene transfer. In Horizontal Gene Transfer: Genomes in Flux, M.B. Gogarten, J.P. Gogarten, and L.C. Olendzenski, eds. (New York: Humana Press), pp. 227–240.

Puigbò, P., Wolf, Y.I., and Koonin, E.V. (2009). Search for a 'Tree of Life' in the thicket of the phylogenetic forest. J. Biol. *8*, 59.

Ragan, M.A., and Beiko, R.G. (2009). Lateral genetic transfer: Open issues. Phil. Trans. R. Soc. B *364*, 2241–2251.

Ragan, M.A., McInerney, J.O., and Lake, J.A. (2009). The network of life: Genome beginnings and evolution. Phil. Trans. R. Soc. B. *364*, 2169–2175.

Reanney, D. (1977). Gene transfer as a mechanism of microbial evolution. BioScience *27*, 340–344.

Richards, T.A., Soanes, D.M., Foster, P.G., Leonard, G., Thornton, C.R., and Talbot, N.J. (2009). Phylogenomic analysis demonstrates a pattern of rare and ancient horizontal gene transfer between plants and fungi. Plant Cell *21*, 1897–1911.

Saccone, C., Gissi, C., Lanave, C., and Pesole, G. (1995). Molecular classification of living organisms. J. Mol. Evol. *40*, 273–279.

Salzberg, S.L., White, O., Peterson, J., and Eisen, J.A. (2001). Microbial genes in the human genome: Lateral transfer or gene loss? Science *292*, 1903–1906.

Sapp, J. (2009). The New Foundations of Evolution: On the Tree of Life (Oxford: Oxford University Press).

Schwartz, R.M., and Dayhoff, M.O. (1978). Origins of prokaryotes, eukaryotes, mitochondria, and chloroplasts: A perspective is derived from protein and nucleic acid sequence data. Science *199*, 395–403.

Sibley, C.G. (1962). The comparative morphology of protein molecules as data for classification. Syst. Zool. *11*, 108–118.

Sidow, A., and Wilson, A.C. (1990). Compositional statistics: An improvement of evolutionary parsiomony and its application to deep branches in the tree of life. J. Mol. Evol. *31*, 51–68.

Snel, B., Huynen, M.A., and Dutilh, B.E. (2005). Genome trees and the nature of genome evolution. Annu. Rev. Microbiol. *59*, 191–209.

Sogin, M.L. 1991. Early evolution and the origin of eukaryotes. Curr. Opin. Genet. Dev. *1*, 457–463.

Sogin, M.L., Hinkle, G., and Leipe, D.D. (1993). Universal tree of life. Nature *362*, 795.

Sprague, G.F. Jr. (1991). Genetic exchange between kingdoms. Curr. Opin. Genet. Dev. *1*, 530–533.

Spratt, B.G., Hanage, W.P., and Feil, E.J. (2001). The relative contributions of recombination and point mutation to the diversification of bacterial clones. Curr. Opin. Microbiol. *4*, 602–606.

Stanhope, M.J., Lupas, A., Italia, M.J., Kovetke, K.K., Volker, C., and Brown, J.R. (2001). Phylogenetic analyses do not support horizontal gene transfers from bacteria to vertebrates. Nature *411*, 940–44.

Stanier, RY., and van Niel, C.B. (1941). The main outlines of bacterial classification. J. Bacteriol. *42*, 437–466.

Stanier, R.Y., and van Niel, C.B. (1962). The concept of a bacterium. Arch. Mikrobiol. *42*, 17–35.

Sterelny, K., and Griffiths, P.E. (1999). Sex and Death: An Introduction to Philosophy of Biology (Chicago: Chicago University Press).

Stevens, P.F. (1984). Metaphors and typology in the development of botanical systematics 1690–1960, or the art of putting new wine in old bottles. Taxon *33*, 169–211.

Swithers, K.S., Gogarten, J.P., and Fournier, G.P. (2009). Trees in the web of life. J. Biol. *8*, 54.

Syvanen, M. (1987). Molecular clocks and evolutionary relationships: Possible distortions due to horizontal gene flow. J. Mol. Evol. *26*, 16–23.

Syvanen, M. (1984). The evolutionary implications of mobile genetic elements. Annu. Rev. Genet. *18*, 271–293.

Tettelin, H., Riley, D., Cattuto, C., and Medini, D. (2008). Comparative genomics: The bacterial pan-genome. Curr. Opin. Microbiol. *12*, 472–477.

Ward, N., and Fraser, C.M. (2005). How genomics has affected the concept of microbiology. Curr. Opin. Microbiol. *8*, 564–571.

Welch, R.A., Burland, V., Plunkett, G., 3rd, Redford, P., Roesch, P., Rasko, D., Buckles, E.L., Liou, S.R., Boutin, A., Hackett, J., Stroud, D., Mayhew, G.F., Rose, D.J., Zhou, S., Schwartz, D.C., Perna, N.T., Mobley, H.L., Donnenberg, M.S., and Blattner, F.R. (2002). Extensive mosaic structure revealed by the complete genome sequence of uropathogenic *Escherichia coli*. Proc. Natl. Acad. Sci. U.S.A. *99*, 17020–17024.

Wheelis, M.L., Kandler, O., and Woese, C.R. (1992). On the nature of global classification. Proc. Natl. Acad. Sci. U.S.A. *89*, 2930–2934.

Whittaker, R.H. (1959). On the broad classification of organisms. Quart. Rev. Biol. *34*, 210–226.

Whittaker, R.H. (1969). New concepts of kingdoms of organisms. Science *163*, 150–160.

Wilson, A.C., Carlson, S.S., and White, T.J. (1977). Biochemical evolution. Annu. Rev. Biochem. *46*, 573–639.

Wilson, A.C., and Sarich, V.M. (1969). A molecular time scale for human evolution. Proc. Natl. Acad. Sci. U.S.A. *63*, 1088–1093.

Woese, C.R. (1987). Bacterial evolution. Microbiol. Rev. *51*, 221–271.

Woese, C.R. (1994a). There must be a prokaryote somewhere: Microbiology's search for itself. Microbiol. Rev. *58*, 1–9.

Woese, C.R. (1994b). Microbiology in transition. Proc. Natl. Acad. Sci. U.S.A. *91*, 1601–1603.

Woese, C.R. (1996). Phylogenetic trees: Whither microbiology? Curr. Biol. *6*, 1060–1063.

Woese, C.R. (2000). Interpreting the universal phylogenetic tree. Proc. Natl. Acad. Sci. U.S.A. *97*, 8392–8396.

Woese, C.R. (2002). On the evolution of cells. Proc. Natl. Acad. Sci. U.S.A. *99*, 8742–8747.

Woese, C.R. (2005). The archaeal concept and the world it lives in: A retrospective. In Discoveries in Photosynthesis, Govindjee, J.T. Beatty, H. Gest, and J.F. Allen, eds. (Dordrecht: Springer), pp. 1109–1120.

Woese, C.R., and Fox, G.E. (1977). Phylogenetic structure of the prokaryotic domain: The primary kingdoms. Proc. Natl. Acad. Sci. U.S.A. *74*, 5088–5090.

Woese, C.R., Gibson, J., and Fox, G.E. (1980). Do genealogical patterns in purple photosynthetic bacteria reflect interspecific gene transfer? Nature *283*, 212–214.

Woese, C.R., Kandler, O., and Wheelis, M.L. (1990). Towards a natural system of organisms: Proposal for the domains Archaea, Bacteria, and Eucarya. Proc. Natl. Acad. Sci. U.S.A. *87*, 4576–4579.

Zhaxybaeva, O. (2009). Detection and quantitative assessment of horizontal gene transfer. In Horizontal

Gene Transfer: Genomes in Flux, M.B. Gogarten, J.P. Gogarten, and L.C. Olendzenski, eds. (New York: Humana Press), pp. 195–213.

Zhaxybayeva, O., Swithers, K.S., Lapierre, P., Fournier, G.P., Bickhart, D.M., DeBoy, R.T., Nelson, K.E., Nesbø, C.L., Doolittle, W.F., Gogarten, J.P., and Noll, K.M. (2009a). On the chimeric nature, thermophilic origin, and phylogenetic placement of the Thermotogales. Proc. Natl. Acad. Sci. U.S.A. *106*: 5865–5870.

Zhaxybayeva, O., Doolittle, W.F., Papke, R.T., and Gogarten, J.P. (2009b). Intertwined evolutionary histories of marine *Synechococcus* and *Prochlorococcus marinus*. Genome Biol. Evol. *1*, 325–329.

Zuckerkandl, E. (1987). On the molecular evolutionary clock. J. Mol. Evol. *26*, 34–46.

Zuckerkandl, E., and Pauling, L. (1965). Molecules as documents of evolutionary history. J. Theor. Biol. *8*, 357–366.

# Horizontal Gene Transfer and the Formation of Groups of Microorganisms

David Williams, Cheryl P. Andam and J. Peter Gogarten


## Abstract

In this chapter, we discuss the impact of gene transfer on the formation of groups of organisms. We begin by discussing the obvious: gene transfer can make it more difficult to define and determine relationships. In those cases where many genes have been transferred between preferred partners, the majority of genes in a genome may reflect gene acquisition, and as a consequence, if a coherent signal is detected, one nevertheless might not be sure that the signal is due to organismal shared ancestry. In the second part of this chapter we will focus on two positive aspects of gene transfer. The presence of a particular transferred gene was shown, in several cases, to constitute a shared derived character useful in classification. Gene transfer can put together new metabolic pathways that open up new ecological niches, and consequently, the transfer of an adaptive gene might create a new group of organisms.


## Problems encountered in phylogenetic reconstruction

### Towards a natural taxonomy

Phylogeny asks the question 'How did higher taxonomic units come into existence?' According to Hennig, a natural taxonomic system aims for a classification that reflects organismal shared ancestry (Hennig, 1966). In such a natural system of classification, the formation of proper taxonomic groups is based on shared derived characters. The distinction between primitive and derived characters is important in a cladistic framework. Only shared derived characters define a monophyletic clade,[1] whereas a primitive character defines a paraphyletic group. A paraphyletic group is derived from a common ancestor, but this common ancestor also gave rise to organisms outside the group. A group defined by a primitive character is paraphyletic because those organisms that possess the derived character state also evolved from the same ancestor as those that retained the primitive character state. In molecular phylogenies, derived and primitive character states usually are not, and often cannot, be distinguished. A tree resulting from sequence-based phylogenetic reconstruction is usually unrooted. The use of outgroups is one way to polarize part of a phylogenetic tree, but the placement of the root should be explicitly discussed. In case of doubt it is preferable to avoid cladistic terminology altogether (Wilkinson *et al.*, 2007). This comment is not trivial. Many data suggest that for many molecular systems the root

1 Sometimes monophyletic clades *sensu* Hennig (1966) are labelled as holophyletic, and the term monophyletic is re-defined to include holo- and paraphyletic groups (Ashlock, 1971). Here we use the terms as defined by Hennig.

of the net[2] of life is located on the bacterial branch (Zhaxybayeva and Gogarten, 2007; Zhaxybayeva *et al.*, 2005), making the Archaea and Eukaryotes sister domains. However, the root of the bacterial domain is uncertain with respect to the different bacterial phyla. Ribosomal data, including both ribosomal proteins and ribosomal RNA, often place the Aquificae as the deepest branching lineage in the bacterial domain (Woese, 1987), but the outgroup is located on a very long branch, and prudence suggests that the point where the outgroup joins the ingroup should be considered as unresolved. In unrooted ribosomal phylogenies, Aquificae and Thermotogales form a strongly supported clan (*sensu* Wilkinson *et al.*, 2007), but if the traditional rRNA phylogeny were correct, these two organisms would not form a clade, but a paraphyletic grouping. The same uncertainty also permeates placing the root of the eukaryotes (we use this term to denote the nucleocytoplasmic component of the eukaryotic cell) and the relationships between Eukaryotes and Archaea (for discussion see Dagan and Martin, 2007b; Kurland *et al.*, 2006; Martin *et al.*, 2007; Penny and Poole, 1999).

## Gene transfer and phylogenetic reconstruction

In addition to the problems generated by the difficulty and uncertainty of phylogenetic reconstruction, horizontal gene transfer (HGT) creates additional complications. Because of the horizontal transfer of genetic information, genes coexisting in a genome frequently have different histories (Doolittle, 1999; Gogarten *et al.*, 2002; Gogarten and Townsend, 2005; Hilario and Gogarten, 1993; Lawrence and Ochman, 1997; Ochman *et al.*, 2000). Based on the frequency with which genes are encountered in related organisms, genomes of microorganisms can be divided into different sets of genes. The extended core includes genes found in all members of the group. The extended bacterial core includes about 250 gene families, i.e. 250 gene families are found represented in nearly all bacterial genomes. In contrast, the pan-genome of a group, such as a species, genus or larger taxonomic unit, is defined as the non-overlapping set of all gene families found in at least one genome. According to current estimations, the pan-genome of many bacterial groups (Tettelin *et al.*, 2005), and of bacteria as a whole (Lapierre and Gogarten, 2009) is open, i.e. each randomly selected bacterial genome that is sequenced will encode many proteins that do not belong to any previously characterized family. At present, the average number of novel genes per newly sequenced genome is estimated to be over a hundred genes per genome, with no levelling off being detected (Lapierre and Gogarten, 2009). The genes that are encountered only once, or very infrequently in other genomes, have been labelled as accessory genes and constitute on average about one-quarter of each bacterial genome. Most of these genes are not well characterized: they are strain specific, frequently acquired through gene transfer, most do not persist in a lineage for long periods of time, and they are more AT-rich and shorter than the average gene (Daubin and Ochman, 2004; Lapierre and Gogarten, 2009; Lawrence and Ochman, 1997). For accessory genes found in closely related strains the rate of non-synonymous to synonymous substitutions reveals a low level of purifying selection (Daubin and Ochman, 2004); however, it was suggested that this low level of selection might be due to mutations generating detrimental properties and not due to an adaptive value of the transferred gene (Gogarten and Townsend, 2005). The majority of recently acquired genes might be neutral or nearly neutral in their effect on the fitness of the recipient.

A corollary to the large number of accessory genes per genome is that the common shared gene pool of strains that belong to the

2 We use the term 'net of life' in place of 'tree of life' to highlight that evolution has been a reticulate process that should not be reduced to a single bifurcating process.

same species is surprisingly small.[3] Welch *et al.* (2002) reported that the common shared gene pool of three *Escherichia coli* genomes was less than 40% of the non-overlapping gene set; in the case of three *Frankia* strains whose small subunit ribosomal RNA showed less than 3% sequence divergence, only 20% of the common shared gene pool had detectable homologues in all three genomes (Normand *et al.*, 2007). The accessory genes illustrate that microbial genomes are changing rapidly, and most of the accessory genes constitute derived characters for the strains in which they are found; however, these genes are not found in other organisms, thus they do not provide a taxonomic marker. In contrast, core genes contain phylogenetic information; however, the debate is ongoing with respect to the presence and meaning of a phylogenetic consensus signal (see Bapteste *et al.* (2009) and Gogarten-Boekels *et al.* (2009) for an overview of recent discussions).

The phylogenetic analysis of extreme thermophilic bacteria provides an excellent illustration of the problems created through highways of gene sharing (Beiko *et al.*, 2005). Ribosomal RNA and ribosomal protein-based phylogenies of Thermotogales (Zhaxybayeva *et al.*, 2009b) and Aquificales (Boussau *et al.*, 2008) place them as early branching groups at the base of the bacterial domain; about 10–20% of the genes in the genomes of Thermotogales appear to have been acquired by horizontal gene transfer from archaeal donors (Nelson *et al.*, 1999; Nesbø *et al.*, 2001; Zhaxybayeva *et al.*, 2009b); however, a majority of their genes reveal affinities of Thermotogales with Clostridia (Gophna *et al.*, 2005; Zhaxybayeva *et al.*, 2009b) and of Aquificales with Epsilonproteobacteria (Boussau *et al.*, 2008). If we assume that the ribosomal phylogeny reflects the organismal phylogeny of the Thermotogales, then the vast majority of genes would have to have been transferred from Clostridia.

The Thermotogales contain three distinct phylogenetic signatures in their genome: archaeal genes acquired through HGT, clostridial genes that likely also were acquired through HGT, and the ribosome that groups them with Aquificales and that may reflect the evolutionary history of the translation machinery. Reducing the phylogeny of these organisms to only one of these components misses a major process; in particular, the plurality signal, also known as emerging consensus (Wolf *et al.*, 2002) or the central trend (Puigbò *et al.*, 2009), at least in the case of the Thermotogales and Aquificales appears to be overwhelmed by a highway of gene sharing and might not represent the history of the ribosomal nor the organismal tree as defined by cell divisions.

The complexity of genome evolution cannot be captured in a single tree-diagram without reticulations, nor by a hierarchical classification system that is based on a strictly furcating process. A hierarchical classification system possibly might be founded on a sub-cellular system such as the ribosome, which consists of many interacting parts that encounter higher barriers to HGT than other genes (Sorek *et al.*, 2007). Such a system would fall short of describing organismal evolution and phylogeny; however, phylogenies for such subsystems promise to constitute a good starting point to reconstruct the reticulated history of life (Gogarten, 1995; Swithers *et al.*, 2009).

## HGT can create patterns indistinguishable from those created through shared ancestry

In the previous section we considered HGT as a process that can give rise to conflicting patterns of phylogenetic information. In this section, we

3 Assume the following thought experiment: three genomes from three different strains of a species each encode 2000 genes. 600 genes per genome, or 30% of the genes per genome are not found in the other two genomes. 1400 genes are present in each of the three genomes, i.e. they belong to the core. Therefore, in total there are 1400 gene families in the core genome, and 3 time 600 = 1800 genes in the accessory pool. The pan-genome defined as the total non-overlapping gene pool is 1800+1400 = 3200 genes, and the common shared gene pool, in this example, is 1400 genes or 44% of the non-overlapping gene pool.

consider HGT as a process that can create phylogenetic patterns that may be indistinguishable from those created through shared ancestry. We suggest that even in the case where only a single phylogenetic pattern is detected, this pattern may have been created or co-created through biased HGT. Relatedness of organisms in phylogenetic reconstruction has always been correlated with a vertical genealogical path originating from a common ancestor. In the same way, taxonomical classification of organisms, including microorganisms, often implies that individuals possess similar properties that they have inherited from their parents. However, if we trace back the history of a species, particularly for prokaryotic evolution, we observe a more complicated story that reveals not just a single evolutionary process. More specifically, the current taxonomic associations of organisms may actually have been shaped by HGT, thus negating the argument that HGT only creates discord in phylogenetic reconstruction.

Fig. 9.1 illustrates how preferential gene transfer can create clusters of organisms possessing similar properties. Within each group, high levels of gene transfer act as a homogenizing force that holds the group together. Under the scenario depicted in Fig. 9.1, molecular phylogenies will reflect these exchange groups, and not shared ancestry of the organisms. An organism's propensity for specific exchange partners may be a consequence of the transcription and translation machineries that resemble its own, facilitating the entry and establishment of novel genes (Gogarten *et al.*, 2002; Lawrence and Hendrickson, 2005; Olendzenski *et al.*, 2001). On the other hand, incompatible genetic apparatus can reduce the likelihood and frequency of successful transfers (Hendrickson and Lawrence, 2006). Ecological isolation of independent lineages can also enhance biased transfer, due to the limited partners available.

## Plurality of pattern and process

If members of the group exchange genes, they become more similar. The more genes they share, the more similar they become, and the more bias may be created for future transfers. Although this may seem circuitous, it only emphasizes the observation that biased transfers can, at the very least, reinforce hierarchical groups formed through vertical inheritance. Later on, transferred genes can lead to the vertical transmission of innovations that were not present in the ancestor of the group (Gogarten and Olendzenski, 1999) and can persist as shared derived characters that unite the descendants of the group (Huang and Gogarten, 2006; Huang *et al.*, 2005). To the extent that a phylogenetic pattern is created through biased HGT, the observed phylogenetic signal has to be considered phenetic and not cladistic. A consensus phylogenetic pattern should not automatically be equated with shared organismal ancestry. Considering that ribosomal RNA genes are also prone to horizontal transfer (Schouls *et al.*, 2003; van Berkum *et al.*, 2003; Wang and Zhang, 2000; Yap *et al.*, 1999), even ribosomal trees, considered to be a reflection of organismal evolution, may well be, in part, a manifestation of HGT bias. Following transfer, ribosomal RNA encoding genes often are integrated into the recipient genome through homologous recombination (Morandi *et al.*, 2005; Yap *et al.*, 1999). In some instances, the ribosomal rRNA operon might be a mosaic that integrates over many gene transfer events between related organisms (Gogarten *et al.*, 2002). Similarity of organisms and their relatedness as identified through molecular phylogenies can be created through two processes: shared ancestry and gene transfer between preferred partners. HGT does create distinct patterns in molecular phylogenies; in addition, biased HGT also provides an additional process that can generate the patterns usually ascribed to shared organismal ancestry.

## A population genetic point of view for higher level taxonomic units

Exchange groups not only exist at the population and species level. Higher-level taxonomic units (such as order and phylum) also display partiality toward certain HGT partners, creating larger groups that exhibit phenotypic coherence. Barriers to gene exchange, such as geographical, genetic and physiological features, can induce higher-level groups to swap genes with certain other assemblages of organisms and not with others. A pattern emerges in which higher taxonomic groups exhibit behaviour that is analogous to that found in populations of a single species. Hence, taxa, at any level, will persist as long as they

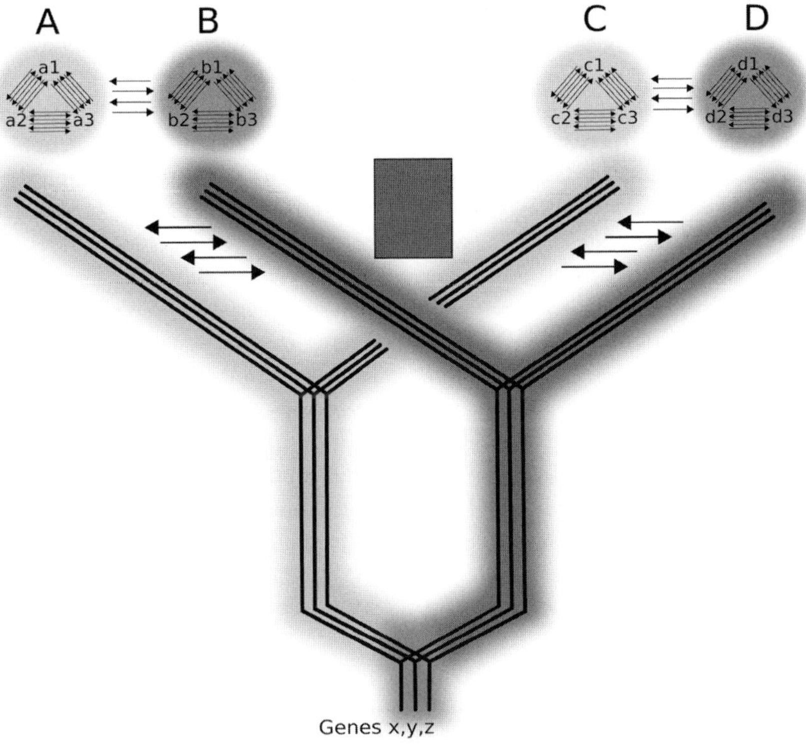


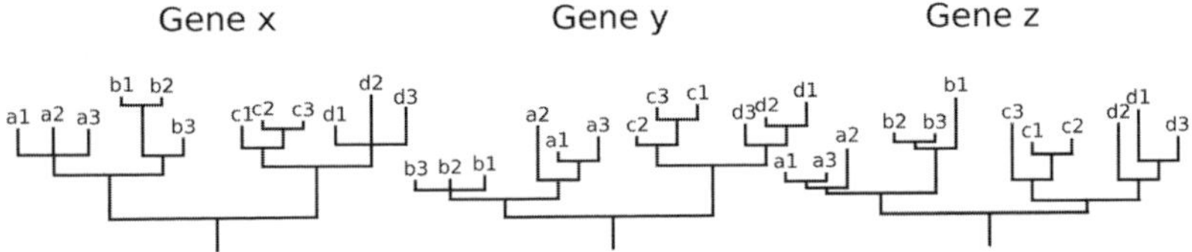


**Figure 9.1** The role of biased gene transfer in the evolution of organismal lineages. Panel A shows four groups of organisms representing lineages, composed of members that frequently swap genes among each other. Within each group, high levels of gene transfer act as a homogenizing force. The organismal tree of these four groups indicates that lineages A and C share a last common ancestor, different from the last ancestor of lineages B and D. However, a barrier to gene transfer (represented by a grey rectangle) between A and B and between C and D is established, halting gene exchange between more closely related organisms. As a consequence, A and B continue to swap genes with each other, and C with D. Without assuming a known organismal phylogeny, we would conclude from the gene phylogenies that lineages A and B (and C and D) are more related to each other and thus will cluster together. Hence, it will be inaccurate to assume that these two larger groups were formed as a result of vertical inheritance; rather these groups were formed as a consequence of high rates of exchange between the lineages. Panel B depicts phylogenetic trees reconstructed for different genes (x, y, z) for groups A–D shown in the top panel. Each gene tree might not necessarily show similar topology to the organismal tree. At the level of lower taxonomic units (a1, a2, a3; ...), constant shuffling of genes occur within each lineage, creating conflicting topological patterns, even though the group as a whole remain consistent. More frequent transfer between specific lineages (between A and B and between C and D) can skew their groupings, reflecting a bias between exchange partners. This bias may not always be attributed to organismal shared ancestry. The figure is based on concepts and diagrams from Gogarten *et al.* (2002) and Olendzenski *et al.* (2001).

maintain their propensity toward their exchange partners. A population genetic model is therefore necessary to account for the processes that underlie the genetic structure of lineages, particularly in the microbial realm.

HGT frequencies differ according to the type of gene, group of organism, and ecosystem (Beiko *et al.*, 2005; Jain *et al.*, 1999; Zhaxybayeva *et al.*, 2007). In addition, methods to detect HGT events have complementary sensitivities with respect to phylogenetic depth of the transfer (Ragan, 2001), and have different error rates for false positives and false negatives in detecting gene transfer events (Poptsova and Gogarten, 2007; Zhaxybayeva, 2009). Dagan and Martin (2007a) reported a minimum rate of 1.1 HGT events per gene family, and that at least 65% of all gene families have been affected by HGT; Zhaxybayeva *et al.* (2006) found traces of gene transfer in more than 50% of gene families in cyanobacterial genomes. In some ecosystems and for some closely related organisms in the process of de-speciation or de-generation, the gene transfer rates likely are much higher (Sheppard *et al.*, 2008; Zhaxybayeva *et al.*, 2009a). The impact of HGT on organismal phylogeny will not be constant throughout life's history.

## HGT and the creation of metabolic pathways

### Categorizing gene transfer

#### *Homologous and non-homologous recombination*

Natural selection requires variation as raw material. Throughout the biosphere, this variation is achieved through mutation of genetic sequences (Chao and Cox, 1983; Kimura, 1967; Levins, 1967) and recombination, the mixing of existing genetic material (Fisher, 1930; Jeffreys and Neumann, 2002; Papke *et al.*, 2007). Numerous mechanisms to increase the amount of both sources of novelty, representing the evolution of evolvability, have been described including mutator phenotypes in bacteria (Gross and Siegel, 1981; Taddei *et al.*, 1997), conjugation in Bacteria (Kuusinen, 1996) and Archaea (Prangishvili *et al.*, 1998; Schleper *et al.*, 1994) as well as the elaborately choreographed homologous recombination between chromosomes in sexually reproducing organisms such as ourselves (Creighton and McClintock, 1931). This last example is found in populations of sexually reproducing individuals among which recombination occurs every generation but which also experience strong barriers to recombination between individuals of different populations. This is the basis for Mayr's Biological Species Definition (Mayr, 1942, 1963) but evidence is building that similar barriers influence diversity within and between microbial lineages (Doolittle and Zhaxybayeva, 2009; Dykhuizen and Green, 1991; Fraser *et al.*, 2009; Papke, 2009). As described above, recombination may occur with sufficient frequency to create patterns resembling those of vertical inheritance at different taxonomic levels, as delineated by the effectiveness of barriers to recombination. Recombination between closely related microorganisms (sometimes referred to as HGT, although in this context has an effect similar to sex) can occur at very high frequencies as in colonizing populations of the human pathogen *Helicobacter pylori* (Suerbaum *et al.*, 1998) and is usually homologous due to high levels of sequence similarity and genome synteny. Between more distant lineages, recombination becomes less likely as sequence similarity and synteny decrease (Gogarten *et al.*, 2002; Majewski and Cohan, 1999; Wolf *et al.*, 2001). Integration of transferred genes by non-homologous recombination, causing a decrease in synteny between closely related lineages, has been postulated as part of the speciation process in some microorganisms (Retchless and Lawrence, 2007).

#### *Effects on fitness*

The evolutionary impact of HGT can, to some extent, be categorized by the frequency of successful HGT events in conjunction with the apparent evolutionary fitness effects on the recipient. One category includes transfers among an exchange group of closely related lineages as defined above. The sufficiently relaxed recombination barriers among lineages result in discrete, random, one-way transfers at a high enough frequency to allow a degree of perceived reciprocity and an allele exchange-like process. Analyses of genetic data have revealed such processes in diverse groups including between pathogenic and commensal strains of the *Neisseria* genus (Feil *et al.*, 1996;

Zhou *et al.*, 1997); between *Agrobacterium* biovar 1 'species' (Costechareyre *et al.*, 2009); and between isolates of the free living haloarchaeal genus *Halorubrum* (Papke *et al.*, 2007). Between a closely related donor and recipient, ecological and intracellular conditions are likely to be similar so that potential fitness gains by genetic transfer will usually be slight, but enough to allow persistence in and potential subsequent transfer from the recipient. However, stronger selective pressures may cause more abrupt changes in the allele-like population dynamics of transferred genetic material, for example the rapid spread of antibiotic resistance genes among phyla of microbial pathogens (Davies, 1994), such as among strains of the *Neisseria* genus (Feil *et al.*, 1996; Zhou *et al.*, 1997).

### *HGT between distant relatives*

Despite their comparative rarity, HGT events between distant lineages have also had profound evolutionary impacts. The success of some very large extant clades of organisms appears to be due to HGT-mediated synergism between pre-existing metabolic machinery and the products of the newly acquired genetic material. Examples include the assembly to oxygen producing photosynthesis in cyanobacteria (also referred to as blue–green algae and Cyanophyta; Oren, 2004) which are among the most abundant organisms known and are found throughout the oceans (Chisholm *et al.*, 1988; Waterbury *et al.*, 1979; Zwirglmaier *et al.*, 2008), rivers and lakes (Zwart *et al.*, 2002), hot springs (Papke *et al.*, 2003), hypersaline pools (Sørensen *et al.*, 2005), and in symbiotic niches (Kuusinen, 1996; Lesser *et al.*, 2004). Their oxygenic photosynthetic metabolic pathway is credited with the oxygenation of the atmosphere starting 2.7 billion years ago based on geological evidence (Buick, 1992). Members of the euryarchaeal order Methanosarcinales are capable of acetoclastic methanogenesis, a process that today produces approximately two-thirds of biogenic methane annually (Ferry, 1992), and are widely distributed in marine and freshwater sediments, soils and the gastrointestinal tracts of animals (Donovan *et al.*, 2004; Koizumi *et al.*, 2003). The enzymes used to initiate this process by members of the genus *Methanosarcina* are coded for by genes atypical in Archaea and phylogenetic evidence reveals that they were acquired from cellulolytic clostridia via HGT (Fournier and Gogarten, 2008).

Rather than modifying an existing metabolic function, this category of HGT event leads to more profound innovation but may be rare for two reasons. Firstly, successful chromosomal integration of genetic material between distant relatives is less likely because of several reasons, including low sequence similarity preventing homology dependent recombination (Gogarten *et al.*, 2002; Majewski and Cohan, 1999; Wolf *et al.*, 2001), non-overlapping specificity of genetic vectors such as phage (Breitbart and Rohwer, 2005), and geographical isolation because of contrasting ecological niches or natural history. Secondly, following chromosomal integration, the fitness effect on the recipient is likely to be near-neutral or negative if the promoter system is incompatible or if the gene product interferes with the existing cell chemistry (which is more likely between distantly related, divergent organisms). A decrease in fitness of the recipient is likely to lead to its extinction and leave no trace in the evolutionary record. The frequency of such events is retrospectively undetectable and the impact negligible when considering evolutionary legacies of HGT. Even if HGT products with near-neutral fitness effects are fixed due to genetic drift, such transfers alone do not hold much promise for the recipient's descendants.

## HGT and the expansion of metabolic networks

However, the acquisition of a gene that allows the recipient to occupy a new, previously unoccupied niche provides an increase in fitness even if the expression and regulation of the gene is not optimally integrated into the recipients metabolic network (Gogarten *et al.*, 2002). Consequently, it is not surprising that many of the genes successfully transferred across domain boundaries encode enzymes that transport metabolizable substrates into the cell (e.g. Noll and Thirangoon, 2009) and that expand the capabilities at the periphery of the recipient's metabolic network (Pal *et al.*, 2005). Much like winning a state lottery, some of the rare successes have been spectacular as demonstrated by the above examples. The key to this success is in the nature of cell metabolism: the energy transductions and matter transitions, mediated by electron transfer among chemical species, necessary for the proliferation of life. This complex process can be considered a network of discrete nodes or modules linked together as a network (Duarte *et al.*, 2004; Reed *et al.*, 2003). Each node is usually a multimeric protein complex catalysing a specific chemical reaction or the transfer of a specific chemical species across a membrane. Reagents are transferred between connected nodes for subsequent reactions. If the products of laterally transferred genetic material happen to facilitate one or more chemical reactions under the recipient's internal conditions, and if that chemical reaction fits into the existing metabolic network, a significant change in eco-physiology can occur; as well as the potential for enhancing the competitiveness in a presently occupied ecological niche. Whole new modes of metabolism may become possible, opening up new niches for the HGT recipient and its descendants (Gogarten *et al.*, 2002; Pal *et al.*, 2005) and even causing globally significant changes in biogeochemistry. The legacies of such rare events include, but are not limited to (Boucher *et al.*, 2003), the widely distributed *Methanosarcina* and their influence on the global methane budget and the even more ubiquitous cyanobacteria as primary producers in many ecosystems and their oxygenation of the atmosphere.

In the following section, we discuss a few selected detailed examples and a fortuitous side effect of such rapidly appearing derived characters with respect to cladistics.

*Oxygen producing photosynthesis*

Bacteria that are able to gain energy through anoxygenic photosynthetic pathways form a diverse polyphyletic group. Some members of the alpha, beta, and gamma proteobacteria use light energy to transfer electrons from sulfide, elemental sulfur or hydrogen to inorganic carbon compounds for subsequent release of energy in anaerobic photoautotrophy (purple bacteria). However, when molecular oxygen is available, heterotrophic oxidation of organic matter is possible by reversing the flow of electrons through cytochrome *b/c* (Blankenship *et al.*, 1995). Chloroflexi (green filamentous bacteria) possess a reaction centre similar to that of the purple bacteria and photosytem II of the cyanobacteria, whilst Chlorobi (green sulfur bacteria) and Heliobacteria (of the Gram-positive phylum Firmicutes) each possesses similar reaction centres that contrast with the former and are similar to photosynthetic reaction centre I in cyanobacteria (Raymond, 2008; Sadekar *et al.*, 2006). The recent discovery of members of other bacterial phyla (Acidobacteria) capable of photoheterotrophy (Bryant *et al.*, 2007) further increases the wide distribution of anoxygenic photosynthetic metabolism (Blankenship, 2001; Raymond, 2008).

The oxygenic photosynthetic machinery of cyanobacteria is more complex combining two distinct reaction centres (photosystems) that work in series when electrons are transported from water to NADP (Ferreira *et al.*, 2004). The photosynthetic machinery is encoded in more than 100 genes, whose products form a tightly interacting machinery (Shi *et al.*, 2005). In addition to their electron transport chain, cyanobacteria have evolved many diverse adaptations to the environments that live in, including complex carbon dioxide concentrating centres (Badger and Price, 2003). Some of these adaptations occurred in the cyanobacterial lineage after some cyanobacteria became endosymbionts that evolved into the plastids of eukaryotes. Structural, functional, and sequence similarity have been noted between each of the cyanobacterial reaction centres and those of the green filamentous bacteria and the purple bacteria respectively (Buttner *et al.*, 1992; Liebl *et al.*, 1993; Sadekar *et al.*, 2006).

Two distinct scenarios have long been discussed (see Woese, 1987, for an early review) on how the more complex electron transport chain came into existence:

1 Internal gene duplications of the reaction centres in the cyanobacterial stem, i.e. the lineage leading to the most recent common ancestor (MRCA) of all extant cyanobacteria, gave rise to the two types of reaction centres now working in series (Mulkidjanian *et al.*, 2006). These reaction centres were then transferred horizontally from the cyanobacteria to other bacterial groups.
2 The different types of reaction centres evolved in different groups of photosynthetic bacteria and were brought together in the

cyanobacterial stem though HGT. In this scenario, the cyanobacterial stem contained photosystem I bacteria that were not able to use water as an electron donor in photosynthesis. Before the successful integration of the two photosytems, the organisms of the cyanobacterial stem group might have used PSI for cyclic electron transport (as in heterocysts and the bundle sheath cells of some $C_4$ plants) and/or they might have used either of the photosystems with another electron donor.

Several observations strongly favour the latter scenario for the creation of oxygenic photosynthesis through HGT:

- Gene transfers, and not internal gene duplications, are the pathways by which extant bacteria expand their metabolic capabilities (Gogarten *et al.*, 2002; Pal *et al.*, 2005; see the following discussion of proton and sodium pumping ATPases for an illustration).
- HGT of genes encoding the photosynthetic machinery, and in particular components of the reaction centre complexes, occur frequently. In some Proteobacteria, the genes encoding the photosynthetic machinery are clustered together in one region of the genome, and phylogenetic analyses revealed that this super cluster was transferred from alpha- to beta-proteobacteria (Igarashi *et al.*, 2001). Marine cyanophages carry genes encoding peptides integral to photosynthetic reaction centres, and the phylogeny of the phage and bacterial version of this enzyme is intertwined suggesting transfer in both directions (Lindell *et al.*, 2004; Mann *et al.*, 2003; Sharon *et al.*, 2009).
- The plurality consensus phylogeny, the phylogeny of the ribosome, and the phylogeny of enzymes encoding the chlorophyll biosynthetic pathway do not agree with each other, revealing that the photosynthetic machinery did not evolve exclusively through vertical inheritance, but included HGT (Raymond *et al.*, 2002, 2003;; Xiong and Bauer, 2002; Xiong *et al.*, 2000; Zhaxybayeva *et al.*, 2004). Furthermore, the two reaction centres in cyanobacteria are not closely related to each other (Sadekar *et al.*, 2006).

While debate continues on how the reaction centre dimers themselves evolved, and in which directions they were transferred, the autochthonous evolution of the two reaction centres within the cyanobacterial stem group without the involvement of HGT is unlikely. The synergistic union in ancestral cyanobacteria of pre-existing metabolic units containing the two reaction centres also is compatible with the rather long branch that connects the cyanobacteria to the other bacterial phyla, a pattern for the number of lineages through time that is not typical for other bacterial phyla (Gogarten and Lapierre, unpublished). The coming together of two different reaction centres in a single cell, and their subsequent evolution to work in series in a single electron transport chain, was a singular event in evolution. After it had occurred, electrons could be transported over larger electrochemical potential differences allowing for the use of water as electron donor. The ability to use a ubiquitous substrate as electron donor made the cyanobacteria an extraordinarily successful and diverse group.

*Acetoclastic methanogenesis in Methanosarcina*

In the case of the acquisition of genes innovating acetoclastic methanogenesis in *Methanosarcina*, the supportive phylogenetic inferences and the direction of the transfer are clearer as methanogenesis is restricted to a relatively narrow group of euryarchaeota (Fournier, 2009). Although this group, consisting of several classes of the phylum Euryarchaeota, is paraphyletic according to phylogenies of genes involved in transcription and translation (Bapteste *et al.*, 2005; Brochier *et al.*, 2005; Woese, 1987), the non-methanogens that would make it monophyletic (Thermoplasmatales, Archaeoglobales and Halobacteriales) are likely to have independently lost the core methanogenic genes. In addition, as these core genes have not been found in other lineages, the evolution of the core methanogenesis pathway appears to have been dominated by vertical descent (Bapteste *et al.*, 2005). In contrast, the history of the enzymes that allow methanogens to use different substrates (methylamines and acetate) involves HGT (Fournier, 2009; Fournier *et al.*, 2009).

*Methanosarcina* perform the transformation of acetate to acetyl-CoA in two steps using acetate kinase and phosphoacetyltransferase. The genes coding for these two enzymes are not homologous to any known archaeal gene, but are widely distributed among bacteria (Fournier and Gogarten, 2008). Phylogenetic analyses of these two enzymes suggest they were acquired by an ancestor of *Methanosarcina* in a single HGT event from a donor within the group of cellulolytic clostridia. HGT between methanogens and clostridia is not unprecedented (Beiko *et al.*, 2005) and modern representatives of the two groups can be found living in close proximity (Stams, 1994).

### *Evolution of $Na^+$ and $H^+$ pumping ATPases/ ATP synthase*

In prokaryotes, genes that are typically considered as paralogues are often created through gene transfer (Jeffery Lawrence, personal communication; Gogarten *et al.*, 2002). *Sensu stricto*, paralogues are genes that at the root of their phylogenetic relationship are related by a gene duplication event (Fitch, 2000); in contrast, genes related through a speciation are labelled as orthologues. One way for two homologous genes with different functions to end up in the same genome is through adaptation to a different function in two distinct lineages, followed by HGT of the then functionally distinct version into the same genome. In this scenario, the enzyme acquires a new function in isolation, thus avoiding problems caused through gene conversion and recombination. At this point the genes can be considered diverging orthologues. After one or both of the diverging genes have acquired a different function, the genes are so different in sequence that following an HGT event, they no longer undergo homologous recombination and, following non-homologous recombination, can co-exist in the same genome.

The ATP synthases/ATPases provide an illustration for this scenario. ATP synthases belong to a large group of multi-subunit enzymes that includes the bacterial coupling factor ATP synthase (also labelled as F-ATPases); the vacuolar ATPase found as an exclusively proton pumping ATPase in Eukaryotes energizing the endomembrane system (e.g. vacuoles and lysosomes), but also on the plasma membrane of specialized cells, such as osteoclasts and in cells of transport epithelia (Gogarten *et al.*, 1992; Harvey and Nelson, 1992); and as coupling factor ATPase in Archaea (Gogarten *et al.*, 1989a; Gogarten *et al.*, 1989b; Gogarten and Taiz, 1992). The latter ATPase type is often labelled as A-type ATPase, because it is mainly found in Archaea, where it has the same function as the F-type ATPase in bacteria (Gogarten *et al.*, 1989a; Gruber and Marshansky, 2008; Ihara and Mukohata, 1991). Both the F- and A-ATPases are reversible, and for both types different versions of the enzymes are known that transport protons or sodium ions (Dimroth *et al.*, 1999; Takase *et al.*, 1993; Yokoyama *et al.*, 1998). However, because A-ATPases are more similar in sequence to the V-ATPases, they are often included under the category of vacuolar type ATPases. Further complicating the terminology is the finding that both the bacterial and the archaeal type ATPase have been horizontally transferred between the domains (Hilario and Gogarten, 1993), e.g. *Methanosarcina* possess genes encoding an F-type ATPase in addition to its A-type ATPase (Sumi *et al.*, 1997); the Deinococcaceae only encode an A-type ATPase in their genome (Lapierre *et al.*, 2006; Olendzenski *et al.*, 2000; Tsutsumi *et al.*, 1991).

Other enzymes homologous to the ATP synthase are ATPases that are part of the type III secretion system, the bacterial flagella assembly machinery (Vogler *et al.*, 1991), and the rho transcription termination factor (Richardson, 2002). The latter functions as a helicase that unwinds the DNA–RNA heteroduplex. All these ATPases appear to function as rotary motors, consisting of a ring of six homologous subunits that rotate a central element, either a protein subunit, a protein substrate, or a nucleic acid strand (Abrahams *et al.*, 1994; Mulkidjanian *et al.*, 2007). Sequence similarity between the ATP-binding subunits is sufficient to establish homology unambiguously, but insufficient phylogenetic information has been retained in the sequences to reliably reconstruct the phylogenetic relations between the different enzymes; however, interesting scenarios have been developed that relate ion pumps and protein translocating ATPases (Kibak *et al.*, 1992; Mulkidjanian *et al.*, 2007).

Mulkidjanian *et al.* (2008) argue, based on phylogenetic reconstruction and protein structure, that the bacterial and archaeal coupling factor-type ATPases first evolved to transport sodium ions, and that the lineages independently evolved specificity for protons (protons usually have a much lower concentration than sodium ions, thus a higher discrimination between substrates is needed for enzymes to become specific for protons). In agreement with this hypothesis, some lineages that often are recovered as deep branching lineages in molecular phylogenies contain sodium ion translocating ATPases; for example, the Thermotogales (Ludwig *et al.*, 1998; Mulkidjanian *et al.*, 2008; Woese, 1987) use sodium ions to energize their plasma membrane (Galperin and Koonin, 1997) and possess a typical F-type ATPase (Ludwig *et al.*, 1998), whereas most Bacteria and Archaea use protons in energy coupling for ATP synthesis.

Some organisms, including the firmicute *Enterococcus hirae*, possess both an A-type ATPase with specificity for sodium (Takase *et al.*, 1994), and an additional F-ATPase that transports protons and is involved in pH regulation (Shibata *et al.*, 1992). A similar situation is found in *Thermotoga neapolitana* (Iida *et al.*, 2002), although the enzymes in *T. neapolitana* have not been characterized as well as in the case of *E. hirae*. Analysis of molecular phylogenies and of gene presence/absence data reveals that these F and A- type ATPase coexisting in the same cell did not evolve by gene duplication in the *E. hirae* or *T. neapolitana* lineages, respectively (Lapierre, 2007; Mulkidjanian *et al.*, 2008). Rather, these enzymes are homologues that have a long independent evolutionary history, and that were brought together in the same lineage through HGT. The presence of two homologous enzymes with different substrate specificity allows the recipient to use one or both of these ATPases for ion transport, and the other for ATP synthesis.

## Transferred genes as shared derived characters

It is perhaps counterintuitive that transferred genes also aid the reconstruction of organismal phylogeny. If a gene is transferred from a divergent donor and incorporated into the recipient in a way that places the gene under purifying selection, then the presence of the transferred gene can be considered a shared derived character for the recipient and its descendent (Huang and Gogarten, 2006). Only a small minority of transferred genes will attain the status of a shared derived character. Many transferred genes will be neutral or nearly neutral in the recipient (Gogarten and Townsend, 2005), or be under selection only for a short period of time (Lawrence and Roth, 1996), and may persist in the recipient lineage for less than a few hundred million years (Lawrence and Ochman, 2002). For a gene to survive in the recipient lineage on the long run, it must become essential for the recipient, either because it provides or contributes to a function that becomes essential for the recipient lineage (e.g. photosynthesis; Huang and Gogarten, 2007), or because it replaces the gene that was originally present in the recipient (e.g. the ATP synthase in Deinococcaceae; Olendzenski *et al.*, 2000). The assembly of new complex traits frequently uses genes that were acquired by gene transfer (Huang and Gogarten, 2008).

### HGT of aminoacyl tRNA synthetases

Aminoacyl tRNA synthetases (aaRS), i.e. the enzymes that charge the tRNAs with their cognate amino acids, often provide useful derived characteristics. These enzymes are essential in all living organisms, they are sufficiently conserved to allow phylogenetic reconstruction at the level of the whole net-of-life (Wolf *et al.*, 1999), and they often exist in different versions that have the same basic function (Farahi *et al.*, 2004), but might differ in sensitivity to antibiotic resistance. The fact that aaRSs only make specific interactions with few other macromolecules (i.e. the tRNAs), and that they are frequent targets of antibiotics (Brown *et al.*, 2003) might make them prone to HGT between divergent organisms (Woese *et al.*, 2000).

A possible complication of using genes as phylogenetically informative characters that resulted from orthologous replacement is that the replacement process is not instantaneous. The transferred gene is so divergent from its orthologue that it cannot be integrated into the recipient genome by homologous recombination.

Consequently, two versions of the gene coexist for some time in the recipient lineage, until one or the other is deleted (Stern *et al.*, in press). The ultimate deletion of one or the other gene likely occurs because functional redundancy reduces the level of purifying selection acting on the individual copy. Lineage splitting events that occurred during the time when both of the genes were still present in the recipient might not be correctly resolved (Huang and Gogarten, 2009).

The presence of a haloarchaeal tyrRS in all opisthokonts supports the animals and fungi as a monophyletic clade (Huang *et al.*, 2005). A thrRS in *Prochlorococcus* that was transferred from within the Gammaproteobacteria, supports the notion of *Prochlorococcus* as a monophyletic group (Huang and Gogarten, 2009; Zhaxybayeva *et al.*, 2006), which is in contrast to many phylogenies that are based on an average of the phylogenetic information contained in the genomes, which surprisingly recover *Prochlorococcus* as paraphyletic (e.g. Beiko *et al.*, 2005; Zhaxybayeva *et al.*, 2006; for further discussion see Zhaxybayeva *et al.*, 2009a). The presence of proline and alanine RSs in diplomonads and parabasalids that group with the homologue from *Nanoarchaeum equitans* suggests a closer than expected relationship between these two protist lineages (Andersson *et al.*, 2005).

## The origin of primary plastids

Chamydial-type genes were detected in several plant and algal lineages (Becker *et al.*, 2008; Brinkman *et al.*, 2002; Huang and Gogarten, 2007; Moustafa *et al.*, 2008). These genes often have functions in plastids (Brinkman *et al.*, 2002). Different scenarios were proposed, when these genes were first identified:

- Everett *et al.* (2005) suggested that these genes might have been acquired by plants through HGT from Chlamydiae via plant feeding insects; however, these genes are also found in red algae, and the split between red and green algae predates the existence insects. Furthermore, the plant homologues group within the Chlamydiae, often specifically with homologues from Protochlamydia, the chlamydial homologues do not group within the plants (Huang and Gogarten, 2007).
- Brinkman *et al.* (2002) proposed that these genes revealed a close relationship between Cyanobacteria and Chlamydiae, and that these genes were contributed to the plant and algal cells via the cyanobacterial endosymbiont that evolved into the plastid. However, when more of the chlamydial-type genes present in plants were analysed, cyanobacterial homologues were identified that did not group with the chlamydial homologues (Huang and Gogarten, 2007), illustrating that the chlamydial and plant homologues are distinct from the cyanobacterial ones.
- Huang and Gogarten (2007) suggested that a chlamydial parasite might have played a crucial role in establishing the cyanobacterial symbiosis that let to the evolution of the primary plastid.

To date, over 50 genes of chlamydial origin have been reported in plants, red algae and glaucocystophytes. In addition to these chlamydial genes, additional bacterial genes were found in red and green algae, plants, and in some instances glaucocystophytes (Huang and Gogarten, 2006, 2008; Tyra *et al.*, 2007). The presence of these transferred genes in all lineages that possess primary plastids supports a single origin of the primary plastid and suggests that these lineages form a monophyletic clade, the Archaeplastida (Adl *et al.*, 2005)

## Correlating evolution across the domains of life

In addition to defining clades in the net of life, genes transferred between divergent organisms also provide a means to correlate evolutionary events that occurred in different groups. The donor and recipient of an HGT event had to live either at the same time, or in the case of intermediate vectors or carriers, the donor might have existed before the recipient. Applied to the HGT of tyrRS from *Haloarcula* to the opisthokonts, we can conclude that the halophilic Archaea had already diversified into different groups, including *Haloarcula* as distinct from *Halobacterium*, before the opisthokont lineage split into animals and fungi. The grouping of mitochondrial and cyanobacterial proteins with the alphaproteobacterial and cyanobacterial lineage (Gray, 1993), respectively, and the presence of chlamydial type genes in all Archaeplastida (Adl *et al.*, 2005) reveals that the bacteria were already diversified into different phyla before the different kingdoms and phyla of eukaryotes evolved.

## Conclusions

Horizontal gene transfer is an innovative force in evolution that has changed the face of planet Earth. HGT impacts the creation and recognition of groups in different ways: gene transfer biased towards more related partners maintains and possibly creates groups of recognizably related organisms, but the same process also has the potential to create a phylogenetic signal that may be different from organismal shared ancestry; gene transfer between divergent organism can create new metabolic pathways and extend existing metabolic networks; and it can provide characters useful for a natural taxonomic classification. The examples discussed in this chapter illustrate that the pathways created through HGT have shaped Earth's biosphere. A more detailed mapping of HGT events will help to unravel the intertwined histories of organismal evolution and the evolution of biochemical and regulatory networks.

### Acknowledgements

The authors thank the National Science Foundation Assembling the Tree of Life program (DEB 0830024) for support.

### References

Abrahams, J.P., Leslie, A.G., Lutter, R., and Walker, J.E. (1994). Structure at 2.8 Å resolution of $F_1$-ATPase from bovine heart mitochondria. Nature *370*, 621–628.

Adl, S.M., Simpson, A., Farmer, M., Andersen, R., Anderson, O., Barta, J., Bowser, S., Brugerolle, G., Fensome, R., Fredericq, S., James, T.Y., Karpov, S., Kugrens, P., Krug, J., Lane, C.E., Lewis, L.A., Lodge, J., Lynn, D.H., Mann, D.G., McCourt, R.M., Mendoza, L., Moestrup, O., Mozley-Standridge, S.E., Nerad, T.A., Shearer, C.A., Smirnov, A.V., Spiegel, F.W., and Taylor, M.F. (2005). The new higher level classification of eukaryotes with emphasis on the taxonomy of protists. J. Eukaryot. Microbiol. 5, 399–451.

Andersson, J.O., Sarchfield, S.W., Roger, A.J., Sjogren, A.M., Davis, L.A., and Embley, T.M. (2005). Gene transfers from Nanoarchaeota to ancestor of Diplomonads and Parabasalids – phylogenetic analyses of diplomonad genes reveal frequent lateral gene transfers affecting eukaryotes. Mol. Biol. Evol. *22*, 85–90.

Ashlock, P.D. (1971). Monophyly and associated terms. Syst. Zool. *20*, 63–69.

Badger, M.R., and Price, G.D. (2003). $CO_2$ concentrating mechanisms in cyanobacteria: Molecular components, their diversity and evolution. J. Exp. Bot. *54*, 609–622.

Bapteste, E., Brochier, C., and Boucher, Y. (2005). Higher-level classification of the Archaea: evolution of methanogenesis and methanogens. Archaea *1*, 353–363.

Bapteste, E., O'Malley, M.A., Beiko, R.G., Ereshefsky, M., Gogarten, J.P., Franklin-Hall, L., Lapointe, F.J., Dupré, J., Dagan, T., Boucher, Y., and Martin, W. (2009). Prokaryotic evolution and the tree of life are two different things. Biol. Direct *4*, 34.

Becker, B., Hoef-Emden, K., and Melkonian, M. (2008). Chlamydial genes shed light on the evolution of photoautotrophic eukaryotes. BMC Evol. Biol. *8*, 203.

Beiko, R.G., Harlow, T.J., and Ragan, M.A. (2005). Highways of gene sharing in prokaryotes. Proc. Natl. Acad. Sci. U.S.A. *102*, 14332–14337.

Blankenship, R.E. (2001). Molecular evidence for the evolution of photosynthesis. Trends Plant Sci. *6*, 4–6.

Blankenship, R.E., Madigan, M.T., and Bauer, C.E. (1995). Anoxygenic Photosynthetic Bacteria (Dordrecht: Kluwer).

Boucher, Y., Douady, C.J., Papke, R.T., Walsh, D.A., Boudreau, M.E., Nesbø, C.L., Case, R.J., and Doolittle, W.F. (2003). Lateral gene transfer and the origins of prokaryotic groups. Annu. Rev. Genet. *37*, 283–328.

Boussau, B., Gueguen, L., and Gouy, M. (2008). Accounting for horizontal gene transfers explains conflicting hypotheses regarding the position of Aquificales in the phylogeny of Bacteria. BMC Evol. Biol. *8*, 272.

Breitbart, M., and Rohwer, F. (2005). Here a virus, there a virus, everywhere the same virus? Trends Microbiol. *13*, 278–284.

Brinkman, F.S., Blanchard, J.L., Cherkasov, A., Av-Gay, Y., Brunham, R.C., Fernandez, R.C., Finlay, B.B., Otto, S.P., Ouellette, B.F., Keeling, P.J., Rose, A.M., Hancock, R.E., Jones, S.J., and Greberg, H. (2002). Evidence that plant-like genes in *Chlamydia* species reflect an ancestral relationship between Chlamydiaceae, cyanobacteria, and the chloroplast. Genome Res. *12*, 1159–1167.

Brochier, C., Forterre, P., and Gribaldo, S. (2005). An emerging phylogenetic core of Archaea: phylogenies of transcription and translation machineries converge following addition of new genome sequences. BMC Evol. Biol. *5*, 36.

Brown, J.R., Gentry, D., Becker, J.A., Ingraham, K., Holmes, D.J., and Stanhope, M.J. (2003). Horizontal transfer of drug-resistant aminoacyl-transfer-RNA synthetases of anthrax and Gram-positive pathogens. EMBO Rep. *4*, 692–698.

Bryant, D.A., Garcia Costas, A.M., Maresca, J.A., Chew, A.G.M., Klatt, C.G., Bateson, M.M., Tallon, L.J., Hostetler, J., Nelson, W.C., Heidelberg, J.F., and Ward, D.M. (2007). *Candidatus* Chloracidobacterium thermophilum: An aerobic phototrophic acidobacterium. Science *317*, 523–526.

Buick, R. (1992). The antiquity of oxygenic photosynthesis: Evidence from stromatolites in sulfate-deficient archaean lakes. Science *255*, 74–77.

Buttner, M., Xie, D.L., Nelson, H., Pinther, W., Hauska, G., and Nelson, N. (1992). Photosynthetic reaction center genes in green sulfur bacteria and in photosystem 1 are related. Proc. Natl. Acad. Sci. U.S.A. *89*, 8135–8139.

Chao, L., and Cox, E.C. (1983). Competition between high and low mutating strains of *Escherichia coli*. Evolution *37*, 125–134.

Chisholm, S.W., Olson, R.J., Zettler, E.R., Goericke, R., Waterbury, J.B., and Welschmeyer, N.A. (1988). A novel free-living prochlorophyte abundant in the oceanic euphotic zone. Nature *334*, 340–343.

Costechareyre, D., Bertolla, F., and Nesme, X. (2009). Homologous recombination in *Agrobacterium*: Potential implications for the genomic species concept in bacteria. Mol. Biol. Evol. *26*, 167–176.

Creighton, H.B., and McClintock, B. (1931). A correlation of cytological and genetical crossing-over in *Zea mays*. Proc. Natl. Acad. Sci. U.S.A. *17*, 492–497.

Dagan, T., and Martin, W. (2007a). Ancestral genome sizes specify the minimum rate of lateral gene transfer during prokaryote evolution. Proc. Natl. Acad. Sci. U.S.A. *104*, 870–875.

Dagan, T., and Martin, W. (2007b). Testing hypotheses without considering predictions. Bioessays *29*, 500–503.

Daubin, V., and Ochman, H. (2004). Bacterial genomes as new gene homes: The genealogy of ORFans in *E. coli*. Genome Res. *14*, 1036–1042.

Davies, J. (1994). Inactivation of antibiotics and the dissemination of resistance genes. Science *264*, 375–382.

Dimroth, P., Wang, H., Grabe, M., and Oster, G. (1999). Energy transduction in the sodium F-ATPase of *Propionigenium modestum*. Proc. Natl. Acad. Sci. U.S.A. *96*, 4924–4929.

Donovan, S.E., Purdy, K.J., Kane, M.D., and Eggleton, P. (2004). Comparison of Euryarchaea strains in the guts and food-soil of the soil-feeding termite *Cubitermes fungifaber* across different soil types. Appl. Environ. Microbiol. *70*, 3884–3892.

Doolittle, W.F., and Zhaxybayeva, O. (2009). On the origin of prokaryotic species. Genome *19*, 744–756.

Doolittle, W.F. (1999). Phylogenetic classification and the universal tree. Science *284*, 2124–2129.

Duarte, N.C., Herrgard, M.J., and Palsson, B.O. (2004). Reconstruction and validation of *Saccharomyces cerevisiae* iND750, a fully compartmentalized genome-scale metabolic model. Genome Res. *14*, 1298–1309.

Dykhuizen, D.E., and Green, L. (1991). Recombination in *Escherichia coli* and the definition of biological species. J. Bacteriol. *173*, 7257–7268.

Everett, K.D., Thao, M., Horn, M., Dyszynski, G.E., and Baumann, P. (2005). Novel chlamydiae in whiteflies and scale insects: endosymbionts *'Candidatus* Fritschea bemisiae' strain Falk and *'Candidatus* Fritschea eriococci' strain Elm. Int. J. Syst. Evol. Microbiol. 55, 1581–1587.

Farahi, K., Pusch, G.D., Overbeek, R., and Whitman, W.B. (2004). Detection of lateral gene transfer events in the prokaryotic tRNA synthetases by the ratios of evolutionary distances method. J. Mol. Evol. *58*, 615–631.

Feil, E., Zhou, J., Smith, J.M., and Spratt, B.G. (1996). A comparison of the nucleotide sequences of the *adk* and *recA* genes of pathogenic and commensal *Neisseria* species: Evidence for extensive interspecies recombination within adk. J. Mol. Evol. *43*, 631–640.

Ferreira, K.N., Iverson, T.M., Maghlaoui, K., Barber, J., and Iwata, S. (2004). Architecture of the photosynthetic oxygen-evolving center. Science *303*, 1831–1838.

Ferry, J.G. (1992). Methane from acetate. J. Bacteriol. *174*, 5489–5495.

Fisher, R.A. (1930). Sexual reproduction and sexual selection. In: The Genetical Theory of Natural Selection (Oxford: Clarendon Press), pp. 121–145.

Fitch, W. (2000). Homology a personal view on some of the problems. Trends Genet. *16*, 227–231.

Fournier, G. (2009). Horizontal gene transfer and the evolution of methanogenic pathways. Meth. Mol. Biol. *532*, 163–179.

Fournier, G.P., and Gogarten, J.P. (2008). Evolution of acetoclastic methanogenesis in *Methanosarcina* via horizontal gene transfer from cellulolytic clostridia. J. Bacteriol. *190*, 1124–1127.

Fournier, G.P., Huang, J., and Gogarten, J.P. (2009). Horizontal gene transfer from extinct and extant lineages: biological innovation and the coral of life. Phil. Trans. R. Soc. Lond. B *364*, 2229–2239.

Fraser, C., Alm, E.J., Polz, M.F., Spratt, B.G., and Hanage, W.P. (2009). The bacterial species challenge: Making sense of genetic and ecological diversity. Science *323*, 741–746.

Galperin, M.Y., and Koonin, E.V. (1997). A diverse superfamily of enzymes with ATP-dependent carboxylate-amine/thiol ligase activity. Protein Sci. *6*, 2639–2643.

Gogarten, J.P. (1995). The early evolution of cellular life. Trends Ecol. Evol. *10*, 147–151.

Gogarten, J.P., and Olendzenski, L. (1999). Orthologs, paralogs and genome comparisons. Curr. Opin. Genet. Dev. *9*, 630–636.

Gogarten, J.P., and Taiz, L. (1992). Evolution of proton pumping ATPases: Rooting the tree of life. Photosynth. Res. *33*, 137–146.

Gogarten, J.P., and Townsend, J.P. (2005). Horizontal gene transfer, genome innovation and evolution. Nature Rev. Microbiol. *3*, 679–687.

Gogarten, J.P, Kibak, H., Dittrich, P., Taiz, L., Bowman, E., Bowman, B., Manolson, M., Poole, R., Date, T., Oshima, T., Konishi, J., Denda, K., and Yoshida, M. (1989a). Evolution of the vacuolar $H^+$-ATPase: implications for the origin of eukaryotes. Proc. Natl. Acad. Sci. U.S.A. *86*, 6661–6665.

Gogarten, J.P., Rausch, T., Bernasconi, P., Kibak, H., and Taiz, L. (1989b). Molecular evolution of $H^+$-ATPases. I. *Methanococcus* and *Sulfolobus* are monophyletic with respect to eukaryotes and eubacteria. Z. Naturforsch. *44*, 641–650.

Gogarten, J.P., Starke, T., Kibak, H., Fishman, J., and Taiz, L. (1992). Evolution and isoforms of V-ATPase subunits. J. Exp. Biol. *172*, 137–147.

Gogarten, J.P., Doolittle, W.F., and Lawrence, J.G. (2002). Prokaryotic evolution in light of gene transfer. Mol. Biol. Evol. *19*, 2226–2238.

Gogarten-Boekels, M., Olendzenski, L., and Gogarten, J.P. (2009). Horizontal Gene Transfer – Genomes in Flux (New York: Humana Press).

Gophna, U., Doolittle, W.F., and Charlebois, R.L. (2005). Weighted genome trees: refinements and applications. J. Bacteriol. *187*, 1305–1316.

Gray, M.W. (1993). Origin and evolution of organelle genomes. Curr. Opin. Genet. Dev. *3*, 884–890.

Gross, M.D., and Siegel, E.C. (1981). Incidence of mutator strains in *Escherichia coli* and coliforms in nature. Mutation Res. *91*, 107–110.

Gruber, G., and Marshansky, V. (2008). New insights into structure-function relationships between archaeal ATP synthase ($A_1A_0$) and vacuolar type ATPase ($V_1V_0$). Bioessays *30*, 1096–1109.

Harvey, W.R., and Nelson, N. (1992). V-ATPases. J. Exp. Biol., Vol 172 (Cambridge, U.K.: The Company of Biologists).

Hendrickson, H., and Lawrence, J.G. (2006). Selection for chromosome architecture in bacteria. J. Mol. Evol. *62*, 615–629.

Hennig, W. (1966). Phylogenetic Systematics (Urbana: University of Illinois Press).

Hilario, E., and Gogarten, J.P. (1993). Horizontal transfer of ATPase genes – the tree of life becomes a net of life. Biosystems *31*, 111–119.

Huang, J., and Gogarten, J.P. (2006). Ancient horizontal gene transfer can benefit phylogenetic reconstruction. Trends Genet. *22*, 361–366.

Huang, J., and Gogarten, J.P. (2007). Did an ancient chlamydial endosymbiosis facilitate the establishment of primary plastids? Genome Biol. *8*, R99.

Huang, J., and Gogarten, J.P. (2008). Concerted gene recruitment in early plant evolution. Genome Biol. *9*, R109.

Huang, J., and Gogarten, J.P. (2009). Ancient gene transfer as a tool in phylogenetic reconstruction. Meth. Mol. Biol. *532*, 127–139.

Huang, J., Xu, Y., and Gogarten, J.P. (2005). The presence of a haloarchaeal type tyrosyl-tRNA synthetase marks the opisthokonts as monophyletic. Mol. Biol. Evol. *22*, 2142–2146.

Igarashi, N., Harada, J., Nagashima, S., Matsuura, K., Shimada, K., and Nagashima, K.V. (2001). Horizontal transfer of the photosynthesis gene cluster and operon rearrangement in purple bacteria. J. Mol. Evol. *52*, 333–341.

Ihara, K., and Mukohata, Y. (1991). The ATP synthase of *Halobacterium salinarium* (*halobium*) is an archaebacterial type as revealed from the amino acid sequences of its two major subunits. Arch. Biochem. Biophys. *286*, 111–116.

Iida, T., Inatomi, K., Kamagata, Y., and Maruyama, T. (2002). F- and V-type ATPases in the hyperthermophilic bacterium *Thermotoga neapolitana*. Extremophiles *6*, 369–375.

Jain, R., Rivera, M.C., and Lake, J.A. (1999). Horizontal gene transfer among genomes: the complexity hypothesis. Proc. Natl. Acad. Sci. U.S.A. *96*, 3801–3806.

Jeffreys, A.J., and Neumann, R. (2002). Reciprocal crossover asymmetry and meiotic drive in a human recombination hot spot. Nature Genet. *31*, 267–271.

Kibak, H., Taiz, L., Starke, T., Bernasconi, P., and Gogarten, J.P. (1992). Evolution of structure and function of V-ATPases. J. Bioenerg. Biomembr. *24*, 415–424.

Kimura, M. (1967). On the evolutionary adjustment of spontaneous mutation rates. Genet. Res. *9*, 23–24.

Koizumi, Y., Takii, S., Nishino, M., and Nakajima, T. (2003). Vertical distributions of sulfate-reducing bacteria and methane-producing archaea quantified by oligonucleotide probe hybridization in the profundal sediment of a mesotrophic lake. FEMS Microbiol. Ecol. *44*, 101–108.

Kurland, C.G., Collins, L.J., and Penny, D. (2006). Genomics and the irreducible nature of eukaryote cells. Science *312*, 1011–1014.

Kuusinen, M. (1996). Cyanobacterial macrolichens on *Populus tremula* as indicators of forest continuity in Finland. Biol. Conserv. *75*, 43–49.

Lapierre, P. (2007). The Impact of Horizontal Gene Transfers on Prokaryotic Genome Evolution. PhD Thesis (Storrs: University of Connecticut).

Lapierre, P., and Gogarten, J.P. (2009). Estimating the size of the bacterial pan-genome. Trends Genet. *25*, 107–110.

Lapierre, P., Shial, R., and Gogarten, J.P. (2006). Distribution of F- and A/V-type ATPases in *Thermus scotoductus* and other closely related species. Syst. Appl. Microbiol. *29*, 15–23.

Lawrence, J.G., and Hendrickson, H. (2005). Genome evolution in bacteria: order beneath chaos. Curr. Opin. Microbiol. *8*, 572–578.

Lawrence, J.G., and Ochman, H. (1997). Amelioration of bacterial genomes: rates of change and exchange. J. Mol. Evol. *44*, 383–397.

Lawrence, J.G., and Ochman, H. (2002). Reconciling the many faces of lateral gene transfer. Trends Microbiol. *10*, 1–4.

Lawrence, J.G., and Roth, J.R. (1996). Selfish operons: horizontal transfer may drive the evolution of gene clusters. Genetics *143*, 1843–1860.

Lesser, M.P., Mazel, C.H., Gorbunov, M.Y., and Falkowski, P.G. (2004). Discovery of symbiotic nitrogen-fixing cyanobacteria in corals. Science *305*, 997–1000.

Levins, R. (1967). Theory of fitness in a heterogenous environment. VI. The adaptive significance of mutation. Genetics *56*, 163–178.

Liebl, U., Mockensturm-Wilson, M., Trost, J.T., Brune, D.C., Blankenship, R.E., and Vermaas, W. (1993). Single core polypeptide in the reaction center of the photosynthetic bacterium *Heliobacillus mobilis*: Structural implications and relations to other photosystems. Proc. Natl. Acad. Sci. U.S.A. *90*, 7124–7128.

Lindell, D., Sullivan, M.B., Johnson, Z.I., Tolonen, A.C., Rohwer, F., and Chisholm, S.W. (2004). Transfer of photosynthesis genes to and from *Prochlorococcus* viruses. Proc. Natl. Acad. Sci. U.S.A. *101*, 11013–11018.

Ludwig, W., Strunk, O., Klugbauer, S., Klugbauer, N., Weizenegger, M., Neumaier, J., Bachleitner, M., and

Schleifer, K.H. (1998). Bacterial phylogeny based on comparative sequence analysis. Electrophoresis *19*, 554–568.

Majewski, J., and Cohan, F.M. (1999). DNA sequence similarity requirements for interspecific recombination in *Bacillus*. Genetics *153*, 1525–1533.

Mann, N.H., Cook, A., Millard, A., Bailey, S., and Clokie, M. (2003). Bacterial photosynthesis genes in a virus. Nature *424*, 741.

Martin, W., Dagan, T., Koonin, E.V., Dipippo, J.L., Gogarten, J.P., and Lake, J.A. (2007). The evolution of eukaryotes. Science *316*, 542–543; author reply: Science *316*, 542–543.

Mayr, E. (1942). Systematics and the Origin of Species (New York: Columbia University Press).

Mayr, E. (1963). Animal species and evolution (Cambridge, MA: Harvard University Press).

Morandi, A., Zhaxybayeva, O., Gogarten, J.P., and Graf, J. (2005). Evolutionary and diagnostic implications of intragenomic heterogeneity in the 16S rRNA gene in *Aeromonas* strains. J. Bacteriol. *187*, 6561–6564.

Moustafa, A., Reyes-Prieto, A., and Bhattacharya, D. (2008). Chlamydiae has contributed at least 55 genes to Plantae with predominantly plastid functions. PLoS One *3*, e2205.

Mulkidjanian, A.Y., Koonin, E.V., Makarova, K.S., Mekhedov, S.L., Sorokin, A., Wolf, Y.I., Dufresne, A., Partensky, F., Burd, H., Kaznadzey, D., Haselkorn, R., and Galperin, M.Y. (2006). The cyanobacterial genome core and the origin of photosynthesis. Proc. Natl. Acad. Sci. U.S.A. *103*, 13126–13131.

Mulkidjanian, A.Y., Makarova, K.S., Galperin, M.Y., and Koonin, E.V. (2007). Inventing the dynamo machine: the evolution of the F-type and V-type ATPases. Nature Rev. Microbiol. *5*, 892–899.

Mulkidjanian, A.Y., Galperin, M.Y., Makarova, K.S., Wolf, Y.I., and Koonin, E.V. (2008). Evolutionary primacy of sodium bioenergetics. Biology Direct *3*, 13.

Nelson, K.E., Clayton, R.A., Gill, S.R., Gwinn, M.L., Dodson, R.J., Haft, D.H., Hickey, E.K., Peterson, J.D., Nelson, W.C., Ketchum, K.A., McDonald, L., Utterback T.R., Malek, J.A., Linher, K.D., Garrett, M.M., Stewart, A.M., Cotton, M.D., Pratt, M.S., Phillips, C.A., Richardson, D., Heidelberg, J., Sutton, G.G., Fleischmann, R.D., Eisen, J.A., White, O., Salzberg, S.L., Smith, H.O., Venter, J.C., and Fraser, C.M. (1999). Evidence for lateral gene transfer between archaea and bacteria from genome sequence of *Thermotoga maritima*. Nature *399*, 323–329.

Nesbø, C.L., L'Haridon, S., Stetter, K.O., and Doolittle, W.F. (2001). Phylogenetic analyses of two "archaeal" genes in *Thermotoga maritima* reveal multiple transfers between archaea and bacteria. Mol. Biol. Evol. *18*, 362–375.

Noll, K.M., and Thirangoon, K. (2009). Interdomain transfers of sugar transporters overcome barriers to gene expression. Meth. Mol. Biol. *532*, 309–322.

Normand, P., Lapierre, P., Tisa, L.S., Gogarten, J.P., Alloisio, N., Bagnarol, E., Bassi, C.A., Berry, A.M., Bickhart, D.M., Choisne, N., Couloux, A., Cournoyer, B., Cruveiller, S., Daubin, V., Demange, N., Francino, M.P., Goltsman, E., Huang, Y., Kopp, O.R., Labarre, L., Lapidus, A., Lavire, C., Marechal, J., Martinez, M., Mastronunzio, J.E., Mullin, B.C., Niemann, J., Pujic, P., Rawnsley, T., Rouy, Z., Schenowitz, C., Sellstedt, A., Tavares, F., Tomkins, J.P., Vallenet, D., Valverde, C., Wall, L.G., Wang, Y., Medigue, C., and Benson, D.R. (2007). Genome characteristics of facultatively symbiotic *Frankia* sp. strains reflect host range and host plant biogeography. Genome Res. *17*, 7–15.

Ochman, H., Lawrence, J.G., and Groisman, E.A. (2000). Lateral gene transfer and the nature of bacterial innovation. Nature *405*, 299–304.

Olendzenski, L., Liu, L., Zhaxybayeva, O., Murphey, R., Shin, D.G., and Gogarten, J.P. (2000). Horizontal transfer of archaeal genes into the Deinococcaceae: Detection by molecular and computer-based approaches. J. Mol. Evol. *51*, 587–599.

Olendzenski, L., Zhaxybayeva, O., and Gogarten, J. (2001). What's in a tree? Does horizontal gene transfer determine microbial taxonomy? In: Cellular Origin and Life in Extreme Habitats, J. Seckbach, ed. (Dordrecht: Kluwer Academic Publishers), pp. 67–78.

Oren, A. (2004). A proposal for further integration of the cyanobacteria under the bacteriological code. Int. J. Syst. Evol. Microbiol. *54*, 1895–1902.

Pal, C., Papp, B., and Lercher, M.J. (2005). Adaptive evolution of bacterial metabolic networks by horizontal gene transfer. Nature Genet. *37*, 1372–1375.

Papke, R.T. (2009). A critique of prokaryotic species concepts. Meth. Mol. Biol. *532*, 379–395.

Papke, R.T., Ramsing, N., Bateson, M., and Ward, D.M. (2003). Geographical isolation in hot spring cyanobacteria. Environ. Microbiol. *5*, 650–659.

Papke, R.T., Zhaxybayeva, O., Feil, E.J., Sommerfeld, K., Muise, D., and Doolittle, W.F. (2007). Searching for species in haloarchaea. Proc. Natl. Acad. Sci. U.S.A. *104*, 14092–14097.

Penny, D., and Poole, A. (1999). The nature of the last universal common ancestor. Curr. Opin. Genet. Dev. *9*, 672–677.

Poptsova, M.S., and Gogarten, J.P. (2007). The power of phylogenetic approaches to detect horizontally transferred genes. BMC Evol. Biol. *7*, 45.

Prangishvili, D., Albers, S.V., Holz, I., Arnold, H.P., Stedman, K., Klein, T., Singh, H., Hiort, J., Schweier, A., Kristjansson, J.K., and Zillig, W. (1998). Conjugation in Archaea: frequent occurrence of conjugative plasmids in *Sulfolobus*. Plasmid *40*, 190–202.

Puigbo, P., Wolf, Y.I., and Koonin, E.V. (2009). Search for a 'Tree of Life' in the thicket of the phylogenetic forest. J. Biol. *8*, 59.

Ragan, M.A. (2001). On surrogate methods for detecting lateral gene transfer. FEMS Microbiol. Lett. *201*, 187–191.

Raymond, J. (2008). Coloring in the tree of life. Trends Microbiol. *16*, 41–43.

Raymond, J., Zhaxybayeva, O., Gogarten, J.P., Gerdes, S.Y., and Blankenship, R.E. (2002). Whole-genome analysis of photosynthetic prokaryotes. Science *298*, 1616–1620.

Raymond, J., Zhaxybayeva, O., Gogarten, J.P., and Blankenship, R.E. (2003). Evolution of photosynthetic prokaryotes: a maximum-likelihood mapping approach. Phil. Trans. R. Soc. Lond. B *358*, 223–230.

Reed, J.L., Vo, T.D., Schilling, C.H., and Palsson, B.O. (2003). An expanded genome-scale model of *Escherichia coli* K-12 (iJR904 GSM/GPR). Genome Biol. *4*, R54

Retchless, A.C., and Lawrence, J.G. (2007). Temporal fragmentation of speciation in bacteria. Science *317*, 1093–1096.

Richardson, J.P. (2002). Rho-dependent termination and ATPases in transcript termination. Biochim. Biophys. Acta *1577*, 251–260.

Sadekar, S., Raymond, J., and Blankenship, R.E. (2006). Conservation of distantly related membrane proteins: photosynthetic reaction centers share a common structural core. Mol. Biol. Evol. *23*, 2001–2007.

Schleper, C., Roder, R., Singer, T., and Zillig, W. (1994). An insertion element of the extremely thermophilic archaeon *Sulfolobus solfataricus* transposes into the endogenous galactosidase gene. Mol. Gen. Genet. *243*, 91–96.

Schouls, L.M., Schot, C.S., and Jacobs, J.A. (2003). Horizontal transfer of segments of the 16S rRNA genes between species of the *Streptococcus anginosus* group. J. Bacteriol. *185*, 7241–7246.

Sharon, I., Alperovitch, A., Rohwer, F., Haynes, M., Glaser, F., Atamna-Ismaeel, N., Pinter, R.Y., Partensky, F., Koonin, E.V., Wolf, Y.I., Nelson, N., and Béjà, O. (2009). Photosystem I gene cassettes are present in marine virus genomes. Nature *461*, 258–262.

Sheppard, S.K., McCarthy, N.D., Falush, D., and Maiden, M.C. (2008). Convergence of *Campylobacter* species: implications for bacterial evolution. Science *320*, 237–239.

Shi, T., Bibby, T.S., Jiang, L., Irwin, A.J., and Falkowski, P.G. (2005). Protein interactions limit the rate of evolution of photosynthetic genes in cyanobacteria. Mol. Biol. Evol. *22*, 2179–2189.

Shibata, C., Ehara, T., Tomura, K., Igarashi, K., and Kobayashi, H. (1992). Gene structure of *Enterococcus hirae* (*Streptococcus faecalis*) $F_1F_0$-ATPase, which functions as a regulator of cytoplasmic pH. J. Bacteriol. *174*, 6117–6124.

Sørensen, K., Canfield, D., Teske, A., and Oren, A. (2005). Community composition of a hypersaline endoevaporitic microbial mat. Appl. Environ. Microbiol. *71*, 7352–7365.

Sorek, R., Zhu, Y., Creevey, C.J., Francino, M.P., Bork, P., and Rubin, E.M. (2007). Genome-wide experimental determination of barriers to horizontal gene transfer. Science *318*, 1449–1452.

Stams, A.J. (1994). Metabolic interactions between anaerobic bacteria in methanogenic environments. Antonie van Leeuwenhoek *66*, 271–294.

Stern, A., Mayrose, I., Penn, O., Shaul, S., Gophna, U., and Pupko, T. (In press). An evolutionary analysis of lateral gene transfer in thymidylate synthase enzymes. Syst. Biol.

Suerbaum, S., Maynard Smith, J., Bapumia, K., Morelli, G., Smith, N.H., Kunstmann, E., Dyrek, I., and Achtman, M. (1998). Free recombination within *Helicobacter pylori*. Proc. Natl. Acad. Sci. U.S.A. *95*, 12619–12624.

Sumi, M., Yohda, M., Koga, Y., and Yoshida, M. (1997). $F_0F_1$-ATPase genes from an archaebacterium, *Methanosarcina barkeri*. Biochem. Biophys. Res. Commun. *241*, 427–433.

Swithers, K.S., Gogarten, J.P., and Fournier, G.P. (2009). Trees in the web of life. J. Biol. *8*, 54.

Taddei, F., Radman, M., Maynard-Smith, J., Toupance, B., Gouyon, P.H., and Godelle, B. (1997). Role of mutator alleles in adaptive evolution. Nature *387*, 700–702.

Takase, K., Yamato, I., and Kakinuma, Y. (1993). Cloning and sequencing of the genes coding for the A and B subunits of vacuolar-type $Na^+$-ATPase from *Enterococcus hirae*. Coexistence of vacuolar- and $F_0F_1$-type ATPases in one bacterial cell. J. Biol. Chem. *268*, 11610–11616.

Takase, K., Kakinuma, S., Yamato, I., Konishi, K., Igarashi, K., and Kakinuma, Y. (1994). Sequencing and characterization of the *ntp* gene cluster for vacuolar-type $Na^+$-translocating ATPase of *Enterococcus hirae*. J. Biol. Chem. *269*, 11037–11044.

Tettelin, H., Masignani, V., Cieslewicz, M.J., Donati, C., Medini, D., Ward, N.L., Angiuoli, S.V., Crabtree, J., Jones, A.L., and Durkin, A.S. (2005). Genome analysis of multiple pathogenic isolates of *Streptococcus agalactiae*: implications for the microbial "pan-genome". Proc. Natl. Acad. Sci. U.S.A. *102*, 13950–13955.

Tsutsumi, S., Denda, K., Yokoyama, K., Oshima, T., Date, T., and Yoshida, M. (1991). Molecular cloning of genes encoding major two subunits of a eubacterial V-type ATPase from *Thermus thermophilus*. Biochim. Biophys. Acta *1098*, 13–20.

Tyra, H.M., Linka, M., Weber, A.P., and Bhattacharya, D. (2007). Host origin of plastid solute transporters in the first photosynthetic eukaryotes. Genome Biol. *8*, R212.

van Berkum, P., Terefework, Z., Paulin, L., Suomalainen, S., Lindstrom, K., and Eardly, B.D. (2003). Discordant phylogenies within the rrn loci of rhizobia. J. Bacteriol. *185*, 2988–2998.

Vogler, A.P., Homma, M., Irikura, V.M., and Macnab, R.M. (1991). *Salmonella typhimurium* mutants defective in flagellar filament regrowth and sequence similarity of FliI to $F_0F_1$, vacuolar, and archaebacterial ATPase subunits. J. Bacteriol. *173*, 3564–3572.

Wang, Y., and Zhang, Z. (2000). Comparative sequence analyses reveal frequent occurrence of short segments containing an abnormally high number of non-random base variations in bacterial rRNA genes. Microbiology *146*, 2845–2854.

Waterbury, J.B., Watson, S.W., Guillard, R.R.L., and Brand, L.E. (1979). Widespread occurrence of a unicellular, marine, planktonic, cyanobacterium. Nature *277*, 293–294.

Welch, R.A., Burland, V., Plunkett, G., 3rd, Redford, P., Roesch, P., Rasko, D., Buckles, E.L., Liou, S.R., Boutin, A., Hackett, J., Stroud, D., Mayhew, G.F., Rose, D.J., Zhou, S., Schwartz, D.C., Perna, N.T., Mobley, H.L., Donnenberg, M.S., and Blattner, F.R. (2002). Extensive mosaic structure revealed by the complete genome sequence of uropathogenic *Escherichia coli*. Proc. Natl. Acad. Sci. U.S.A. *99*, 17020–17024.

Wilkinson, M., McInerney, J.O., Hirt, R.P., Foster, P.G., and Embley, T.M. (2007). Of clades and clans: terms for phylogenetic relationships in unrooted trees. Trends Ecol. Evol. *22*, 114–115.

Woese, C.R. (1987). Bacterial evolution. Microbiol. Rev. *51*, 221–271.

Woese, C.R., Olsen, G.J., Ibba, M., and Söll, D. (2000). Aminoacyl-tRNA synthetases, the genetic code, and the evolutionary process. Microbiol. Mol. Biol. Rev. *64*, 202–236.

Wolf, Y.I., Aravind, L., Grishin, N.V., and Koonin, E.V. (1999). Evolution of aminoacyl-tRNA synthetases – analysis of unique domain architectures and phylogenetic trees reveals a complex history of horizontal gene transfer events. Genome Res. *9*, 689–710.

Wolf, Y.I., Rogozin, I.B., Kondrashov, A.S., and Koonin, E.V. (2001). Genome alignment, evolution of prokaryotic genome organization, and prediction of gene function using genomic context. Genome Res. *11*, 356–372.

Wolf, Y.I., Rogozin, I.B., Grishin, N.V., and Koonin, E.V. (2002). Genome trees and the tree of life. Trends Genet. *18*, 472–479.

Xiong, J., and Bauer, C.E. (2002). Complex evolution of photosynthesis. Annu. Rev. Plant. Biol. *53*, 503–521.

Xiong, J., Fischer, W.M., Inoue, K., Nakahara, M., and Bauer, C.E. (2000). Molecular evidence for the early evolution of photosynthesis. Science *289*, 1724–1730.

Yap, W.H., Zhang, Z., and Wang, Y. (1999). Distinct types of rRNA operons exist in the genome of the actinomycete *Thermomonospora chromogena* and evidence for horizontal transfer of an entire rRNA operon. J. Bacteriol. *181*, 5201–5209.

Yokoyama, K., Muneyuki, E., Amano, T., Mizutani, S., Yoshida, M., Ishida, M., and Ohkuma, S. (1998). V-ATPase of *Thermus thermophilus* is inactivated during ATP hydrolysis but can synthesize ATP. J. Biol. Chem. *273*, 20504–20510.

Zhaxybayeva, O. (2009). Detection and quantitative assessment of horizontal gene transfer. Meth. Mol. Biol. *532*, 195–213.

Zhaxybayeva, O., and Gogarten, J.P. (2007). Horizontal gene transfer, gene histories and the root of the tree of life. In: Astrobiology and the Origins of Life, R.E. Pudritz, P.G. Higgs, and J. Stone, eds. (Cambridge, UK: Cambridge University Press).

Zhaxybayeva, O., Hamel, L., Raymond, J., and Gogarten, J. (2004). Visualization of the phylogenetic content of five genomes using dekapentagonal maps. Genome Biol. *5*, R20.

Zhaxybayeva, O., Lapierre, P., and Gogarten, J.P. (2005). Ancient gene duplications and the root(s) of the tree of life. Protoplasma *227*, 53–64.

Zhaxybayeva, O., Gogarten, J.P., Charlebois, R.L., Doolittle, W.F., and Papke, R.T. (2006). Phylogenetic analyses of cyanobacterial genomes: Quantification of horizontal gene transfer events. Genome Res. *16*, 1099–1108.

Zhaxybayeva, O., Nesbø, C., and Doolittle, W.F. (2007). Systematic overestimation of gene gain through false diagnosis of gene absence. Genome Biol. *8*, 402.

Zhaxybayeva, O., Doolittle, W.F., Papke, R.T., and Gogarten, J.P. (2009a). Intertwined evolutionary histories of marine *Synechococcus* and *Prochlorococcus marinus*. Genome Biol. Evol. *1*, 325–339.

Zhaxybayeva, O., Swithers, K., Lapierre, P., Fournier, G., Bickhart, D., DeBoy, R., Nelson, K., Nesbo, C., Doolittle, W., Gogarten, J., and Noll, K.M. (2009b). On the chimeric nature, thermophilic origin, and phylogenetic placement of the Thermotogales. Proc. Natl. Acad. Sci. U.S.A. *106*, 5865–5870.

Zhou, J., Bowler, L.D., and Spratt, B.G. (1997). Interspecies recombination, and phylogenetic distortions, within the glutamine synthetase and shikimate dehydrogenase genes of *Neisseria meningitidis* and commensal *Neisseria* species. Mol. Microbiol. *23*, 799–812.

Zwart, G., Crump, B.C., Kamst-van Agterveld, M.P., Hagen, F., and Han, S.K. (2002). Typical freshwater bacteria: An analysis of available 16S rRNA gene sequences from plankton of lakes and rivers. Aquat. Microb. Ecol. *28*, 141–155.

Zwirglmaier, K., Jardillier, L., Ostrowski, M., Mazard, S., Garczarek, L., Vaulot, D., Not, F., Massana, R., Ulloa, O., and Scanlan, D.J. (2008). Global phylogeography of marine *Synechococcus* and *Prochlorococcus* reveals a distinct partitioning of lineages among oceanic biomes. Environ. Microbiol. *10*, 147–161.

# Endosymbiosis and the Evolution of Plastids

# 10

Christopher E. Lane and Dion G. Durnford

Abstract

Photosynthesis is one of the most successful energy production strategies on the planet and has been co-opted numerous times throughout evolutionary history via the uptake and retention of photosynthetic cells by non-photosynthetic eukaryotic heterotrophs. Whereas the result of this process is clear, what is not settled is the mode and tempo of plastid movement among eukaryotes, particularly plastids of red algal derivation. Recent changes in our understanding of the relationships between eukaryotic supergroups have only served to complicate the picture further. In this chapter we summarize current information on the evolution of plastids with a particular focus on relationships among photosynthetic eukaryotes, the process of endosymbiogenesis and the variation in ways plastids have been modified to suit the light harvesting needs of their hosts. The understanding of all of these factors is an active field of continued research that will undoubtedly lead to further discoveries in the coming years.

## Introduction

The ability of organisms to harvest sunlight for energy is undoubtedly one of the most important evolutionary landmarks in the history of our planet. Not only was the evolution of photosynthesis the arrival of a novel life strategy, but it played a central role in transforming the planet's atmosphere to an oxygen-rich environment over 3 billion years ago, spurring the processes that have resulted in the diversity of life seen today. When most people think of photosynthesis, they immediately associate the term with land plants and the chloroplast. However, single-celled organisms, on land and in aquatic environments, perform the majority of eukaryotic photosynthesis using a diversity of photosynthetic organelles more broadly referred to as 'plastids'. Whereas all plastids are derived from a single ancient fusion of a cyanobacterium-like prokaryote and a eukaryotic cell, plastids have evolved via a contorted path among eukaryotes in a tale of enslavement, theft and transfer.

There are many examples of endosymbiotic associations where hosts have acquired photosynthetic partners that are useful in providing examples of the metabolic dependency of the different systems. There are even curious examples of 'kleptoplasty' where hosts have internalized only the plastids in an example of partial digestion following feeding (Rumpho *et al.*, 2006). This was observed in a variety of protist systems (Gustafson *et al.*, 2000), sea slugs (Green *et al.*, 2000), and even benthic foraminiferans that exist below the photic zone (Bernhard and Bowser, 1999; Grzymski *et al.*, 2002). In the latter case, though the plastids are photosynthetically active in the light, it seems unlikely that plastid retention is for the purpose of photosynthesis. Grzymski *et al.* (2002) suggested that inorganic nitrogen assimilation may be an important role. Thus, the metabolic rationale for establishing endosymbioses is not necessarily limited to the milking of carbon metabolites from photosynthesis though it is a significant outcome.

## Endosymbiogenesis: primary versus secondary plastids

### Primary plastids

Originally a prokaryotic innovation, the ability to harvest solar light energy in a usable form for the cell, has been adopted by eukaryotic cells. Rather than reinvent the machinery *de novo*, photosynthesis was acquired by eukaryotes via the enslavement of a Gram-negative cyanobacterial cell (Cavalier-Smith and Lee, 1985; Cavalier-Smith, 2000). Undoubtedly, cyanobacteria were a major food source of early eukaryotic cells, but in what must have been an exceedingly rare event, an endosymbiotic relationship between a photosynthetic prokaryote and a eukaryote developed. The uptake and retention of a cyanobacterium by a heterotrophic eukaryote, termed Primary Endosymbiosis (Fig. 10.1), established photosynthesis in eukaryotes and led to the diversification of the three lineages: green algae (including land plants), red algae and the less widely known glacophytes (Rodriguez-Ezpeleta *et al.*, 2005; Deschamps *et al.*, 2008). All three lineages contain plastids characterized by two membranes, derived from the plasma membrane and outer membrane of the cyanobacterium (Cavalier-Smith and Lee, 1985; Cavalier-Smith, 2000).

Based on phylogenetic evidence, glacophytes most likely diverged as an independent lineage before red and green algae, roughly 550 million years ago (mya) (Yoon *et al.*, 2004). This hypothesis is also supported by the similarity of glaucophyte plastids (sometimes called cyanelles) to extant cyanobacteria. Among the three primary plastid-containing lineages, only glaucophytes produce a peptidoglycan layer between the two plastid membranes (Bhattacharya and Schmidt, 1997). Both glaucophytes and red algae retain phycobilisomes, which can be found in free-living

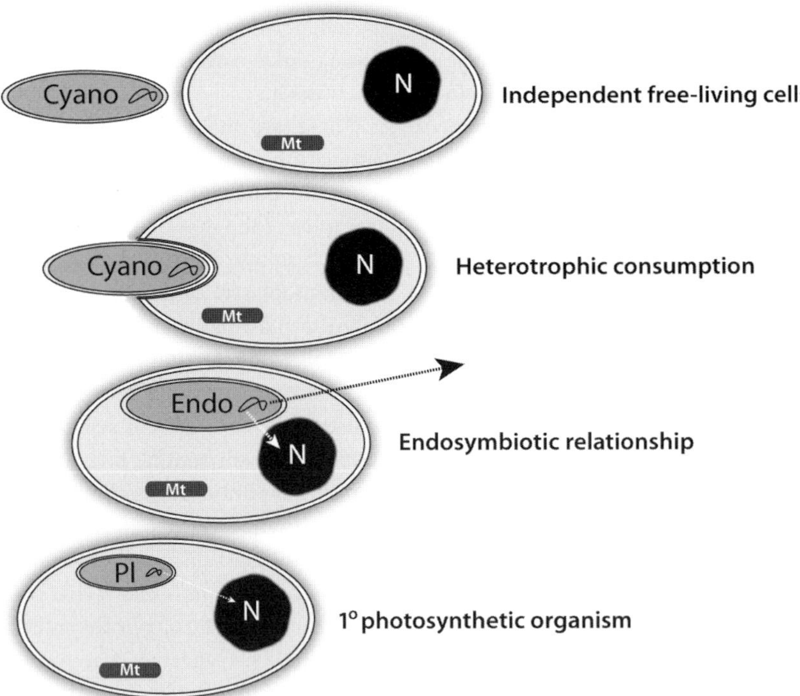


**Figure 10.1** Primary endosymbiosis is the process by which a cyanobacterial cell (Cyano) is taken up by a heterotrophic eukaryotic cell, but not digested. Over time, the cells form an endosymbiotic relationship and the genome of the endosymbiont both transfers genes to the host (white arrow) and loses DNA (black arrow). Once the endosymbiont is fully integrated into the cell as the plastid, DNA transfer continues at a reduced rate.

cyanobacteria (Steiner *et al.*, 2005) but green algae and land plants have a more derived and complex light harvesting apparatus (Durnford *et al.*, 1999).

Evidence for a single acquisition of the primary plastid has been demonstrated using a variety of molecular phylogenetic studies from both nuclear (e.g. Hackett *et al.*, 2007) and plastid genes (e.g. Rodriguez-Ezpeleta *et al.*, 2005), as well as components of the Calvin cycle machinery (Reyes-Prieto and Bhattacharya, 2007). The weight of evidence falls very strongly in favour of a single primary plastid acquisition, but there are alternative viewpoints (Stiller *et al.*, 2003), whose proponents argue that multiple origins should not be entirely discounted. Due to the roughly 1–1.5 billion years since the primary plastid was first captured and the exchange of genes that has occurred between plastid and host, a definitive answer has been elusive. The consensus view, however, supports a single origin (Delwiche, 1999; Palmer, 2003; Keeling, 2004; Deschamps *et al.*, 2008; Archibald, 2009) with the exception of the case of *Paulinella chromatophora*.

*P. chromatophora* is the only photosynthetic species among the filose amoeba (Burki *et al.*, 2009), but its photosynthetic organelle (chromatophore) resembles a modern *Prochlorococcus*-like cell more than it does a modern plastid (Marin *et al.*, 2005; Yoon *et al.*, 2006b). Although the chromatophore carries out similar functions for the cell as the typical plastid, the genome of the chromatophore is only marginally reduced compared to modern cyanobacteria (Yoon *et al.*, 2006b) and was clearly derived from a separate and much more recent event than the primary plastid of glacophytes, red and green algae. The fact that *Paulinella ovalis*, the sister species of *P. chromatophora*, is non-photosynthetic but a voracious predator of cyanobacteria (Johnson *et al.*, 1988), lends further credence to a modern primary endosymbiosis in the making.

### Secondary plastids

The lineages that were spawned following the primary endosymbiosis do not account for the extent of photosynthetic taxa on Earth. Plastids have been laterally transferred among eukaryotic lineages by way of a heterotrophic eukaryote engulfing a photosynthetic eukaryote, without ingesting it (Fig. 10.2; Palmer, 2003; Archibald and Keeling, 2005). Remarkably, the incorporation of one eukaryotic cell into another has happened multiple times in the course of evolution, with both red and green algae having been utilized at least once to turn a heterotrophic lineage into a photosynthetic one (Fig. 10.3). The question of how many times a primary alga has been converted from free-living cell into a secondary plastid is heavily debated (e.g. Bhattacharya *et al.*, 2004; Archibald and Keeling, 2005; Bodyl, 2005; Reyes-Prieto *et al.*, 2007; Archibald, 2009; Bodyl *et al.*, 2009; Keeling, 2009; Lane and Archibald, 2009; Stiller *et al.*, 2009), but there is consensus that green algae have been taken in as plastids at least twice and that red algae have been converted to a plastid at least once. The debate surrounding the timing and number of secondary plastid acquisitions centres on the difficulty in understanding the relationships between the host lineages.

## Red 2° plastids – the evolving Chromalveolata

The evolution of red algal-derived secondary plastids is far more controversial than their green algal cousins. The most influential hypothesis describing the process that led to the pattern of red secondary plastids is the 'chromalveolate hypothesis', advanced by Cavalier-Smith (1999b). In essence, he argues that all lineages that have a red secondary plastid, and their immediate non-photosynthetic relatives, are derived from a common ancestor. At its inception, Chromalveolata was organized into two main groups, alveolates (apicomplexans, ciliates and dinoflagellates) and Chromista (cryptophytes, haptophytes and stramenopiles [= heterokonts]) (Cavalier-Smith, 1999c, 2003). This grouping of taxa has remained one of the most controversial eukaryotic 'supergroups' since it was first described, primarily based on lack of molecular evidence from phylogenetic studies using nuclear genes. It is possible to construct an overlapping character set that supports the group (Keeling, 2009), but no single character or molecular phylogeny based on nuclear genes has convincingly done so.

The justification for this hypothesis was based mainly on the theoretical difficulty of acquiring a plastid versus losing one. As is the case for the

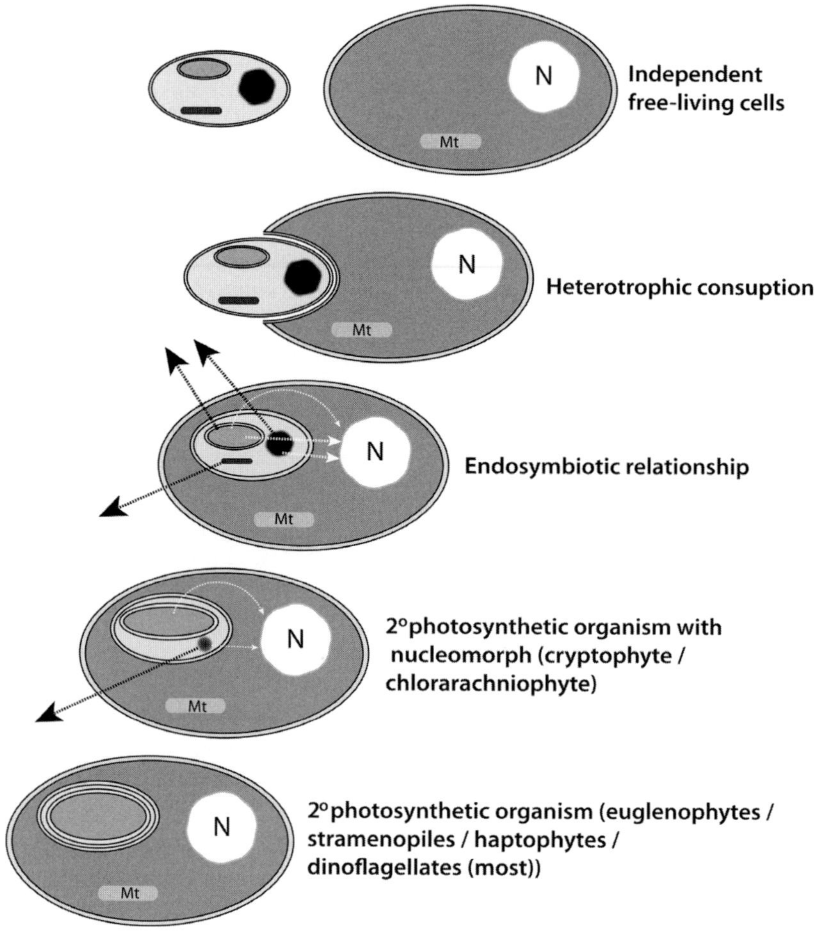

**Figure 10.2** Secondary endosymbiosis is the fusion of two eukaryotic cells that results from a heterotrophic cell taking in a photosynthetic one that escapes digestion. Like primary endosymbiosis (Figure 10.1), the relationship between the two cells results in massive DNA loss (black arrows) and transfer to the host (white arrows). In the case of secondary endosymbionts, some DNA transferred to the secondary host may be the result of transfer from the primary plastid to the primary nucleus. Both cryptophytes and chlorarachniophytes maintain a relic algal nucleus in the form of the nucleomorph.

primary plastids discussed above, the conversion of a free-living organism into an organelle is a complex problem, made even more challenging in the case of secondary plastids by the complexity of eukaryotic cells. Whereas the primary plastid event consisted of the reduction of a single genome and the transfer of thousands of genes to the primary host nucleus, secondary plastids are the remnants of eukaryotic cells, which already contained a nucleus, plastid and mitochondrion. In all cases, the formerly red algal mitochondrion has been lost, and in most cases (discussed below), so has the entire red algal nucleus with its full complement of nuclear genes. Many of those formerly red algal genes encode proteins required for plastid function, meaning that they had to be replaced or transferred to the secondary host nucleus before the red algal nucleus could be eliminated. Additionally, a membrane transport system had to be invented to navigate host nucleus-encoded proteins through the three or four membranes of secondary plastid and into the appropriate compartments of the endosymbiont (discussed below). The complexity of this process, compared with the relative theoretical ease of losing a plastid and returning to a heterotrophic

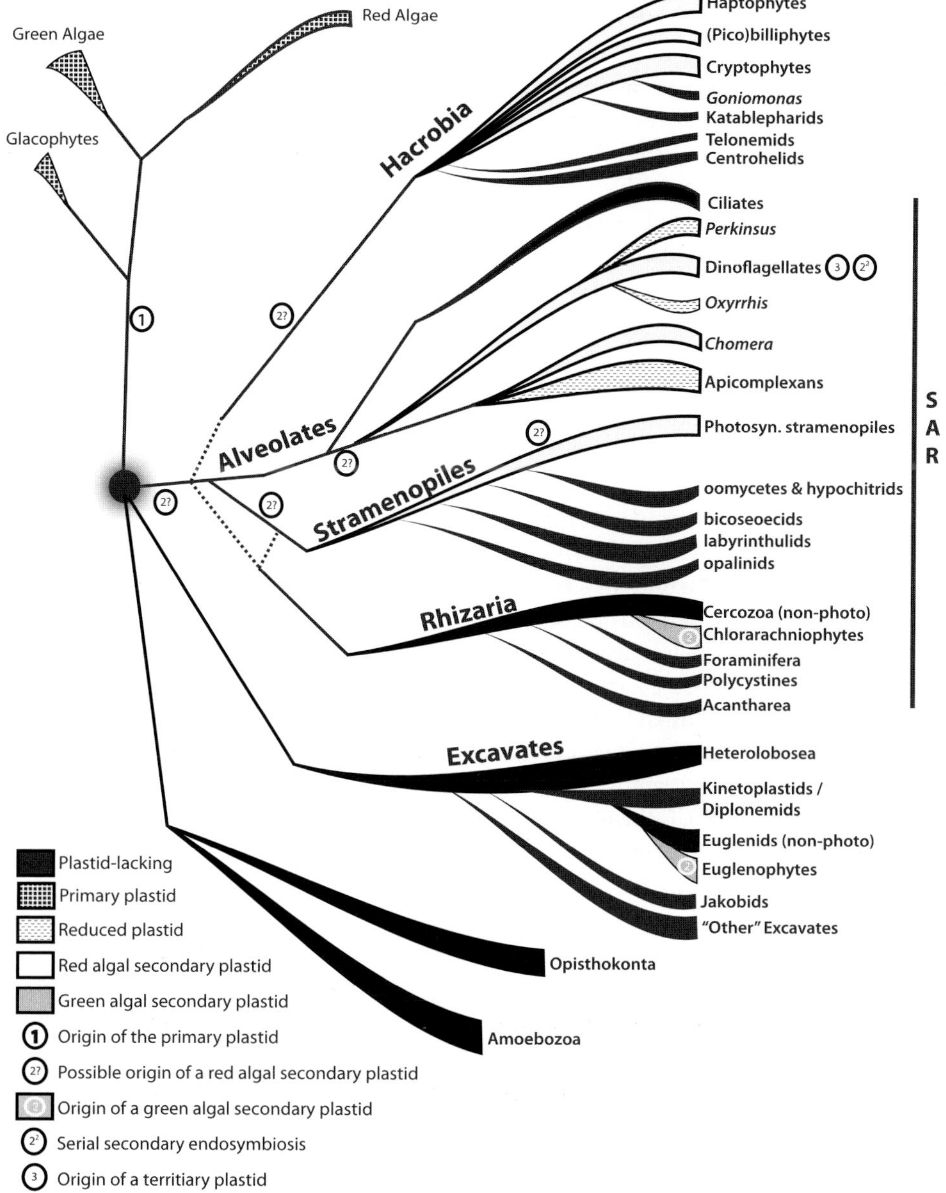


**Figure 10.3** An illustration of phylogenetic relationships between major eukaryotic lineages. The dashed lines represent ambiguous phylogenetic placement. Possible origins of red secondary plastids are indicated, with the origin closest to the eukaryotic root (black circle) representing the event proposed by the chromalveolate hypothesis.

life style, was the primary justification for the chromalveolate group.

In 2007, three independent papers added a new wrinkle to the chromalveolate story by uniting stramenopiles with the Alveolata, to the exclusion of haptophytes and cryptophtes, which formed an independent clade (Burki *et al.*, 2007; Hackett *et al.*, 2007; Patron *et al.*, 2007). Far more shocking, however, was the resolution of the supergroup Rhizaria nested within the chromalveolates (Burki *et al.*, 2007; Hackett *et al.*, 2007), forming the 'SAR' clade of stramenopiles, alveolates and Rhizaria (discussed below). Prior to these studies, the Rhizaria had been an independent lineage of mostly non-photosynthetic single-celled organisms, with the one plastid-bearing group being the chlorarachniophytes. This novel topology has subsequently been resolved by additional data (Burki *et al.*, 2009; Hampl *et al.*, 2009), significantly complicating the plastid evolution picture. In the following sections we will describe the chromalveolate lineages and the arguments for and against their common ancestry. Before delving into the changing theoretical framework, however, it is important to introduce the players.

## Hacrobia

The recently proposed group Hacrobia, is comprised of two well-known photosynthetic lineages (cryptophytes and haptophytes) and several relatively unstudied heterotrophic lineages (kataplepharids, centrohelid heliozoa and telonemids) (Burki *et al.*, 2009; Okamoto *et al.*, 2009). An additional photosynthetic lineage that was not formally included in Hacrobia, but has been shown to be related to the cryptophyte/haptophyte clade (Not *et al.*, 2007), is the biliphytes.

Molecular evidence for a cryptophyte/haptophyte clade began to appear in 2007 (Burki *et al.*, 2007; Hackett *et al.*, 2007; Patron *et al.*, 2007) and has continued to be a feature of large-scale eukaryotic phylogenies since that time (Burki *et al.*, 2009; Hampl *et al.*, 2009). Even prior to multigene phylogenies resolving the sister relationship between haptophytes and cryptophytes, a relationship between the lineages was suggested based on the unique shared presence of a bacterial-derived rpl36 encoded in the plastid (Rice and Palmer, 2006). The bacterial-type rpl36 in cryptophytes and haptophytes has apparently replaced the canonical copy in the plastids of these groups, providing evidence (independent of phylogenetic support) for a common origin of their plastids.

### *Cryptophytes*

Cryptophytes are single-celled organisms that inhabit both marine and freshwater environments and can be readily identified using light microscopy by their asymmetric shape, two flagella of unequal length and prominent ejectosomes, which are extrusive organelles that line the gut furrow (Butcher, 1967). The majority of cryptophytes are photosynthetic, but some members of the genus *Cryptomonas* contain a non-photosynthetic plastid (leucoplast) (Sepsenwol, 1973) and the earliest-diverging members of the lineage, *Goniomonas* spp., show no evidence of a plastid remnant (Mignot, 1965). Species of *Goniomonas* are also unlike other cryptophytes in their locomotion, which tends to be gliding along surfaces rather than the more typical spiral swimming style of canonical cryptophytes.

Photosynthetic cryptophytes contain one or two plastids that are typically parietal and can contain a central pyrenoid. The plastid is surrounded by four membranes, the outer most of which is continuous with the nuclear membrane and endoplasmic reticulum (Cavalier-Smith, 1999a). The two outer plastid membranes are derived from the plasma membrane of the secondary endosymbiont, whereas the inner membrane pair are derived from the cyanobacterial progenitor of the plastid (Palmer, 2003; Archibald and Keeling, 2005; Archibald, 2009). Between these two pairs of membranes is, arguably, the most concrete evidence for secondary endosymbiosis — the nucleomorph.

### *The cryptophyte nucleomorph*

The discovery of the nucleomorph (Greenwood, 1974; Greenwood *et al.*, 1977) was critical in validating the theory of secondary plastid acquisition proposed in the 1970s (Taylor, 1974; Gibbs, 1978). The space between the second and third membranes of the four membrane-bound cryptophyte plastid (Fig. 10.2) corresponds to the former cytosolic space of the red alga (McFadden, 1993) that was taken up and transformed over time into the 2° plastid. Within that space resides an organelle, which is the remnant of the red algal nucleus. This miniaturized nucleus retains all the characteristics of a canonical nucleus (Archibald, 2007), but is heavily reduced in size and coding capacity. Much like the reduction of the cyanobacterially derived genome that we recognize as the plastid genome, the nuclear genome of the organisms that were the progenitors of secondary plastids have been reduced, and in most cases, lost entirely. Of all the secondary plastid-containing lineages, evidence for an extant nucleomorph exists for only cryptophytes and chlorarachniophytes (discussed below). There is evidence that the recently described picobiliphytes may also contain a nucleomorph (Not *et al.*, 2007), but this remains unconfirmed.

Nucleomorph genomes are the smallest nuclear-derived genomes known, with sizes ranging from 450 to 845 kbp (Archibald, 2007). Curiously, all nucleomorph genomes are arranged in three chromosomes (Rensing *et al.*, 1994; Douglas *et al.*, 2001; Lane and Archibald, 2006; Lane *et al.*, 2006, 2007; Archibald, 2007), but the significance of this observation is unknown. A common feature of nucleomorph chromosomes is that ribosomal DNA (rDNA) cistrons are found at each end of all three chromosomes (Archibald, 2007), with the exception of nucleomorphs in the genus *Hemiselmis* (Lane *et al.*, 2007; Lane and Archibald, 2008b). Cryptophyte nucleomorph genomes are on the larger side of the size range, but have gene densities similar to those of prokaryotic genomes, typically ~1 gene/kbp (Archibald, 2007). Intergenic space is so reduced in the nucleomorph of *Guillardia theta* that some genes overlap (Douglas *et al.*, 2001). In addition to reduced intergenic space, gene sequences are also reduced when compared to orthologues in red algal nuclei (Lane *et al.*, 2007). This phenomenon appears to be the result of both 3′ and 5′ truncation as well as the loss of DNA between functional domains (Lane *et al.*, 2007).

Genome compaction has not only affected nucleomorph gene size and density, but also the size and number of introns (Gilson, 2001; Archibald, 2007). In the cryptophyte *G. theta*, only 17 introns persist in the nucleomorph and none is larger than 52 bp (Douglas *et al.* 2001; Archibald 2007). Surprisingly, in the distantly related *Hemiselmis andersenii*, introns have been lost entirely from the nucleomorph genome, along with the machinery to process them (Lane *et al.*, 2007). The *H. andersenii* nucleomorph is the only known instance of introns being lost from a nuclear genome.

In terms of gene content, nucleomorph genomes consist of mostly 'housekeeping' genes necessary for the persistence and function of the nucleomorph (Gilson, 2001). The gene content is not enough, however, for the nucleomorph to be self-sufficient and a large number of host nuclear-encoded proteins are imported from the cytosol (Gould *et al.*, 2007). From this perspective, the continued existence of the cryptophyte nucleomorph seems counter-intuitive, particularly given that all other plastids derived from a red algal secondary endosymbiosis have lost their nucleomorphs. The hypothesis for why nucleomorphs still exist revolves around the 30 nucleomorph-encoded genes whose products are targeted to the plastid (Gilson, 2001). As long as these genes that encode essential plastid proteins are not transferred to the nucleus or functionally replaced by nuclear genes, the nucleomorph's role is critical for the cell.

*Haptophytes*

Haptophytes (= Coccolithophorids) are united by the presence of a unique structure, termed the haptonema (Andersen, 2004). From a gross morphology perspective, although it has been secondarily lost in some genera, the haptonema appears very similar to the two unequal flagella of haptophytes, but the microtubules are arranged as 6 or 7 single microtubules surrounded by three concentric membranes, unlike the typical 9 + 2 structure found in most flagella. The function of the haptonema is also dissimilar to that of flagella, in that it appears to have a role in attachment and feeding, rather than locomotion (Manton, 1964) and may be coiled when the cell is motile.

The haptophytes are environmentally important because they construct scales of calcium carbonate that surround their cell walls. The scales are routinely shed, and coupled with cell death, are a significant component of deep-sea $CaCO_3$ deposits. The famous 'White Cliffs of Dover' are principally made of $CaCO_3$ chalk resulting from the settling of coccolithophores. In the right environmental conditions, haptophytes can produce blooms on a spatial and density scale that rival those of 'red tide' producing dinoflagellates (Jordan and Chamberlain, 1997). Haptophytes are an exceptionally diverse and abundant lineage in marine environments that may be essential drivers of the global carbon pump (Liu *et al.*, 2009).

*Recent additions*

The remaining lineages in the Hacrobia are poorly studied and until very recently (Burki *et al.*, 2009; Okamoto *et al.*, 2009), the phylogenetic placement of each was unclear. In fact the biliphytes (= picobiliphytes) are only known from sequences resulting from environmental studies and fluorescence in situ hybridization (FISH) images (Not *et al.*, 2007). The scarcity of biliphyte cells in the environment and their minute size (2 x 6 μm) has kept them from being cultured or studied more intently, but the limited evidence available suggests they may contain a cryptophyte-like nucleomorph (Not *et al.*, 2007). The phylogenetic position of telonemids was conflicting in several studies (Klaveness *et al.*, 2005; Shalchian-Tabrizi *et al.*, 2006, 2007), and the centrohelids were never placed with any certainty, despite some morphological characters linking them to haptophytes (Nikolaev *et al.*, 2004; Cavalier-Smith and von der Heyden, 2007). Phylogenies based on over one hundred nuclear genes have brought some resolution to the position of these enigmatic taxa (Burki *et al.*, 2009; Okamoto *et al.*, 2009), but it may be too early to say for certain in these under-sampled groups.

## Alveolates

The Alveolata was first proposed in 1991 (Cavalier-Smith, 1991) and encompasses the apicomplexans, dinoflagellates and ciliates. Like all members of the Chromalveolata, the group was proposed to have a photosynthetic ancestor despite the lack of a plastid in many of its members, including the entire ciliate lineage. Despite the diversity of form and function of these cells, evidence for the monophyly of Alveolata based on cellular features (such as 'alveoli', membranous sacs found beneath the plasma membrane and the feature from which the group derives its name) and nuclear gene phylogenies is strong (Fast *et al.*, 2002; Harper *et al.*, 2005; Burki *et al.*, 2007). Phylogenies also indicate a closer relationship between apicomplexans and dinoflagellates, to the exclusion of ciliates, and the recent placement of early diverging members of the first two groups has revealed a number of characters bridging them (see below).

### *Apicomplexans*

This lineage is a large and diverse group made up alsmost entirely of unicellular obligate parasites, the most well-known and -studied being *Plasmodium falciparum*, the parasite responsible for malaria. Little is known about the evolutionary relationships within Apicomplexa (Leander, 2007; Morrison, 2009) because classification of these parasites was intended to be utilitarian rather than reflective of their evolutionary relationships (Morrison, 2009). Two features that all apicomplexans share, however, is an unusual cell invasion structure known as the 'apical complex' (from which the group derives its name) and the recently described cortical proteins, dubbed 'alveolins' (Gould *et al.*, 2008). Though the apical complex is shared by the parasitic relatives of apicomplexans, such as the colpodellids and perkinsids, the cytoskeletal structure of the apical complex is entirely closed in the apicomplexans and not in the other groups (Perkins *et al.*, 2002).

Apicomplexans harbour a second unusual feature as well: the apicoplast. This organelle has been shown to be a relic plastid, derived from secondary endosymbiosis (Marechal and Cesbron-Delauw, 2001; Ralph *et al.*, 2001) that is no longer photosynthetic. Instead, the apicoplast has a role in fatty acid and heme biosynthesis and is critical for host invasion (Ralph *et al.*, 2001). The genome of the apicoplast is highly reduced, having lost numerous genes for photosynthesis (Wilson *et al.*, 1996). However, the apicoplast is surrounded by four membranes, typical for secondarily derived plastids. The discovery of the photosynthetic sister taxon to apicomplexans, *Chromera velia* (Moore *et al.*, 2008), has provided a critical link in connecting the apicoplast to other 2° plastids of red algal origin and particularly to dinoflagellates (Keeling, 2008).

### *Dinoflagellates*

From an evolutionary perspective, dinoflagellates seem to know no boundaries. Their diversity of form and function spans an enormous range and the have evolved complex structures that are analogous to features of animals (e.g. highly derived photoreception and defence structures that function like nematocysts). At the molecular level, the nucleus of dinoflagellates divides by a unique form of mitosis (Bhaud *et al.*, 2000) and gene products are processed via the *trans*-splicing of a splice leader to all mRNA products (Lidie and Van Dolah, 2007). The nuclear genomes of dinoflagellates are some of the largest known, lack a canonical histone structure and are permanently condensed in a liquid crystal structure (Wong and Kwok, 2005). In contrast, the mitochondrial and plastid genomes are highly reduced in gene content and arranged on hundreds of single-gene circular chromosomes (Jackson *et al.*, 2007; Waller and Jackson, 2009).

The diversity of plastid types found among dinoflagellates is also unusual and unparalleled. Roughly half of dinoflagellates are photosynthetic, although almost all of the plastid-bearing species are mixotrophic. Inasmuch as the word 'typical' can be used in reference to dinoflagellates, the typical plastid found in this group is a peridinin-containing plastid. Peridinin is a diatom-specific carotenoid pigment, but additional pigments, such as chlorophylls *a* and $c_2$, are also present. Despite a 2° origin of the peridinin plastid, three membranes surround the organelle instead of the four membranes found surrounding other plastids of secondary red algal origin. Furthermore, dinoflagellates are apparently able to swap out their peridinin plastids for other plastid types.

Four different dinoflagellate genera contain plastids acquired either by tertiary uptake (from a secondary alga) or serial secondary transfer (replacement of the peridinin plastid with that of a green alga). The dinoflagellate *Karenia brevis* houses a plastid derived from a haptophyte (Ishida and Green, 2002), *Dinophysis* has taken on a cryptophyte plastid (Hackett *et al.*, 2003) and *Karlodinium micrum* harbours a plastid of heterokont origin (Yoon *et al.*, 2002). In contrast to taking up other secondary alga, *Lepidodinium* has incorporated a green algal symbiont (Schnepf and Elbrachter, 1999). There are no other described cases of these phenomenon, although some models of plastid evolution invoke these mechanisms to explain the presence of a red algal plastid in 'chromalveolate' taxa without assuming a common ancestor of the group (Bodyl, 2005). Dinoflagellates appear to be unique, however, in their ability to change plastid types and use existing cellular machinery to run the novel plastid. Perhaps the highly reduced nature of the peridinin plastid genome and the fact that the vast majority of plastid function genes are encoded in the dinoflagellate nucleus allows dinoflagellates to exchange plastids more easily than other taxa.

*Emerging plastids*

Among the most persuasive arguments for the chromalveolate hypothesis are the new discoveries that continue to expand the range of chromalveolate taxa with a 2° red plastid (or at least a small number of 2° plastid acquisitions), strengthening the proposal of a common origin over multiple gains. In this sense, the description of the plastid-containing relative of the apicomplexans, *Chromera velia*, formed a critical link between apicomplexans and dinoflagellates (Moore *et al.*, 2008).

The connection between the canonical dinoflagellate plastid and the apicoplast was difficult to make based on gene sequences, owing to their independent specializations. There are no shared genes between the highly reduced genome of the non-photosynthetic apicoplast and the small complement of genes present on the minicircles that make up the dinoflagellate plastid genome, so direct phylogenetic comparisons can not be made. However, the *C. velia* plastid genome contains a typical set of plastid genes, providing data that independently overlaps with both the peridinin plastid and apicoplast gene complement, thus making it possible to compare the unusual plastids of dinoflagellates and apicomplexans (Keeling, 2008). These data provide the first unequivocal link between the plastids of these two groups of unusual organisms, indicating that their divergent plastids do indeed share a common ancestor.

At the same time that *C. velia* was being described, evidence for plastids in early-diverging dinoflagellates was also accumulating. Two independently diverging lineages of non-photosynthetic dinoflagellates make up the genera, *Oxyrrhis* and *Perkinsus*. Species of *Oxyrrhis* are voracious marine planktonic consumers and perkinsids are parasites of shellfish that annually cause millions of dollars worth of loss, in the U.S. alone, for the aquaculture industry. No evidence of a plastid was known for either group, and only their phylogenetic location (Fig. 10.3) provided any indication that a plastid might exist in these organisms. A putative plastid was first identified in *Perkinsus alanticus* (Teles-Grilo *et al.*, 2007a,b) based on electron micrographs and growth inhibition studies. Plastid biosynthesis pathways were later identified in *P. marinus*, coupled with the localization of the enzyme, 1-deoxy-D-xylulose 5-phosphate reductoisomerase, to a small organelle associated with the mitochondria (Matsuzaki *et al.*, 2008). DNA appears to be lacking from this organelle (Matsuzaki *et al.*, 2009). Likewise, a genome-containing plastid in *Oxyrrhis* has not been identified, but genes derived from a plastid possessing N-terminal 5′ targeting signals typical of the plastid-localized products have been found (Slamovits and Keeling, 2008). Combined, the available data are strong evidence for a plastid in the common ancestor of apicomplexans and dinoflagellates.

### *Ciliates*

Ciliates are non-photosynthetic unicellular organisms that are typically covered in cilia, from which they derive their name. They are omnipresent in aquatic environments, often harbour ecto- or endosymbionts and many are opportunistic pathogens, some of which infect humans. Research on ciliates, *Tetrahymena thermophila* in particular, has contributed enormously to scientific progress. A few of the important discoveries that have resulted from research on *T. thermophila* include the discovery of telomeres and telomerase, cell cycle control mechanisms, purification of the cytoskeleton motor protein dynein and Nobel prize-winning research on catalytic RNA (see Collins and Gorovsky, 2005).

Ciliates are also known for the complexity of their nuclear genomes. Two nuclei exist in their cells, a larger and transcriptionally active somatic 'macronucleus' and the condensed germline micronucleus. During cell division the polyploid macronucleus dissolves and the daughter cell nuclei are formed by karyogamy of the 'micronucleus'. The macronucleus is then reconstructed by genome amplification resulting in 60–1000 copies of each chromosome (Parfrey *et al.*, 2008), though chromosome fragmentation is prevalent (Katz, 2001). During sexual reproduction ciliates undergo meiosis, but they commonly reproduce asexually via an amitotic process that does not involve spindle formation. The result of this process, coupled with enormous levels of polyploidy is that macronuclear content in asexually reproducing strains can vary widely between individuals producing non-mendelian inheritance (Robinson and Katz, 2007) and elevated evolutionary rates of proteins (Zufall *et al.*, 2006).

## Stramenopiles (Heterokonta)

The most diverse lineages from the perspective of form and function, stramenopiles are united by their two, unequal flagella and the presence of the pigments chlorophyll *a* and *c* (Lee, 2008). The anterior-directed flagellum is typically lined with tripartite projections called mastigonemes, whereas the posterior flagellum is usually lined with less complex tubular hairs and used in a whiplash-like motion. There is little debate that the group is monophyletic, but relationships between stramenopile lineages are not as well resolved (Andersen, 2004; Cavalier-Smith and Chao, 2006). Both plastid-bearing and plastid-lacking taxa are encompassed within the group, and the non-photosynthetic lineages are believed to have diverged earlier than those with a plastid.

With the exception of the oomycetes (a group that includes pathogens which case diseases such the Irish Potato Blight and Sudden Oak Death), little is known about the plastid-lacking stamenopiles. Labyrinthulids are marine fungus-like decomposers, a handful of which are opportunistic or obligate pathogens (Leander and Porter, 2001). Opalanids live exclusively in the hindguts of amphibians and appear to be parasitic or commensal with their host (Delvinquier and Patterson, 1993), though this relationship is not well understood. Similarly, *Blastocystis* has been linked to some intestinal disorders in humans, but is not clearly the causative agent (Stenzel and Boreham, 1996). A common thread among these groups, however, is their parasitic or saprobic lifestyle.

The photosynthetic stramenopiles include more familiar groups, such as the ubiquitous diatoms and brown algae, which populate every marine coastal area from pole to pole. Brown algae have diversified into a broad array of multicellular morphologies, from simple filaments to 50 m kelp, with complex nutrient transport cells analogous to phloem in land plants (Phillips *et al.*, 2008). With the exception of two genera of xanthophytes (the sister group to brown algae), however, all other photosynthetic stramenopiles are single celled organisms with few external morphological features to differentiate them (Andersen,

2004). For instance, the classes Pelagophyceae (Andersen *et al.*, 1993) and Pinguiophyceae (Kawachi *et al.*, 2002) were described only after molecular sequencing techniques gave some phylogenetic structure to the myriad of 'little brown balls' in culture at various institutions.

### D) Rhizaria

The Rhizaria were initially proposed based on the majority of members possessing either reticulopodia or ventral, branched pseudopods (Cavalier-Smith, 2002). Other than these characters, few cellular features unite these diverse organisms that include cercozoans, foraminiferans, polycistines and chlorarachniophytes. Nevertheless, molecular phylogenies consistently confirm the monophyly of the lineage (Burki and Pawlowski, 2006). The only photosynthetic groups within Rhizaria are *Paulinella chromatophora* (an emerging plastid; see above) and the chlorarachniophytes, which have obtained their plastid via the secondary endosymbiosis involving a green alga (discussed below). However, this group has taken on new significance in regard to the evolution of plastids in recent years as its phylogenetic position has been refined.

#### *The 'SAR' clade*

In 2007, three papers were concurrently published by different labs that caused a dogmatic shift in eukaryotic phylogenetics. All three robustly demonstrated (Burki *et al.*, 2007; Hackett *et al.*, 2007; Patron *et al.*, 2007), based on multigene phylogenetic analysis, the relationship between the cryptophytes and haptophytes to the exclusion of other chromalveolate taxa. This relationship forms the basis of the Hacrobia group described above and had not previously been strongly resolved. More surprisingly, however, the two analyses that included members of Rhizaria (Burki *et al.*, 2007; Hackett *et al.*, 2007) produced support for the Rhizaria nested within the chromalveolate group with the stramenopiles and Alveolates, forming the SAR clade (Burki *et al.*, 2007). Not only was this relationship never considered based on the morphology of the cells, but it has major consequences for the interpretation of plastid gains and losses across eukaryotic lineages. Based on the almost entire lack of plastids within the group, the Rhizaria is hypothesized to be ancestrally plastid-lacking, and if the group has diverged from within the Chromalveolata, would represent another instance of plastid loss within the expanded chromalveolate lineage. The significance of this has yet to be determined because the prevalence of plastid loss in nature is not clear.

## Green 2° plastids

Secondary plastids derived from green algae are less prevalent than their red cousins and their number of origins less contentious. Only two unrelated lineages, a subclade of the Euglenozoa and the chlorarachniophytes, contain green 2° plastids. These are clearly derived from independent events based on their distant phylogenetic placement in the Excavata and Rhizaria, respectively (e.g. Keeling *et al.*, 2005).

### Euglenophytes

This subclade of Euglenozoa is a monophyletic assemblage of 2° photosynthetic organisms that have clearly derived from a phagotrophic ancestor (Leander, 2004). The green algal origin of the euglenid plastid is well established, both biochemically and based on sequence data (Gibbs, 1978; Hallick *et al.*, 1993) and is surrounded by three membranes. The photosynthetic euglenophytes share several derived cellular features, in comparison to their non-photosynthetic relatives, which are putatively the result of their mode of nutrition (Leander, 2004). Reduced feeding apparatus, flagellar adjustments that allow for swimming instead of gliding and convergence on a hardened pellicle all appear to be adaptations related to the acquisition of the plastid (Leander, 2004).

Euglenophytes typically contain multiple plastids with prominent pyrenoids and the plastid membrane is not continuous with the nuclear membrane (van Dooren *et al.*, 2001). The primary photosynthetic pigments in euglenophytes are chlorophylls *a* and *b* and phototaxis is achieved using a carotenoid-based stigma that shades a photoreceptive swelling, located at the base of the dorsal flagellum (Kuznicki *et al.*, 1990).

### Chlorarachniophytes

Chlorarachniophytes are single-celled photosynthetic organisms that take on a variety of forms. Their morphology can range from flagellated coccoid cells to the amoeboid form from which their name is derived (Moestrup and Sengco, 2001). In amoeboid form, the cells extend multiple cytoplasmic projections that serve to capture and ingest prey in a web-like fashion. Despite adaptations to a heterotrophic lifestyle, the chlorarachniophytes are the only photosynthetic lineage among the diverse Rhizaria group (with the exception of *Paulinella chromatophora*). The chlorarachniophyte plastid is surrounded by four membranes and is derived from a green alga (McFadden *et al.*, 1994). Much like the red alga-derived plastid of cryptophytes, the chlorarachniophyte plastid contains a relic nucleus within the space that was once the green algal cytosol (Gilson, 2001; Archibald, 2007).

The genome of the chlorarachniophyte nucleomorph has been sequenced from *Bigellowiella natans* and shows both amazing convergence with the unrelated cryptophyte nucleomorph and many unique features. Both types of nucleomorph have three similarly sized chromosomes, almost all of which are capped by ribosomal cistrons (Douglas *et al.*, 2001; Gilson *et al.*, 2006; Archibald, 2007; Lane *et al.*, 2007). Unlike the cryptophyte nucleomorph that contains a small number of introns (Douglas *et al.*, 2001), or none at all (Lane *et al.*, 2007) the chlorarachniophyte nucleomorph genome is littered with miniaturized introns between 17 and 22 base pairs in size (Gilson *et al.*, 2006; Archibald, 2007). In a gene-by-gene comparison, the two nucleomorph types encode only a handful of shared homologues, but possess genes of similar functional categories that appear to persist for the purpose of expressing the nucleomorph genes that are plastid-targeted (Gilson *et al.*, 2006). Why these genes have not been transferred or replaced by nuclear genes and the nucleomorph lost entirely remains a mystery, but it has been hypothesized that it is merely a matter of time before nucleomorphs are eliminated in cryptophytes and chlorarachniophytes, as they have been in all other lineages that contain a secondary plastid (Gilson *et al.*, 2006).

## Endosymbiosis and its genomic effects

Implicit in all of the evolutionary scenarios listed above and any other model of plastid transfer among eukaryotes, is the genetic transfer that would accompany such a transfer. As previously discussed for both 1° and 2° plastids, the coordination of the host and symbiont requires massive loss or transfer of genetic material. As demonstrated in *Arabidopsis* (Martin *et al.*, 2002), the extent of endosymbiotic gene transfer (EGT) can be large (18% of the host genome) and transferred genes can take on a wide array of functions. In some cases, transferred genes can even replace the host copy entirely, as has been shown for phosphoglycerate-kinase (PGK) in land plants (Brinkmann and Martin, 1996). Two copies of PGK exist in land plants; one that is plastid-encoded and functions in the Calvin cycle and another that is nuclear-encoded and is involved in cytosolic glycolysis. Phylogenetic analysis of the two PGK copies indicates that both are derived from cyanobacteria, indicating that the cyanobacterial copy has been transferred to the host and replaced the native copy (Brinkmann and Martin, 1996).

In the case of endosymbiotic replacement of PGK, the identification of the transfer is fairly straightforward because prokaryotic genes can retain the phylogenetic signal of their origin for a long period of time since the divergence between donor and host is so great. But what of gene transfers and replacements between two eukaryotic nuclei? In cases of 2° plastids, the endosymbiont was subsumed, nucleus and all. As the endosymbiont nucleus was reduced to extinction (except for nucleomorphs), the opportunities for the transfer of red or green algal genes to the host nucleus would have been enormous. In particular, genes whose products were targeted to the 1° plastid in red or green algae would already have an N-terminal signal, making transfer from the 1° host nucleus to the 2° host nucleus and targeting back to the (now) 2° plastid potentially easier.

Reports of red algal-derived gene transfers to the host nucleus do exist for multiple lineages of chromalveolates (Armbrust *et al.*, 2004; Waller *et al.*, 2006; Nosenko and Bhattacharya, 2007; Bhattacharya and Nosenko, 2008) and endosymbiotic gene transfer has been invoked to explain these. Based on the chromalveolate hypothesis, the nuclear genomes of plastid-lacking lineages, such as ciliates and oomycetes, have been searched for algal-derived genes (Tyler *et al.*, 2006; Reyes-Prieto *et al.*, 2008). The assumption is that endosymbiotic gene transfer should be detectable in lineages that have a photosynthetic ancestry, even if present day species lack a plastid. These studies have turned up algal genes, but the significance of these examples is not clear. In fact, in the case of oomycetes, there is a question as to whether the 'algal' genes detected in the *Phytophthora* genomes (Tyler *et al.*, 2006) are significant at all. Stiller *et al.* (2009) used contingency tests to determine whether the genes reported by Tyler *et al.* (2006) represent an unusual contribution from a red alga, using diatoms as a control group, and determined that alternative explanations such as horizontal transfer were more plausible.

Although the Stiller *et al.* study is the most statistically rigorous examination of the 'algal' genes in a plastid-lacking member of the hotly debated chromalveolate group and forces the reconsideration of the *a posteriori* assumption that any algal genes in oomycetes or ciliates are evidence of a former plastid, the debate is far from settled. Taxonomic sampling of both chromalveolate taxa (especially plastid-lacking members) and red algae remains weak. The only example of a ciliate genome is that of *Tetrahymena brucei* (Eisen *et al.*, 2006) and the examples of oomycete genomes are limited to the derived genus *Phytophthora* (Tyler *et al.*, 2006; Haas *et al.*, 2009), despite oomycetes being an exceedingly diverse lineage. The situation on the red algal side is worse (Lane and Archibald, 2008a). Nuclear data from two red algal genomes exists, *Galdieria sulphuraria* (see the *Galdieria sulphuraria* Genome Project, http://genomics.msu.edu/galdieria/about.html) and *Cyanidioschyzon merolae* (Matsuzaki *et al.*, 2004), but the red algae chosen for sequencing were selected for the size of their genomes rather than their appropriateness as representatives of the lineage. In fact, both organisms belong to the subphylum Cyanidophytina (Yoon *et al.*, 2006a), an anciently diverging and poorly studied lineage with only four known species. Both taxa are adapted to extreme environments and have unusual highly reduced genomes. The more complete of the two, *Cyanidioschyzon merolae* encodes only ~5600 genes (Matsuzaki *et al.*, 2004) – a fraction of most eukaryotic genomes. Therefore, the identification of transferred red algal genes in plastid-lacking members of the chromalveolate group is hindered by poor taxon sampling of both host and endosymbiont groups and further complicated by lineage-specific variation in evolutionary rates (Rodriguez-Ezpeleta *et al.*, 2007).

## The green genes

An added wrinkle to the search for endosymbiotically transferred genes is the presence, and in many cases dominance, of chromalveolate nuclear genes that show preferential phylogenetic affinity to green algae. The *T. pseudonana* genome was the first indication of this phenomenon (Armbrust *et al.*, 2004; Fig. 10.3), where it was reported that between 182 and 254 'red' genes were identified, but between 865 and 2026 'green' genes were present. The second diatom genome, *Phaeodactylum tricornutum* (Bowler *et al.*, 2008) revealed 170 'red' genes shared between the two diatoms, whereas 1700 'green' genes were subsequently documented (Moustafa *et al.*, 2009). Many of these genes can also be found in other chromalveolate taxa, leading to the suggestion that they are the result of a cryptic endosymbiosis, involving a transient green plastid in the history of the chromalveolate lineage, prior to the event that captured the red algal plastid (Moustafa *et al.*, 2009). If this interpretation is correct, then the gain and loss of plastids may be easier than assumed.

Alternative explanations do exist, however. Among them are the taxonomic sampling discussed above and the unknown relationships between major eukaryotic lineages (Dagan and Martin, 2009; Elias and Archibald, 2009). To account for the potential of the reduced *C. merolae* genome to bias the results in favour of more identifiable green genes simply because red homologues do not exist in the genome, Moustafa *et al.* (2009) examined the sub-group of genes that occur in both the *C. merolae* and *Chlamydomonas* genomes. Even in this subset, the number of genes with green algal affinities was substantially higher than those that grouped with a red algal gene, suggesting that the green signal is more than a simple sampling issue.

What cannot be accounted for, however, is the possibility that the chromalveolate group shares a common ancestry with Archaeplastida. As discussed above, the relationships between the eukaryotic supergroups remain contentious, with several major challenges to the previously accepted 'Tree of Life' having been proposed in just the last few years (Burki *et al.*, 2007, 2008; Hackett *et al.*, 2007; Patron *et al.*, 2007; Rodriguez-Ezpeleta *et al.*, 2007; Hampl *et al.*, 2009). Recent studies are converging on a possible relationship between the expanded chromalveolate lineage ('SAR' + Harcobia) and Archaeplastida (Burki *et al.*, 2008) and the (unlikely, in these authors' opinion) possibility has been raised that chromalveolates as a whole, may nest within Archaeplastida (Nozaki *et al.*, 2009). The fact of the matter is that basal eukaryotic relationships remain unresolved, despite growing data sets and more complex methodologies. If the chromalveolate lineages do indeed nest with Archaeplastida, specifically sister to the green algae/plant lineage, then the 'green' genes found within their nuclear genomes would be expected. However, in the absence of this relationship, we are left to search for a mechanism by which genes of green algal origin outnumber those derived from red algae in the nuclei of chromalveolate taxa.

### The consequences for phylogenetics

There are many reasons why basal eukaryotic relationships continue to be a source of debate and remain difficult to decipher. The combination of time, variation in evolutionary rate between both genes and lineages, and problems with our ability to accurately model molecular evolution over time all create phylogenetic artifacts that work against determining a single and correct 'Tree of Life'. Even our efforts to do so operate under the assumption that the evolution of extant life can be depicted in a tree-like fashion – an assumption that has its challenges (Doolittle and Bapteste, 2007; Lopez and Bapteste, 2009). But, assuming this is possible, an often under-appreciated factor in deep eukaryotic phylogenetics is the chimeric nature of many eukaryotic genomes. Using the diatoms as an example, Moustafa *et al.* (2009) found that 16% of the *P. tricornutum* nuclear genes may have green algal origins. Add in the red algal-derived genes and that number increases to close to 20%. Ignoring the probability that additional genes have been contributed to the genome over time in a non-vertical manner, at least one in five diatom genes could be expected to produce a phylogenetic signal at odds with the genes vertically inherited from the original eukaryote.

Large-scale phylogenetic studies have countered this problem in two ways, either by removing genes that produce unusual tree topologies or making the argument that 'core' genes have a low likelihood of replacement by an endosymbiont gene. Of course, removal of genes that do not reproduce the expected or majority topology introduces the potential of biasing the outcome based on *a priori* assumptions. Using diatoms as an example again, the removal of the 20% of genes that show affinities to red and green algae ensures that a topology showing chromalveolates nested within Archaeplastida will not be recovered. Is that removing, or introducing, phylogenetic artifact? Core eukaryotic genes have also been shown to be affected by endosymbiotic gene replacement. In a study of the 2° photosynthetic excavate *Euglena gracilis*, Ahmadinejad *et al.* (2007) demonstrated that 22% of a set of 259 globally distributed eukaryotic genes showed phylogenetic affinity to 1° plastid lineages. These globally distributed genes are exactly the type that is used for broad-scale eukaryotic phylogenetics because they can be compared across eukaryotic lineages.

## Functional integration and diversification of plastids following endosymbiosis

### Symbiont-host gene transfer and protein targeting to plastids

#### *Barriers to endosymbiogenesis*

The transformation of an independent endosymbiont into an organelle faces tremendous barriers. The hurdles include the transfer of much of the endosymbionts genetic material to the host's nucleus, the acquisition of the proper regulatory and targeting sequences to ensure that the transferred gene not only is expressed but possesses the correct targeting information to redirect it into the plastid. This, of course, leads us to the significant challenge of acquiring the appropriate protein import apparatus to ensure that the targeted proteins are properly imported and sorted within the organelle. There is also the issue of integrating and regulating metabolic pathways

(Gould *et al.*, 2008). This would necessitate the redistribution of the appropriate membrane transporters to allow the effective exchange of metabolites between the plastid and the host. It seems amazing that such an event occurred more than once given the apparent complexity of the task, but it has occurred on multiple independent occasions involving several different partners. We have examples of plastid genesis from endosymbiotic prokaryotes (primary) and eukaryotes (secondary/tertiary) and the task of establishing an organelle from either of these systems is conceptually different. In primary endosymbiosis, gene transfer and functional integration would seem to be particularly challenging given the eukaryotic-prokaryotic dichotomy and the inherent differences in transcriptional/translational apparatuses, and perhaps this accounts for the greater number of endosymbioses involving two eukaryotes. Evolution of a functional protein import apparatus, de novo, would also be a significant barrier to primary endosymbiosis, but less challenging in secondary plastids that would have a system to cross two membranes already.

From a genomic perspective, both endosymbiont and host have undergone massive changes in the transition from free-living cyanobacterium and heterotrophic eukaryote, to obligate symbiotic partners, and ultimately, organelle and host. The most obvious change can be seen in the plastid, where the plastid genome now encodes a mere fraction of the genes it once did as a cyanobacterial genome. The loss of genes from the plastid most likely occurred before the divergence of the three primary plastid lineages, as their plastid genomes have a fairly similar gene complement (Martin *et al.*, 1998). Obviously, genes required for an autonomous life-style are no longer required by the plastid, but even the most gene-rich plastid genome sequenced to date from the red algal *Porphyra purpurea* (Reith and Munholland, 1995) encodes about only 259 single-copy genes. Conversely, even the smallest known cyanobacterial genome encodes over 2000 genes (Kettler *et al.*, 2007). Surely the discrepancy between the two numbers does not represent only genes required for a free-living lifestyle, but clearly many genes have been lost from the plastid over time.

In fact, it has been demonstrated that the number of proteins needed for the plastid to function numbers over one thousand and many of these proteins are not encoded on the plastid genome (Zybailov *et al.*, 2008). The majority of the plastid proteome is nucleus encoded – the result of either the transfer of genes from the plastid to the nucleus, the recruitment of genes by horizontal gene transfer (HGT) or the adaptation of nuclear genes for plastid functions. In this way, the nuclear genome of the host lineage has been significantly impacted by the expansion (rather than reduction) of its coding capacity. Several studies of primary plastid-containing lineages have revealed surprisingly high numbers of genes putatively derived from cyanobacteria in the host nucleus, including the plant *Arabidopsis thaliana* (18%) (Martin *et al.*, 2002), the green alga *Chlamydomonas reinhardtii* (6%) (Moustafa and Bhattacharya, 2008), the red alga *Cyanidioschyzon merolae* (12.7%) (Sato *et al.*, 2005) and the glaucophyte alga *Cyanophora paradoxa* (10.8%) (Reyes-Prieto *et al.*, 2006). However, in *Arabidopsis* (the only organism in which it has been actively investigated) over half of the genes that were predicted to be derived from cyanobacteria were also predicted not to function in the plastid (Martin *et al.*, 2002). That is to say that the majority of the prokaryotic genes transferred to the eukaryotic nucleus have taken on new functions within the cell, unrelated to the compartment from which they came. This represents a substantial resource of novel genetic information for photosynthetic organisms.

The transfer of genetic material from the endosymbiont to the host nucleus and its integration into the genome is likely not uncommon (Martin, 2003; Timmis *et al.*, 2004). From genome sequencing efforts, large chunks of mitochondrial and plastid DNA is identifiable in the nucleus of *Arabidopsis* (Lin *et al.*, 1999). Organelle-to-nucleus gene transfers have been observed experimentally in different plant systems (Stegemann *et al.*, 2003; Huang *et al.*, 2003). Stegemann and Brock (2006) even observed functional activation of a gene transferred to the nucleus from the plastid genome, indicating that such transfers are surprisingly frequent. The mechanism of these transfer events is not well understood but the extranuclear DNA seems to integrate at sites of double stranded DNA breaks (Decottignies, 2005). Of course, we presume that there is an inherent advantage to the transfer of DNA from the endosymbiont/plastid to the nucleus, the dominant one having to do with small population size and asexual reproduction that can lead to the accumulation of deleterious mutations, what is described as Muller's Rachet (Rispe and Moran, 2000), thus favouring the transfer of genes from the endosymbiont to the nucleus. We have to keep in mind, however, that with phagotrophic hosts' gene transfer into the nucleus may not be limited to the endosymbiont that gave rise to the plastid. It would be difficult to rule out gene transfer from vacuoles during degradation of captured prey in addition to the transfer from any additional endosymbionts (Berg and Kurland, 2000). This type of gene transfer may be suspected if gene phylogenies are in discord with known or predicted organismal phylogenies. The same data may also suggest 'cryptic' endosymbiosis, or an endosymbiosis that did not give rise to an organelle (Henze *et al.*, 1995; Moustafa *et al.*, 2009), which could also explain the number of reported cases of lateral gene transfer (Rogers *et al.*, 2007).

Transfer and integration of extra-chromosomal DNA into the nucleus is but the start of a complex process. A second major hurdle is acquiring the appropriate regulatory regions such that the transferred gene is actually transcribed and ultimately translated in the cytosol. While it is plausible that a transferred gene may simply acquire sufficient regulatory sequences to initiate a minimal level of transcription, it seems more likely that transferred genes acquired a functional promoter by insertion into a pre-existing gene, essentially co-opting its promoter. Alternatively, chromosomal rearrangements in the vicinity of the transferred gene could move a promoter in the proximity of the transferred gene, allowing some level of expression.

### *Evolution of protein import in diverse plastids*

Getting a transferred gene to be expressed under the appropriate conditions is only part of the solution, and this process would ultimately have to co-evolve with a mechanism to allow the protein product to be imported back into the organelle of origin. Protein targeting to organelles is mediated by N-terminal targeting sequences that direct newly made proteins to their proper location. Although the N-terminal plastid-targeting signal shares little conservation between groups of photosynthetic organisms, common features can be found (Patron and Waller 2007). This targeting information is necessary and sufficient for protein import into the plastid, which is mediated by a complex series of proteins in the plastid envelope that recognize these sequences and mediate import. For primary plastids, one would presume that these targeting sequences would evolve *de novo*, which is not hard to imagine given the degeneracy of targeting sequences. Reumann *et al.* (2005) also proposed that transit peptides evolved from a portion of the cyanobacterial OMP85 protein adjacent to the signal sequence that targets it to the outer membrane. This region was found to resemble transit peptides and that may have a role in assembly of the complex. It also seems plausible that the mitochondrial transit peptides could have been co-opted in some cases. This seems probable given the propensity of some proteins for dual targeting to both organelles (Peeters and Small, 2001). Either way, the acquisition of transit peptides for plastid targeting in organelle-transferred genes could occur by integrating into pre-existing genes, whereby the gene is inserted downstream of the targeting region (Adams *et al.*, 2001; Arimura *et al.*, 1999), through exon shuffling (Gantt *et al.*, 1991; Durnford and Gray, 2006; Tonkin *et al.*, 2008a), or via a variety of chromosomal rearrangements.

The problem of protein import across the plastid membrane is complex and we observe considerable variation in requirements depending on the plastid type, of which we can functionally separate depending on the number of membranes surrounding the plastid. Plastids derived through a primary endosymbiosis have two surrounding plastid membranes and are represented by the members of the Plantae: green algae/plants, red algae, and glaucophytes. Plastids derived secondarily have three (dinoflagellates, euglenophytes) or four (stramenopiles, Apicomplexa, Hacrobia, chlorarachniophytes) plastid membranes. The differences in plastid ultrastructure significantly alter the translocation mechanism in each of these cases that we will briefly consider.

While a detailed description of the import apparatuses in diverse organisms is beyond the scope of this chapter, there are several excellent reviews (Reumann *et al.*, 2005; Kessler and Schnell, 2006; Bolte *et al.*, 2009). The protein import apparatus has been deciphered primarily through extensive research on plants and green algae where a collection of subunits of an import translocon are located both in the outer chloroplast membrane (TOC) and the inner chloroplast membrane (TIC), and these interact to facilitate protein translocation (Kessler and Schnell, 2006). The similarities in the TIC/TOC apparatuses between reds and greens support a common origin for the two (McFadden and van Dooren, 2004) and it seems that the composition of the translocation components evolved from components from the endosymbiont, the host, or evolved *de novo* during endosymbiogenesis (Reumann *et al.*, 2005).

In secondary plastids, however, it would seem that many of the more significant hurdles have built in solutions that just need to be modified, such as the general import mechanism, perhaps explaining the apparent greater frequency of secondary endosymbiotic events over primary. With secondary endosymbioses, the plastid is effectively 'outside the cell' with the outer one or two membranes being part of the endomembrane system. Since the plastid is enclosed in the ER, then targeting of transferred plastid genes would have to acquire a signal sequence for targeting to start the process of reinsertion into the plastid (Bolte *et al.*, 2009; Patron and Waller, 2007). One significant step would be in redirecting the plastid-bound proteins once inserted into the endomembrane system and traversing the third membrane (those with 4 membranes around the chloroplast); however, at least the passage through the inner two membranes of the chloroplast envelope would presumably be similar.

Of the major clades that acquired plastids secondarily, two of them have three membranes around the chloroplast: the dinoflagellates and the euglenophytes. While the 'typical' peridinin-containing dinoflagellate acquired a plastid from a red alga, euglenophytes likely acquired a plastid from a green alga. Nevertheless, they have uncanny resemblances in plastid ultrastructure and protein targeting mechanisms. The third membrane around the plastids lack bound ribosomes but are presumably derived from the endomembrane system. It was a surprise when precursors for the light-harvesting complexes were detected in the Golgi apparatus using immunogold labelling (Osafune *et al.*, 1991), which was the first indication of a significantly different plastid targeting mechanism in *Euglena*. Routing through the Golgi apparatus has also been shown in dinoflagellates, suggesting a similar import mechanism for this group (Nassoury *et al.*, 2003). The similarity of the N-terminal targeting sequences in *Euglena* and dinoflagellates is striking (see Durnford and Gray, 2006; Nassoury *et al.*, 2003; Patron *et al.*, 2005). The most common type includes a signal sequence for targeting to the ER, a transit peptide for targeting to the plastid and this is followed by a hydrophobic domain that functions as a stop-transfer sequence, preventing complete passage into the ER (Sulli and Schwartzbach, 1996; Sláviková *et al.*, 2005). In both groups, a smaller percentage of the targeting regions lack stop-transfer sequences, suggesting that these soluble precursors are insert into the ER lumen and are subsequently targeted to the plastid (Durnford and Gray, 2006; Patron *et al.*, 2005).

The remaining organisms with secondary plastids have four surrounding membranes, the outer membrane sometimes continuous with the ER and often with the presence of bound ribosomes on the outer surface. Targeting to the plastid in this type of plastid thus requires an N-terminal signal sequence, for co-translational import into the ER, followed by a transit peptide for subsequent translocation through the additional three membranes (Apt *et al.*, 2002). Similar N-terminal plastid targeting sequences are apparent for proteins targeted to the plastid of chlorarachniophytes and the non-photosynthetic apicoplast of the Apicocomplexa, whose outer plastid membrane does not bind ribosomes but is part of the endomembrane system (Rogers *et al.*, 2004; Harb *et al.*, 2004). The mechanism of targeting and passage through the next (third) membrane is unknown but may involve a vesicular transport mechanism (Gibbs, 1979). Presumably, targeting to and passage through the last two membranes would proceed in a fashion similar to the mechanism in primary plastids (see reviews by Bolte *et al.*, 2009, Tonkin *et al.*, 2008b). The lack of related

TOC complexes in diatoms, however, is intriguing and requires further investigation (McFadden and van Dooren, 2004).

## Antenna evolution and photosynthetic diversification

While the core oxygenic photosynthetic apparatus remains more or less conserved across the eukaryotic tree of life, there has been significant diversification of antenna systems, presumably due to large variations in light quality, light intensity and the differing requirement for the ability to dissipate the energy from light in a controlled fashion (Fig. 10.4). Antenna structure and composition can provide information of the evolutionary processes in these distinct eukaryotic groups and act as a marker for organismal divergence. Early algal classification schemes, for instance, were based on pigment composition that logistically divided the photosynthetic eukaryotes into three large divisions: the rhodophytes having chlorophyll *a* and possessing phycobiliosmes; the chlorophytes/plants characterized by the presence of chlorophylls *a* and *b*; and the catch-all group termed the chromophytes that typically contained chorophylls *a* and *c* plus a variety of distinctive carotenoids. While this is not an accurate phylogenetic history of the organisms in the different supergroups, it does provide a reasonable framework to discuss the different antenna systems.

With the exception of the chlorophyll *a* antenna surrounding photosystem II, consisting of the CP43 and CP47 proteins, and the

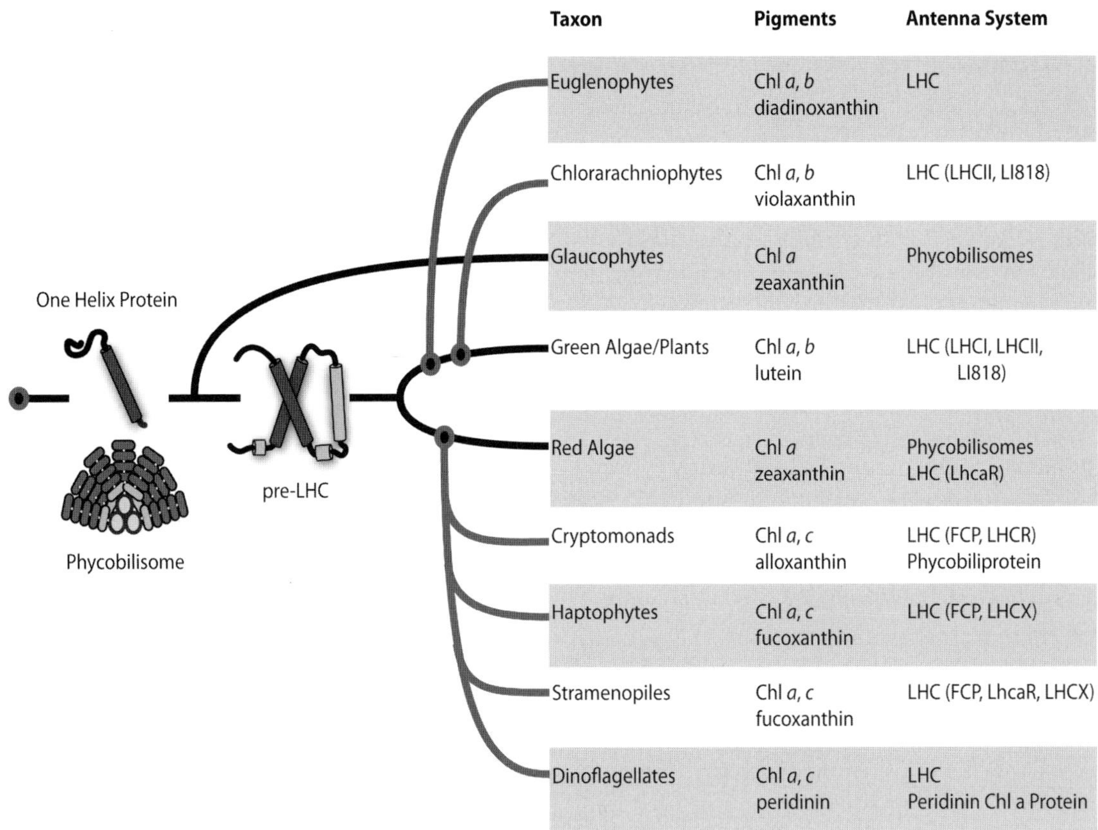


**Figure 10.4** Distribution of pigments and antenna systems in diverse photosynthetic eukaryotes. Figure depicts the presence of the one-helix proteins (LIL2 and 6) and phycobilisomes that were likely present following evolution of the primary plastid. Subsequent divergence and lateral transfer of plastids via secondary endosymbiogenesis gave rise to the diverse antenna systems and pigment compositions. Only the dominant pigments in the main taxa of each group are indicated for simplicity.

built-in chlorophyll *a* antenna of photosystem I core complexes that function to increase the absorption cross section of each reaction centre, for much of the oxygenic photosynthetic life on earth, the phycobilisomes were the dominant light-harvesting structure. This was presumably the first type of accessory light-harvesting system. These soluble protein complexes are abundant in cyanobacteria but are not limited to prokaryotes. Phycobilisomes remain important light-harvesting systems in the Glaucophyceae and the Rhodophyceae and similarities with the cyanobacterial phycobilisome are reflective of the cyanobacterial origin of the plastid.

*Phycobiliproteins*

Phycobilisomes are large, water soluble, multisubunit complexes that are anchored to the photosynthetic membrane (Gantt, 1981). Generally, phycobilisomes are composed of phycobiliproteins covalently attached to phycobilins (chromophores) through thioether linkages. The chromophores themselves are linear tetrapyrroles and synthesized via the heme biosynthesis pathway and structurally similar to the chromophore in phytochrome that is involved in many plant light responses. The number of chromophores attached to each subunit is variable, typically ranging from 1–3. Functionally, the phycobiliproteins and bound phycobilins absorb light optimally in the 450–650 nm range, the region where chlorophyll absorption is weakest.

Each phycobiliprotein assembles as a heterodimer ($\alpha$ and $\beta$) but the basic building block of the phycobilisome appears to be a trimeric unit $(\alpha\beta)_3$. Phylogentic analysis of the different phycobiliproteins has indicated that the $\alpha$ and $\beta$ subunits form separate branches and are likely derived from a common ancestral gene that was subjected to a tandem duplication (Apt *et al.*, 1995). Further gene duplications gave rise to the several different varieties of phycobiliproteins, which consequently modified the absorption and emission characteristics of the functional antenna and a great example of spectral tuning to optimize photosynthesis. All cyanobacteria, rhodophytes and glaucophytes have the cylindrical phycobilisome core composed of allophycocyanin surrounded by assemblages of phycocyanin, phycoerythrin or phycoerythocyanin, as radiating rods attached to the core and fastened with non-pigmented linker proteins (Gantt, 1981). Though there are structural and compositional differences in the phycobilisomes from diverse groups, the components are evolutionarily related.

*Transitioning to a membrane integral antenna system*

Chloroplasts evolved from a cyanobacteria-like organism containing phycobilisomes as the main antenna system in cyanobacteria (Gray and Doolittle, 1982) and that there was a single endosymbiotic event that gave rise to the glaucophytes, red and green algae (Rodriguez-Ezpeleta *et al.*, 2005). Thus phycobilisome presence is ancestral (Fig. 10.4). Of the three primary lineages, the green algae/plants lack phycobilisomes and instead have the membrane integral antenna system called light-harvesting complexes (LHCs, below) that use chlorophyll and carotenoids to absorb light and funnel this excitation energy into the photosystems. Despite sequencing dozens of cyanobacterial genomes, an LHC homologue has not been discovered, suggesting that the LHCs evolved in eukaryotes (Durnford *et al.*, 1999). However, sampling in the cyanobacteria is not sufficient to conclusive say they do not (let alone 'did not') exist. Nevertheless, assuming a monophyletic origin of the primary plastid-containing taxa, the green algae must then have secondarily lost phycobilisomes in favour of the LHCs. Loss of phycobilisomes has also occurred at least twice in eukaryotes following the origin of the plastid, including in algae with plastids derived secondarily from red algae. Phycobilisomes were lost once in the lineage leading to the green algae and at least once in the 'chromalveolate' line. The cryptomonads are the only group in the latter that still retain remnants of the phycobilisome complex, though the structure and localization differ considerable from red algae.

It is difficult to rationalize the selective pressures that may have favoured this replacement in the greens, and their descendants – the land plants – that could have led to improved fitness. Bryant (1992) proposed that phycobiliproteins were 'biosynthetically extravagant' based on the number of amino acids per chromophore in either antennae system, and that nitrogen or carbon deficiency could, in essence, tip the balance in favour of the LHCs. This seems like a reasonable hypothesis, though it does not fully evaluate the abundance of each antenna system nor the machinery required to assemble it. Based on the phycobilisome composition and abundance in the red alga *Porphyridium* (Cunningham *et al.*, 1990; 2200 chromophores/phycobilisomes (PBS); $1.2 \times 10^5$ PBS/cell with a PE, PC, APC hexamer ratio of 58:12:8 per PBS) and the composition and abundance of the LHC-antenna system from the green alga *Chlamydomonas* (assuming LHCII and LHCI account for 50 and 18% of total chlorophyll and that there are 14 chlorophylls per LHC) we can calculate the total cost in terms of amino acid requirement. In this case, the amount of amino acids committed to each antenna system is about the same in each organism (ca. 3 and 2 $\times 10^{10}$ amino acids per cell for LHC vs. PBS) though the actual chromophore content for the LHC antenna in *Chlamydomonas* is about 5 times larger than the phycobilisome in *Porphyridium*. However, these numbers are quite plastic and depend on the growth conditions. Nevertheless, it does not appear likely that nitrogen or carbon demand would be a major factor in driving phycobilisome loss.

So the question remains, what was the reason for the replacement of phycobilisomes with LHCs in most photosynthetic eukaryotes? We suspect that there must be functional considerations tied to the replacement. There are many differences between the different antenna systems, the main ones being that phycobilisomes are large protein complexes attached to the stromal side to the membrane while the LHCs are integral thylakoid membrane proteins. Some consequences of these different characteristics include membrane appression, which is prevented in organisms with phycobilisomes. This may relate to changes in lateral heterogeneity of the thylakoid membrane and separation of PSI and PSII complexes. There would presumably be differences in the ability to directly sense excitation pressure via the magnitude of the pH gradient. LHCs, for instance, have loop regions in the lumen, that are predicted to become protonated as the pH decreases in the light, potentially affecting their structure and function (Horton *et al.*, 1994). Thus, the consequences of these structural features may revolve around photoprotection and the efficiency of dealing with excess light energy and the rapid optimization of light harvesting efficiency. Whatever the cause, the LHC system has been tremendously successful in photosynthetic eukaryotes.

### *Origin of the light-harvesting complexes*

Despite the trend of phycobilisome replacement with an LHC antenna, they are not mutually exclusive (Fig. 10.4). In red algae, distinct LHCs were detected that specifically associated with PSI (Wolfe *et al.*, 1994). Since there is currently no evidence for LHCs in glaucophytes, this implies that LHCs appeared in a lineage after the diversification of the glaucophytes but prior to the separation of red and green algae. LHCs then spread laterally to the chlorophyll a/c-containing groups via a secondary endosymbiosis (Durnford *et al.*, 1999). The origin of the LHCs has been the topic of much speculation (see Montane and Kloppstech, 2000), and it seems likely that the LHCs ultimately evolved from cyanobacterial proteins that share a limited similarity in the chlorophyll-binding domain called high light-inducible proteins (HLIPS, Dolganov *et al.*, 1995). Since the HLIPs possess a single membrane-spanning region (MSR) and the LHCs have three, they would have evolved through a series of gene duplications and fusions (Dolganov *et al.*, 1995; Green and Kuhlbrandt, 1995).

It turns out that there are a variety of structurally unique Light-Harvesting Complex-like (LIL) proteins in photosynthetic eukaryotes, all identified by the presence of at least one chlorophyll-binding motif (Jansson, 1999). Often these are characterized based on the number of predicted MSRs from one to three. Those with one MSR are called one-helix proteins, of which there are two distinctive types (LIL2, LIL6). This class of LIL is a near-universal feature of eukaryotic photosynthesis (Neilson and Durnford, 2010). The second class of LIL proteins have two predicted MSRs and are called stress-enhanced proteins (Heddad and Adamska, 2000; SEP – LIL4, LIL5). An additional two MSR-possessing LIL is simply known as LIL3 (Jansson, 1999). LIL4 and 5 are orthologues and are typically limited to land plants, and perhaps green algae. LIL3, however, is a very well-conserved member of the LIL protein family that is specific to green algal-land plant lineages (Neilson and Durnford, 2010). LIL3 is induced in etiolated plants, can bind chlorophyll and is proposed to function in the assembly of chlorophyll binding protein or act as a transient chlorophyll storage site (Reisinger *et al.*, 2008). The third class of LIL proteins possesses three predicted MSRs, two of which have a recognizable chlorophyll motif (Adamska, 2001). These LIL proteins (also called early light-inducible proteins, ELIPs) are limited to the chlorophyll *a*/*b*-containing lineages and whose function has not yet been conclusively determined. An ELIP-like protein with four MSRs has also been detected in a prasinophyte, again demonstrating the potential for expansion of this complex gene family (Neilson and Durnford, 2010).

In terms of the evolution of the LHC antenna system, there are no obvious transitional states represented in the different LILs. In many cases, it seems that these proteins arose independently from the single helix family so a specific 'path' to the LHCs cannot be discerned. The propensity of the chlorophyll-binding motif to be shuffled about on different proteins suggests the utility of the domain in creating novel proteins (Kilian *et al.*, 2008) and suggests that evolution of LHCs from small precursors is indeed a likely event.

## The LHCs complexity and diversity

The LHCs possess three membrane-spanning regions that are integrated into the thylakoid membrane where they associate with one of the photosystems. All LHCs bind chlorophyll *a* and one or more types of carotenoids that collectively function in both light harvesting and energy dissipation (see review by Barros and Kuhlbrandt, 2009). Many of the differences in the LHC superfamily between diverse groups is in the type of accessory chlorophylls and the type of carotenoids it binds. Nevertheless, despite these differences, all LHCs from the green, red and chromalveolate appear to have originated from a common ancestral LHC precursor and undergone significant diversification in pigment binding and functional capabilities (Durnford *et al.*, 1999).

### *Chlorophyll a antenna: red algae*

The LHC antenna system in the red algal *Porphyridium* is limited to its association with Photosystem I (PSI) and binds chlorophyll *a*, zeaxanthin and β-carotene (Wolfe *et al.*, 1994). It is notable that there are no accessory chlorophylls, such as chlorophyll *b* or *c*, associated with the red algal LHCs as you find in most other photosynthetic groups. Like in other eukaryotes, the red algal LHC family is composed of several related members, in this case ranging from 20–24 kDa (Tan *et al.*, 1997). LHCs specifically associated with PSI have also been detected in *Galderia sulphuraria* (Marquardt and Rhiel, 1997), suggesting PSI association is a common feature in red algae, presumably while the phycobilisomes primarily donate excitation energy to PSII.

### *Chlorophyll a/b-binding antenna: green algae, Euglenophyta, Chlorarachniophyta*

Following the divergence of green and red algae, and coinciding with the loss of phycobilisomes, the antenna proteins in green algae underwent significant diversification as they functionally specialized to fulfil specific tasks in light absorption. At some point, this lineage acquired the ability to synthesize chlorophyll *b* as an accessory pigment, the difference being an aldehyde (CHO) group on C-7 of the chlorin ring rather than a methyl. This ultimately shifts the light absorption profile of chlorophyll *b* such that it broadens the absorption spectra. In plants, the main components of the Photosystem II antenna system (LHCII) binds eight chlorophyll *a* and six chlorophyll *b* plus several carotenoids, including two centrally located luteins, a neoxanthin and an additional xanthophyll presumably involved in the xanthophyll cycle (Liu *et al.*, 2004; Standfuss *et al.*, 2005). Of course, there are a variety of LHCs with different functions and variations in number of chlorophyll and carotenoid binding sites (Barros and Kuhlbrandt, 2009).

The LHC antenna has diversified significantly, one of the main structural distinctions being which reaction centre complex (PSI or PSII) each individual LHC associates. In plants, there are four distinct LHC complexes (LHCI) that specifically form a belt on one side of PSI (Ben-Shem *et al.*, 2003). In a variety of green algae the PSI antenna is larger and there is generally only one very well conserved LHCI protein compared with plants (Koziol *et al.*, 2007). This indicates considerable differences in how the antenna system specifically organizes around PSI (Stauber *et al.*, 2009). The antenna complexes associated with PSII (LHCII) can be divided into the minor antenna, which are more closely associated with the PSII (CP29, CP26 and CP24), plus the major LHCII antenna complexes that typically form trimers (Yakushevska *et al.*, 2003). Because of the variety of EST and genome projects, we can follow the diversification of the antenna system through the chlorophyll *a*/*b*-containing lineages.

It is clear that in all organisms, gene duplication and divergence of LHC-genes has been vital in altering the subunit composition of the PSI and PSII antenna system (Six *et al.*, 2005; Koziol *et al.*, 2007). This is also true for organisms that acquired plastids from green algae secondarily, namely the euglenophyta and chlorarachniophytes. Surprisingly, in these secondary plastids, the PSI antenna system seems to have been significantly modified as evidenced by the lack of homologues compared with extant green algae, unlike the situation with LCHII. In the chlorarachniophyte *Bigelowiella natans*, for instance, our EST sampling failed to identify a single LHCI homologue while there were plenty of LHC sequences related to the PSII antenna system (Koziol *et al.*, 2007). The functional implications of these changes are unknown, but presumably it is related to spectral tuning to optimize light absorption or related to regulating the collection and dissipation of light energy.

### *Chlorophyll a/c-binding antenna: Chromalveolates*

While the chlorophyll *a*/*c*-binding and the chlorophyll *a*/*b*-binding proteins are part of a large super-family of related proteins (Durnford *et al.*, 1999), they have substantially diverged. When comparing the protein sequences of these different clades, there is little sequence similarity outside the main chlorophyll-binding domains in the membrane-spanning regions and these diverse proteins are difficult to align. Not unexpectedly, there are also considerable differences in chromophore binding. In fact, the main LHCII proteins in the different chromealveolate lineages have greater carotenoid–chlorophyll ratios producing plastids with colours ranging from brown to red. In diatoms, brown algae and raphidophytes, the dominant carotenoid is fucoxanthin, a brown-coloured pigment, while in many dinoflagellates it is peridinin, a brick-red coloured pigment (Macpherson and Hiller, 2003). Thus, with the chlorophyll *a*/*c*-binding proteins, carotenoids (such as fucoxanthin) have a more important role in broadening out the absorption spectrum in marine environments where there is lower penetration red light, thus leading to a light environment biased for blue–green wavelengths, that carotenoids capture efficiently (Falkowski and Raven, 1997). In addition to a variety of carotenoids, there is often chlorophyll *c* in addition to chlorophyll *a*. chlorophyll *c* differs in that it lacks a phytol tail in addition to the presence of a double bond between C-17 and C-18 on the chlorin ring, leading to a blue-shifted 'red' absorption peak.

The family of chlorophyll *a*/*c*-binding proteins is becoming increasingly complex as more data streams in from a variety of genome and EST projects. Though a full analysis of the available data has not yet been published, three major clades are emerging. The first is a chlorophyll *a*/*c* clade containing the predominate LHCs (fucoxanthin chlorophyll proteins-FCP) from a variety of stramenopiles such as diatoms, brown algae and raphidophytes (Durnford *et al.*, 1999; Green, 2003; Eppard *et al.*, 2000; Pearson *et al.*, 2009). The second is a clade related to the red algal LhcaR clade that contains LHCs from cryptomonads, and stramenopiles (Koziol *et al.*, 2007; Green, 2003; Eppard *et al.*, 2000). This makes sense since the plastids from these organisms are derived secondarily from red algae and one would predict that these proteins may localized to PSI as they do in red algae. The third group is the LHCX clade that is related to a group of LHC-like proteins in green algae called LI818 (discussed below). Though the functional distinction between these different types is not clear, there is evidence for clear types of FCPs associated with PSI or PSII (Veith *et al.*, 2009), suggesting equivalent complexity as seen in plants.

### *LI818 (LHCSR): an LHC homologue shared by most photosynthetic eukaryotes*

The discovery of an LHC-relative called LI818 (recently renamed LHCSR) was particularly interesting as it was more closely related to LHCs from specific haptophytes and diatoms than to the typical green algal LHCs (Eppard *et al.*, 2000; Savard *et al.*, 1996; Richard *et al.*, 2000). Further sampling confirmed that these LI818 proteins were indeed found in a variety of diverse organisms and were the only clade of LHC proteins shared by the major photosynthetic taxa. This suggests that LI818 appeared before green and red algal separation (Koziol *et al.*, 2007; Richard *et al.*, 2000), although acquisition from a cryptic endosymbiosis with a green alga is a possibility (Moustafa *et al.*, 2009). There are also indications that the role of LI818 may be more related to photoprotection than light-harvesting per se. In *Chlamydomonas*, for instance, LI818 is induced by high light (Richard *et al.*, 2000), and a similar high-light induction of the LI818-like genes (Lhcx) was observed in diatoms (Oeltjen *et al.*, 2002; Zhu and Green, 2008; Nymark *et al.*, 2009). In *Chlamydomonas*, a non-photochemical quenching mutant was deficient in LI818, indicating a role in photoprotection (Peers *et al.*, 2009). The role of this interesting antennae protein needs further investigation, but it is intriguing that this clade of proteins has similar excess light-related expression in diverse groups, potentially indicating a core role in photosynthetic acclimation. Interestingly, LI818, though present in the moss *Physcomitrella* (Nishiyama *et al.*, 2003), is absent in more advanced land plants.

*Pigments and diversification*

The vast array of pigment combinations present in plastids is incredible and the differences often attributed to adaptation driven by the optimization for specific light qualities, which seems like a reasonable hypothesis. It turns out, however, that different LHC apoproteins are not too fussy about the pigments they bind *in vitro*, which led Green (2001) to describe pigment binding as 'molecular opportunism'. In a series of reconstitution experiments, Grabowski *et al.* (2001) were able to assemble functional antenna proteins using a red algal LHCI apoprotein and a variety of pigments not normally present in red algae. Clearly, in this case, the LHCI protein was able to bind a larger complement of pigments than would be expected. This flexibility may have been important during diversification of antenna systems as modifications to the chlorophyll and carotenoid biosynthetic pathways occurred to allow changes in the accessory pigments. Tracking the evolution of pigment biosynthetic enzymes may be valuable in understanding the different strategies in light harvesting (Frommolt *et al.*, 2008).

### D) Novel antennae

There doesn't seem to have been a shortage in novel ways to harvest light during the diversification of photosynthesis and plastid evolution. While much of the diversity stems from modification of the structural features and pigment-binding capabilities of the LHCs, the peridinin-chlorophyll complexes of dinoflagellates represents an entirely new way to harvest light energy. While the peridinin-containing dinoflagellates also have membrane localized LHCs, these peridinin-chlorophyll *a* antennae are soluble pigment-protein complexes that exist as a monomeric 32-kDa protein or a dimer of 15-kDa proteins, depending on the species. In the former case, there is an internal similarity suggesting a gene duplication and fusion (Macpherson and Hiller, 2000). The peridinin-chlorophyll *a* protein (PCP) was crystallized and the protein forms a basket-shaped structure into which sits 8 peridinin and 2 chlorophyll *a* molecules (Hofmann *et al.*, 1996). Based on potential N-terminal targeting information, it is predicted that the PCPs are present in the thylakoid lumen (Norris and Miller, 1994), where they presumably transfer excitation energy into one of the reaction centres. Interestingly, there is nothing similar to the PCPs in Genbank, thus their evolutionary origin remains obscure.

Cryptomonads possess phycobiliproteins, like red algae, but they don't form phycobilisomes and the protein is located within the thylakoid lumen rather than being attached to the stromal side of the thylakoid membrane (Ludwig and Gibbs, 1989). Cryptomonads also possess a single type of phycobiliprotein, either phycoerythrin (PE) or phycocyanin (PC), of which there are two subunits ($\alpha$ and $\beta$). Unlike in red algae, the subunits arrange as a dimer $(\alpha\beta)_2$ rather than the trimeric form $(\alpha\beta)_3$ described above (Wilk *et al.* 1999). What's interesting, is that while the $\beta$ subunit of PE or PC in cryptomonads is related to the $\alpha/\beta$ subunits of red algae, the nucleus-encoded $\alpha$

subunit is unique in sequence (Apt *et al.*, 1995; Sidler *et al.*, 1990). The exact arrangement of the phycobiliproteins and how they are attached to the thylakoid membrane to interact with the reaction centres is unclear. The lack of a similar antenna system in cyanobacteria or red algae argues for its appearance after plastid acquisition in this group, but functionally there remain a number of important questions with respect to the structure and function of this antenna system.

## Conclusion

Endosymbiosis has been a significant mechanism for the divergence of eukaryotic organisms and this was never more obvious than when one considers the spread of photoautotrophy across the tree of life. There are also many fascinating questions related to these evolutionary events that lead to the creation of new organelles and imparting novel metabolic capabilities to distinct groups of eukaryotes. While the number and types of endosymbioses that gave rise to different plastids is inherently fascinating and helps to reveal the mechanisms of endosymbiogenesis, ultimately the metabolic pathways of these different organisms have to be integrated and regulated and understanding the events and mechanisms of endosymbiogenesis is vital for comprehending eukaryotic diversity. Since it is 150 years since Darwin published *On the Origin of Species* we relish the thought of Darwin's amazement if he were able to study such a novel evolutionary mechanism as endosymbiogenesis.

## References


Adams K.L., Rosenblueth M., Qiu Y.L., and Palmer J.D. (2001). Multiple losses and transfers to the nucleus of two mitochondrial succinate dehydrogenase genes during angiosperm evolution. Genetics *158*, 1289–1300.

Adamska, I. (2001). The Elip family of stress proteins in the thylakoid membranes of pro- and eukaryota. In Advances in Photosynthesis and Respiration, E.M. Aro and B. Andersson, eds. (Dordrecht: Kluwer Academic Publishers), pp. 487–505.

Ahmadinejad, N., Dagan, T., and Martin, W. (2007). Genome history in the symbiotic hybrid *Euglena gracilis*. Gene *402*, 35–39.

Andersen, R.A. (2004). Biology and systematics of heterokont and haptophyte algae. Am. J. Bot. *91*, 1508–1522.

Andersen, R.A., Saunders, G.W., Paskind, M.P., and Sexton. J.P. (1993). Ultrastructure and 18S rRNA gene sequence for *Pelagomonas calceolata* gen. et sp. nov. and the description of a new algal class, the Pelagophyceae classis nov. J. Phycol. *29*, 701–715.

Apt, K.E., Collier, J.L., and Grossman, A.R. (1995). Evolution of the phycobiliproteins. J. Mol. Biol. *248*, 79–96.

Apt, K.E., Zaslavkaia, L., Lippmeier, J.C., Lang, M., Kilian, O., Wetherbee, R., Grossman, A.R., and Kroth, P.G. (2002). In vivo characterization of diatom multipartite plastid targeting signals. J. Cell Sci. *115*, 4061–4069.

Archibald, J.M. (2007). Nucleomorph genomes: structure, function, origin and evolution. Bioessays *29*, 1–11.

Archibald, J.M. (2009). The puzzle of plastid evolution. Curr. Biol. *19*, R81-R88.

Archibald, J.M., and Keeling, P.J. (2005). On the origin and evolution of plastids. In Microbial Phylogeny and Evolution, J. Sapp, ed. (New York, Oxford University Press), pp. 238–260.

Arimura, S., Takusagawa, S., Hatano, S., Nakazono, M., Hirai, A., and Tsutsumi, N. (1999). A novel plant nuclear gene encoding chloroplast ribosomal protein S9 has a transit peptide related to that of rice chloroplast ribosomal protein L12. FEBS Lett. *450*, 231–234.

Armbrust, E.V., Berges, J.A., Bowler, C., Green, B.R., Martinez, D., Putnam, N.H., Zhou, S., Allen, A.E., Apt, K.E., Bechner, M., Brzezinski, M.A., Chaal, B.K., Chiovitti, A., Davis, A.K., Demarest,M.S., Detter, J.C., Glavina, T., Goodstein, D., Hadi, M.Z., Hellsten, U., Hildebrand, M., Jenkins, B.D., Jurka, J., Kapitonov, V.V., Kroger, N., Lau, W.W., Lane, T.W., Larimer, F.W., Lippmeier, J.C., Lucas, S., Medina, M., Montsant, M., Obornik, M., Parker, M.S., Palenik, B., Pazour, G.J., Richardson, P.M., Rynearson, T.A., Saito, M.A., Schwartz, D.C., Thamatrakoln, K., Valentin, K., Vardi, A., Wilkerson, F.P., and Rokhsar, D.S. (2004). The genome of the diatom *Thalassiosira pseudonana*: ecology, evolution, and metabolism. Science *306*, 79–86.

Barros, T., and Kühlbrandt, W. (2009). Crystallization, structure and function of plant light-harvesting complex II. Biochim. Biophys. Acta *1787*, 753–772.

Ben-Shem, A., Frolow, F., and Nelson, N. (2003). Crystal structure of plant Photosystem I. Nature *426*, 630–635.

Berg, O.G., and Kurland, C.G. (2000). Why mitochondrial genes are most often found in nuclei. Mol. Biol. Evol. *17*, 951–961.

Bernhard, J.M., and Bowser, S.S. (1999). Benthic foraminifera of dysoxic sediments: Chloroplast sequestration and functional morphology. Earth Sci. Rev. *46*, 149–165.

Bhattacharya, D., and Nosenko. T. (2008). Endosymbiotic and horizntal gene transfer in chromalveolates. J. Phycol. *44*, 7–10.

Bhattacharya, D., and Schmidt, H.A. (1997). Division Glaucocystophyta. In Origin of Algae and their Plastids, D. Bhattacharya, ed. (Vienna: Springer-Verlag), pp. 139–148.

Bhattacharya, D., Yoon, H.S., and Hackett, J.D. (2004). Photosynthetic eukaryotes unite: endosymbiosis connects the dots. Bioessays *26*, 50–60.

Bhaud, Y., Guillebault, D., Lennon, J.F., Defacque, H., Soyer-Gobillard, M.O., and Moreau, H. (2000). Morphology and behaviour of dinoflagellate chromosomes during the cell cycle and mitosis. J. Cell Sci. *113*, 1231–1239.

Bodyl, A. (2005). Do plastid-related characters support the chromalveolate hypothesis? J. Phycol. *41*, 712–719.

Bodyl, A., Stiller, J.W., and Mackiewicz, P. (2009). Chromalveolate plastids: direct descent or multiple endosymbioses? Trends Ecol. Evol. *24*, 119–121.

Bolte, K., Bullmann, L., Hempel, F., Bozarth, A., Zauner, S., and Maier, U.G. (2009). Protein targeting into secondary plastids. J. Eukaryot. Microbiol. *56*, 9–15.

Bowler, C., Allen, A.E., Badger, J.H., Grimwood, J., Jabbari, K., Kuo, A., Maheswari, U., Martens, C., Maumus, F., Otillar, R.P., Rayko, E., Salamov, A., Vandepoele, K., Beszteri, B., Gruber, A., Heijde, M., Katinka, M., Mock, T., Valentin, K., Verret, F., Berges, J.A., Brownlee, C., Cadoret, J.P., Chiovitti, A., Choi, C.J., Coesel, S., De Martino, A., Detter, J.C., Durkin, C., Falciatore, A., Fournet, J., Haruta, M., Huysman, M.J.J., Jenkins, B.D., Jiroutova, K., Jorgensen, R.E., Joubert, Y., Kaplan, A., Kröger, N., Kroth, P.G., La Roche, J., Lindquist, E., Lommer, M., Martin-Jezequel, V., Lopez, P.J., Lucas, S., Mangogna, M., McGinnis, K., Medlin, L.K., Montsant, A., Oudot-Le Secq, M.P., Napoli, C., Obornik, M., Parker, M.S., Petit, J.L., Porcel, B.M., Poulsen, N., Robison, M., Rychlewski, L., Rynearson, T.A., Schmutz, J., Shapiro, H., Siaut, M., Stanley, M., Sussman, M.R., Taylor, A.R., Vardi, A., von Dassow, P., Vyverman, W., Willis, A., Wyrwicz, L.S., Rokhsar, D.S., Weissenbach, J., Armbrust, E.V., Green, B.R., Van De Peer, Y., and Grigoriev, I.V. (2008). The *Phaeodactylum* genome reveals the evolutionary history of diatom genomes. Nature *456*, 239–244.

Brinkmann, H., and Martin, W. (1996). Higher-plant chloroplast and cytosolic 3-phosphoglycerate kinases: a case of endosymbiotic gene replacement. Plant Mol. Biol. *30*, 65–75.

Bryant, D.A. (1992). Puzzles of chloroplast ancestry. Curr. Biol. *2*, 240–242.

Burki, F., and Pawlowski, J. (2006). Monophyly of Rhizaria and multigene phylogeny of unicellular bikonts. Mol. Biol. Evol. *23*, 1922–1930.

Burki, F., Shalchian-Tabrizi, K., Minge, M., Skjæveland, A., Nikolaev, S.I., Jakobsen, K.S., and Pawlowski, J. (2007). Phylogenomics reshuffles the eukaryotic supergroups. PLoS One *2*, e790.

Burki, F., Shalchian-Tabrizi, K., and Pawlowski, J. (2008). Phylogenomics reveals a new 'megagroup' including most photosynthetic eukaryotes. Biol. Lett. *4*, 366–369.

Burki, F., Inagaki, Y., Brate, J., Archibald, J.M., Keeling, P.J., Cavalier-Smith, T., Sakaguchi, M., Hashimoto, T., Horak, A., Klaveness, D., Jokobsen, K.S., Pawlowski, J., and Shalchian-Tabrizi, K. (2009). Massive sequencing and phylogenomics of two enigmatic protist lineages reveal Telonemia and Centrohelizoa are related to photosynthetic chromalveolates. Genome Biol. Evol. *1*, 231–238.

Butcher, R.W. (1967). An introductory account of the smaller algae of British costal waters. Fishery Investigations *Ser. IV*, 1–54.

Cavalier-Smith, T. (1991). Cell diversification in heterotrophic flagellates. In The Biology of Free-living Heterotrophic Flagellates, D.J. Patterson and J. Larsen, eds. (Oxford: Oxford University Press), pp. 113–131.

Cavalier-Smith, T. (1999a). Principles of protein and lipid targeting in secondary symbiogenesis: Euglenoid, dinoflagellate, and sporozoan plastid origins and the eukaryote family tree. J. Eukaryot. Microbiol. *46*, 347–366.

Cavalier-Smith, T. (1999b). Principles of protein and lipid targeting in secondary symbiogenesis: euglenoid, dinoflagellate, and sporozoan plastid origins and the eukaryote family tree. J. Eukaryot. Microbiol. *46*, 347–366.

Cavalier-Smith, T. (1999c). Zooflagellate phylogeny and the systematics of Protozoa. Biol. Bull. *196*, 393–395.

Cavalier-Smith, T. (2000). Membrane heredity and early chloroplast evolution. Trends Plant Sci. *5*, 174–182.

Cavalier-Smith, T. (2002). The phagotrophic origin of eukaryotes and phylogenetic classification of protozoa. Int. J. Syst. Evol. Microbiol. *52*, 297–354.

Cavalier-Smith, T. (2003). Protist phylogeny and the high-level classification of Protozoa. Eur. J. Protistol. *39*, 338–348.

Cavalier-Smith, T., and Chao, E.E.Y. (2006). Phylogeny and megasystematics of phagotrophic heterokonts (kingdom Chromista). J. Mol. Evol. *62*, 388–420.

Cavalier-Smith, T., and Lee, J.J. (1985). Protozoa as hosts for endosymbioses and the conversion of symbionts into organelles. J. Protozool. *32*, 376–379.

Cavalier-Smith, T., and von der Heyden, S. (2007). Molecular phylogeny, scale evolution and taxonomy of centrohelid heliozoa. Mol. Phylog. Evol. *44*, 1186–1203.

Collins, K., and Gorovsky, M.A. (2005). *Tetrahymena thermophila*. Curr. Biol. *15*, R317–318.

Cunningham, F.X., Dennenberg, R.J., Jursinic, P.A., and Gantt, E. (1990). Growth under red light enhances Photosystem II relative to Photosystem I and phycobilisomes in the red alga *Porphyridium cruentum*. Plant Physiol. *93*, 888–895.

Dagan, T., and Martin, W. (2009). Microbiology. Seeing green and red in diatom genomes. Science *324*, 1651–1652.

Decottignies, A. (2005). Capture of extranuclear DNA at fission yeast double-strand breaks. Genetics *171*, 1535–1548.

Delvinquier, B., and Patterson, D. (1993). The Opalines. In Parasitic Protozoa, J.P. Kreier, ed. (New York: Academic Press), pp. 247–325.

Delwiche, C.F. (1999). Tracing the thread of plastid diversity through the tapestry of life. Am. Natur. *154 Suppl.*, S164-S177.

Deschamps, P., Colleoni, C., Nakamura, Y., Suzuki, E., Putaux, J.L., Buleon, A., Haebel, S., Ritte, G., Steup, M., Falcon, L.I., Moreira, D., Löffelhardt, W., Raj, J.N., Plancke, C., d'Hulst, C., Dauvillee, D., and Ball, S. (2008). Metabolic symbiosis and the birth of the plant kingdom. Mol. Biol. Evol. *25*, 536–548.

Dolganov, N.A., Bhaya, D., and Grossman, A.R. (1995). Cyanobacterial protein with similarity to the chlorophyll *a*/*b* binding proteins of higher plants: Evolution and regulation. Proc. Natl. Acad. Sci. U.S.A. *92*, 636–640.

Doolittle, W.F., and Bapteste, E. (2007). Pattern pluralism and the Tree of Life hypothesis. Proc. Natl. Acad. Sci. U.S.A. *104*, 2043–2049.

Douglas, S., Zauner, S., Fraunholz, M., Beaton, M., Penny, S., Deng, L.T., Wu, X.N., Reith, M., Cavalier-Smith, T., and Maier, U.G. (2001). The highly reduced genome of an enslaved algal nucleus. Nature *410*, 1091–1096.

Durnford, D.G., Deane, J.A., Tan, S., McFadden, G.I., Gantt, E., and Green, B.R. (1999). A phylogenetic assessment of the eukaryotic light-harvesting antenna proteins, with implications for plastid evolution. J. Mol. Evol. *48*, 59–68.

Durnford, D.G., and Gray, M.W. (2006). Analysis of *Euglena gracilis* plastid-targeted proteins reveals different classes of transit sequences. Eukar. Cell *5*, 2079–2091.

Eisen, J.A., Coyne, R.S., Wu, M., Wu, D., Thiagarajan, M., Wortman, J.R., Badger, J.H., Ren, Q., Amedeo, P., Jones, K.M., Tallon, L.J., Delcher, A.L., Salzberg, S.L., Silva, J.C., Haas, B.J., Majoros, W.H., Farzad, M., Carlton, J.M., Smith, R.K., Jr., Garg, J., Pearlman, R.E., Karrer, K.M., Sun, L., Manning, G., Elde, N.C., Turkewitz, A.P., Asai, D.J., Wilkes, D.E., Wang, Y., Cai, H., Collins, K., Stewart, B.A., Lee, S.R., Wilamowska, K., Weinberg, Z., Ruzzo, W.L., Wloga, D., Gaertig, J., Frankel, J., Tsao, C.C., Gorovsky, M.A., Keeling, P.J., Waller, R.F., Patron, N.J., Cherry, J.M., Stover, N.A., Krieger, C.J., del Toro, C., Ryder, H.F., Williamson, S.C., Barbeau, R.A., Hamilton, E.P., and Orias, E. (2006). Macronuclear genome sequence of the ciliate *Tetrahymena thermophila*, a model eukaryote. PLoS Biol. *4*, e286.

Elias, M., and Archibald, J.M. (2009). Sizing up the genomic footprint of endosymbiosis. Bioessays *31*, 1273–1279.

Eppard, M., Krumbein, W.E., von Haeseler, A., and Rhiel, E. (2000). Characterization of fcp4 and fcp12, two additional genes encoding light harvesting proteins of *Cyclotella cryptica* (Bacillariophyceae) and phylogenetic analysis of this complex gene family. Plant Biol. *2*, 283–289.

Falkowski, P.G., and Raven, J.A. (1997). Aquatic Photosynthesis (Malden, MD: Blackwell Science).

Fast, N.M., Xue, L., Bingham, S., and Keeling, P.J. (2002). Re-examining alveolate evolution using multiple proteinmolecular phylogenies. J. Eukaryot. Microbiol. *49*, 30–37.

Frommolt, R., Werner, S., Paulsen, H., Goss, G., Wilhelm, C., Zauner, S., Maier, U.G., Grossman, A.R., Bhattacharya, D., and Lohr, M. (2008). Ancient recruitment by chromists of green algal genes encoding enzymes for carotenoid biosynthesis. Mol. Biol. Evol. *25*, 2653–2667.

Gantt, E. (1981). Phycobilisomes. Ann. Rev. Plant Physiol. *32*, 327–347.

Gantt, J.S., Baldauf, S.L., Calie, P.J., Weeden, N.F., and Palmer, J.D. (1991). Transfer of rpl22 to the nucleus greatly preceded its loss from the chloroplast and involved the gain of an intron. EMBO J. *10*, 3073–3078.

Gibbs, S.P. (1978). The chloroplasts of *Euglena* may have evolved from symbiotic green algae. Can. J. Bot. *56*, 2883–2889.

Gibbs, S.P. (1979). Route of entry of cytoplasmically synthesized proteins into chloroplasts of algae possessing chloroplast-ER. J. Cell Sci. *35*, 253–266.

Gilson, P.R. (2001). Nucleomorph genomes: much ado about practically nothing. Genome Biol. *2*, reviews1022.1–1022.5.

Gilson, P.R., Su, V., Slamovits, C.H., Reith, M.E., Keeling, P.J., and McFadden, G.I. (2006). Complete nucleotide sequence of the chlorarachniophyte nucleomorph: Nature's smallest nucleus. Proc. Natl. Acad. Sci. U.S.A. *103*, 9566–9571.

Gould, S.B., Fan, E., Hempel, F., Maier, U.G., and Kloesgen, R.B. (2007). Translocation of a phycoerythrin alpha subunit across five biological membranes. J. Biol. Chem. *282*, 30295–30302.

Gould, S.B., Waller, R.F., and McFadden, G.I. (2008). Plastid evolution. Ann. Rev. Plant Biol. *59*, 491–517.

Gould, S.B., Tham, W.H., Cowman, A.F., McFadden, G.I., and Waller, R.F. (2008). Alveolins, a new family of cortical proteins that define the protist infrakingdom Alveolata. Mol. Biol. Evol. *25*, 1219–1230.

Grabowski, B., Cunningham, F.X., and Gantt, E. (2001). Chlorophyll and carotenoid binding in a simple red algal light- harvesting complex crosses phylogenetic lines. Proc. Natl. Acad. Sci. U.S.A. *98*, 2911–2916.

Gray, M.W., and Doolittle, W.F. (1982), Has the endosymbiont hypothesis been proven? Microbiol. Rev. *46*, 1–42.

Green, B.J., Li, W.Y., Manhart, J.R., Fox, T.C., Summer, E.J., Kennedy, R.A., Pierce, S.K., and Rumpho, M.E. (2000). Mollusc-algal chloroplast endosymbiosis. photosynthesis, thylakoid protein maintenance, and chloroplast gene expression continue for many months in the absence of the algal nucleus. Plant Physiol. *124*, 331–342.

Green, B.R. (2001). Was "molecular opportunism" a factor in the evolution of different photosynthetic light-harvesting pigment systems? Proc. Natl. Acad. Sci. U.S.A. *98*, 2119–2121.

Green, B.R. (2003). The evolution of light-harvesting antennas. In Light-Harvesting Antennas in Photosynthesis, B.R. Green and W.W. Parson, eds. (Dordrecht: Kluwer Academic Publishers), pp. 130–168.

Green, B.R., and Kühlbrandt, W. (1995). Sequence conservation of light-harvesting and stress-response proteins in relation to the three-dimensional molecular structure of LHCII. Photosynth. Res. *44*, 139–148.

Greenwood, A.D. (1974). The Cryptophyta in relation to phylogeny and photosynthesis. In Electron Microscopy, J.V. Sanders and D.J. Goodchild, eds. (Canberra: Australian Academy of Sciences), pp. 566–567.

Greenwood, A.D., Griffiths, H.B., and Santore, U.J. (1977). Chloroplasts and cell compartments in Cryptophyceae. Br. Phycol. J. *12*, 119.

Grzymski, J.J., Schofield, O.M., Falkowski, P.G., and Bernhard, J.M. (2002). The function of plastids in the deep-sea benthic foraminifer, *Nonionella stella*. Limnol. Oceanogr. *47*, 1569–1580.

Gustafson, D.E., Jr, Stoecker, D.K., Johnson, M.D., Van Heukelem, W.F., and Sneider, K. (2000). Cryptophyte algae are robbed of their organelles by the marine ciliate *Mesodinium rubrum*. Nature *405*, 1049–1052.

Haas, B.J., Kamoun, S., Zody, M.C., Jiang, R.H.Y., Handsaker, R.E., Cano, L.M., Grabherr, M., Kodira, C.D., Raffaele, S., Torto-Alalibo, T., Bozkurt, T.O., Ah-Fong, A.M.V., Alvarado, L., Anderson, V.L., Armstrong, M.R., Avrova, A., Baxter, L., Beynon, J., Boevink, P.C., Bollmann, S.R., Bos, J.I.B., Bulone, V., Cai, G.H., Cakir, C., Carrington, J.C., Chawner, M., Conti, L., Costanzo, S., Ewan, R., Fahlgren, N., Fischbach, M.A., Fugelstad, J., Gilroy, E.M., Gnerre, S., Green, P.J., Grenville-Briggs, L.J., Griffith, J., Grunwald, N.J., Horn, K., Horner, N.R., Hu, C.H., Huitema, E., Jeong, D.H., Jones, A.M.E., Jones, J.D.G., Jones, R.W., Karlsson, E.K., Kunjeti, S.G., Lamour, K., Liu, Z.Y., Ma, L.J., MacLean, D., Chibucos, M.C., McDonald, H., McWalters, J., Meijer, H.J.G., Morgan, W., Morris, P.F., Munro, C.A., O'Neill, K., Ospina-Giraldo, M., Pinzon, A., Pritchard, L., Ramsahoye, B., Ren, Q.H., Restrepo, S., Roy, S., Sadanandom, A., Savidor, A., Schornack, S., Schwartz, D.C., Schumann, U.D., Schwessinger, B., Seyer, L., Sharpe, T., Silvar, C., Song, J., Studholme, D.J., Sykes, S., Thines, M., van de Vondervoort, P.J.I., Phuntumart, V., Wawra, S., Weide, R., Win, J., Young, C., Zhou, S.G., Fry, W., Meyers, B.C., van West, P., Ristaino, J., Govers, F., Birch, P.R.J., Whisson, S.C., Judelson, H.S., and Nusbaum, C. (2009). Genome sequence and analysis of the Irish potato famine pathogen *Phytophthora infestans*. Nature *461*, 393–398.

Hackett, J.D., Maranda, L., Yoon, H.S., and Bhattacharya, D. (2003). Phylogenetic evidence for the cryptophyte origin of the plastid of *Dinophysis* (Dinophysiales, Dinophyceae). J. Phycol. *39*, 440–448.

Hackett, J., Yoon, H., Li, S., Reyes-Prieto, A., Rümmele, S., and Bhattacharya, D. (2007). Phylogenomic analysis supports the monophyly of cryptophytes and haptophytes and the association of rhizaria with chromalveolates. Mol. Biol. Evol. *24*, 1702–1713.

Hallick, R.B., Hong, L., Drager, R., Favreau, M., Monfort, A., Orsat, B., Spielmann, A., and Stutz, E. (1993). Complete sequence of *Euglena gracilis* chloroplast DNA. Nucl. Acids Res. *21*, 3537–3544.

Hampl, V., Hug, L., Leigh, J.W., Dacks, J.B., Lang, B.F., Simpson, A.G., and Roger, A.J. (2009). Phylogenomic analyses support the monophyly of Excavata and resolve relationships among eukaryotic "supergroups". Proc. Natl. Acad. Sci. U.S.A. *106*, 3859–3864.

Harb, O.S., Chatterjee, B., Fraunholz, M.J., Crawford, M.J., Nishi, M., and Roos, D.S. (2004). Multiple functionally redundant signals mediate targeting to the apicoplast in the apicomplexan parasite *Toxoplasma gondii*. Eukaryot. Cell. *3*, 663–674.

Harper, J.T., Waanders, E., and Keeling, P.J. (2005). On the monophyly of the chromalveolates using a six-protein phylogeny of eukaryotes. Int. J. Syst. Evol. Microbiol. *55*, 487–496.

Heddad, M., and Adamska, I. (2000). Light stress-regulated two-helix proteins in *Arabidopsis thaliana* related to the chlorophyll *a*/*b*-binding gene family. Proc. Natl. Acad. Sci. U.S.A. *97*, 3741–3746.

Henze, K., Badr, A., Wettern, M., Cerff, R., and Martin, W. (1995). A nuclear gene of eubacterial origin in *Euglena gracilis* reflects cryptic endosymbioses during protist evolution. Proc. Natl. Acad. Sci. U.S.A. *92*, 9122–9126.

Hofmann, E., Wrench, P., Sharples, F., Hiller, R., Welte, W., and Diederichs, K. (1996). Structural basis of light harvesting by carotenoids: Peridinin-chlorophyll-protein from *Amphidinium carterae*. Science *272*, 1788–1791.

Horton, P., Ruban, A.V., and Walters, R.G. (1994). Regulation of light harvesting in green plants (indication by nonphotochemical quenching of chlorophyll fluorescence). Plant Physiol. *106*, 415–420.

Huang, C.Y., Ayliffe, M.A., and Timmis, J.N. (2003). Direct measurement of the transfer rate of chloroplast DNA into the nucleus. Nature *422*, 72–76.

Ishida, K., and Green. B.R. (2002). Second- and third-hand chloroplasts in dinoflagellates: Phylogeny of oxygen-evolving enhancer 1 (PsbO) protein reveals replacement of a nuclear-encoded plastid gene by that of a haptophyte tertiary endosymbiont. Proc. Natl. Acad. Sci. U.S.A. *99*, 9294–9299.

Jackson, C.J., Norman, J.E., Schnare, M.N., Gray, M.W., Keeling, P.J., and Waller, R.F. (2007). Broad genomic and transcriptional analysis reveals a highly derived genome in dinoflagellate mitochondria. BMC Biol. *5*: 41.

Jansson, S. (1999). A guide to the Lhc genes and their relatives in *Arabidopsis*. Trends Plant Sci. *4*, 236–240.

Johnson, P.W., Hargraves, P.E., and Sieburth, J.M. (1988). Ultrastructure and ecology of *Calycomonas ovalis* Wulf, 1919 (Chrysophyceae) and its redescription as a testate rhizopod, *Paulinella ovalis* n. comb. (Filosea: Euglyphina). J. Protozool. *35*, 618–626.

Jordan, R.W., and Chamberlain, A.H.L. (1997). Biodiversity among haptophyte algae. Biodiv. Conserv. *6*, 131–152.

Katz, L.A. (2001). Evolution of nuclear dualism in ciliates: a reanalysis in light of recent molecular data. Int. J. Syst. Evol. Microbiol. *51*, 1587–1592.

Kawachi, M., Inouye, I., Honda, D., O'Kelly, C.J., Bailey, C., Bidigare, R.R., and Andersen, R.A. (2002). The Pinguiophyceae classis nova, a new class of photosynthetic stramenopiles whose members produce large amounts of omega-3 fatty acids. Phycol. Res. *50*, 31–47.

Keeling, P.J. (2004). Diversity and evolutionary history of plastids and their hosts. Am. J. Bot. *91*, 1481–1493.

Keeling, P.J. (2008). Evolutionary biology – Bridge over troublesome plastids. Nature *451*, 896–897.

Keeling, P.J. (2009). Chromalveolates and the evolution of plastids by secondary endosymbiosis. J. Eukaryot. Microbiol. *56*, 1–8.

Keeling, P.J., Burger, G., Durnford, D.G., Lang, B.F., Lee, R.W., Pearlman, R.E., Roger, A.J., and Gray, M.W.

(2005). The tree of eukaryotes. Trends Ecol. Evol. *20*, 670–676.

Kessler, F., and Schnell, D.J. (2006). The function and diversity of plastid protein import pathways: A multilane GTPase highway into plastids. Traffic (Copenhagen) *7*, 248–257.

Kettler, G., Martiny, A., Huang, K., Zucker, J., Coleman, M., Rodrigue, S., Chen, F., Lapidus, A., Ferriera, S., Johnson, J., Steglich, C., Church, G.M., Richardson, P., and Chisholm, S.W. (2007). Patterns and implications of gene gain and loss in the evolution of *Prochlorococcus*. PLoS Genet. *3*, e231.

Kilian, O., Steunou, A.S., Grossman, A.R., and Bhaya, D. (2008). A novel two domain-fusion protein in cyanobacteria with similarity to the CAB/ELIP/HLIP superfamily: Evolutionary implications and regulation. Mol. Plant *1*, 155–166.

Klaveness, D., Shalchian-Tabrizi, K., Thomsen, H.A., Eikrem, W., and Jakobsen, K.S. (2005). *Telonema antarcticum* sp nov., a common marine phagotrophic flagellate. Int. J. Syst. Evol. Microbiol. *55*, 2595–2604.

Koziol, A.G., Borza, T., Ishida, K., Keeling, P., Lee R.W., and Durnford, D.G. (2007). Tracing the evolution of the light-harvesting antennae in chlorophyll *a*/*b*-containing organisms. Plant Physiol. *143*, 1802–1816.

Kuznicki, L., Mikolajczyk, E., and Walne, P.L. (1990). Photobehavior of euglenoid flagellates: theoretical and evolutionary perspectives. Plant Sci. *9*: 343–369.

Lane, C.E., and Archibald, J.M. (2008a). The eukaryotic tree of life: endosymbiosis takes its TOL. Trends Ecol. Evol. *23*, 268–275.

Lane, C.E., and Archibald, J.M. (2006). Novel nucleomorph genome architecture in the cryptomonad genus *Hemiselmis*. J. Eukaryot. Microbiol. *53*, 515–521.

Lane, C.E., and Archibald, J.M. (2008b). New marine members of the genus *Hemiselmis* (Cryptomonadales, Cryptophyceae). J. Phycol. *44*, 439–450.

Lane, C.E., and Archibald, J.M. (2009). Reply to Bodyl, Stiller and Mackiewicz: "Chromalveolate plastids: direct descent or multiple endosymbioses?". Trends Ecol. Evol. *24*, 121–122.

Lane, C.E., Khan, H., MacKinnon, M., Fong, A., Theophilou, S., and Archibald, J.M. (2006). Insight into the diversity and evolution of the cryptomonad nucleomorph genome. Mol. Biol. Evol. *23*, 856–865.

Lane, C.E., van den Heuvel, K., Kozera, C., Curtis, B.A., Parsons, B.J., Bowman, S., and Archibald, J.M. (2007). Nucleomorph genome of *Hemiselmis andersenii* reveals complete intron loss and compaction as a driver of protein structure and function. Proc. Natl. Acad. Sci. U.S.A. *104*, 19908–19913.

Leander, B.S. (2004). Did trypanosomatid parasites have photosynthetic ancestors? Trends Microbiol. *12*, 251–258.

Leander, B.S. (2007). Molecular phylogeny and ultrastructure of *Selenidium serpulae* (Apicomplexa, Archigregarinia) from the calcareous tubeworm *Serpula vermicularis* (Annelida, Polychaeta, Sabellida). Zool. Scripta 36, 213–227.

Leander, C.A., and Porter, D. (2001). The Labyrinthulomycota is comprised of three distinct lineages. Mycologia 93, 459–464.

Lee, R.E. (2008). Phycology (Cambridge: Cambridge University Press).

Lidie, K.B., and Van Dolah, F.M. (2007). Spliced leader RNA-mediated trans-splicing in a dinoflagellate, *Karenia brevis*. J. Eukaryot. Microbiol. *54*, 427–435.

Lin, X., Kaul, S., Rounsley, S., Shea, T.P., Benito, M.I., Town, C.D., Fujii, C.Y., Mason, T., Bowman, C.L., Barnstead, M., Feldblyum, T.V., Buell, C.R., Ketchum, K.A., Lee, J., Ronning, C.M., Koo, H.L., Moffat, K.S., Cronin, L.A., Shen, M., Pai, G., Van Aken, S., Umayam, L., Tallon, L.J., Gill, J.E., Adams, M.D., Carrera, A.J., Creasy, T.H., Goodman, H.M., Somerville, C.R., Copenhaver, G.P., Preuss, D., Nierman, W.C., White, O., Eisen, J.A., Salzberg, S.L., Fraser, C.M., and Venter, J.C. (1999) Sequence and analysis of chromosome 2 of the plant *Arabidopsis thaliana*. Nature *402*, 761–768.

Liu, Z., Yan, H., Wang, K., Kuang, T., Zhang, J., Gui, L., An, X., and Chang, W. (2004). Crystal structure of spinach major light-harvesting complex at 2.72 Å resolution. Nature *428*, 287–292.

Liu, H., Probert, I., Uitz, J., Claustre, H., Aris-Brosou, S., Frada, M., Not, F., and de Vargas, C. (2009). Extreme diversity in noncalcifying haptophytes explains a major pigment paradox in open oceans. Proc. Natl. Acad. Sci. U.S.A. *106*, 12803–12808.

Lopez, P., and Bapteste, E. (2009). Molecular phylogeny: reconstructing the forest. C. R. Biol. *332*, 171–182.

Ludwig, M., and Gibbs, S.P. (1989). Localization of phycoerythin at the lumenal surface of the thylakoid membrane in *Rhodomonas lens*. J. Cell Biol. *108*, 875–884.

Macpherson, A.N., and Hiller, R.G. (2003). Light-harvesting systems in chlorophyll *c*-containing algae. In Light Harvesting Antenna in Photosynthesis, B.R. Green and W.W. Parson, eds. (Dordrecht: Kluwer Academic Publishers), pp. 324–352.

Manton, I. (1964). Further observations on the fine structure of the haptonema in *Prymnesium parvum*. Arch. Microbiol. *49*, 315–330.

Marechal, E., and Cesbron-Delauw, M.F. (2001). The apicoplast: a new member of the plastid family. Trends Plant Sci. *6*, 200–205.

Marin, B., Nowack, E.C., and Melkonian, M. (2005). A plastid in the making: evidence for a second primary endosymbiosis. Protist *156*, 425–432.

Marquardt, J., and Rhiel, E. (1997). The membrane-intrinsic light-harvesting complex of the red alga *Galdieria sulphuraria* (formerly *Cyanidium caldarium*): Biochemical and immunochemical characterization. Biochim. Biophys. Acta *1320*, 153–164.

Martin, W. (2003). Gene transfer from organelles to the nucleus: Frequent and in big chunks. Proc. Natl. Acad. Sci. U.S.A. *100*, 8612–8614.

Martin, W., Stoebe, B., Goremykin, V., Hansmann, S., Hasegawa, M., and Kowallik, K.V. (1998). Gene transfer to the nucleus and the evolution of chloroplasts. Nature *393*, 162–165.

Martin, W., Rujan, T., Richly, E., Hansen, A., Cornelsen, S., Lins, T., Leister, D., Stoebe, B., Hasegawa, M., and Penny, D. (2002). Evolutionary analysis of *Arabidopsis*, cyanobacterial, and chloroplast genomes reveals plastid phylogeny and thousands of cyanobacterial genes in the nucleus. Proc. Natl. Acad. Sci. U.S.A. *99*, 12246–12251.

Matsuzaki, M., Misumi, O., Shin-I, T., Maruyama, S., Takahara, M., Miyagishima, S.Y., Mori, T., Nishida, K., Yagisawa, F., Nishida, K., Yoshida, Y., Nishimura, Y., Nakao, S., Kobayashi, T., Momoyama, Y., Higashiyama, T., Minoda, A., Sano, M., Nomoto, H., Oishi, K., Hayashi, H., Ohta, F., Nishizaka, S., Haga, S., Miura, S., Morishita, T., Kabeya, Y., Terasawa, K., Suzuki, Y., Ishii, Y., Asakawa, S., Takano, H., Ohta, N., Kuroiwa, H., Tanaka, K., Shimizu, N., Sugano, S., Sato, N., Nozaki, H., Ogasawara, N., Kohara, Y., and Kuroiwa, T. 2004. Genome sequence of the ultrasmall unicellular red alga *Cyanidioschyzon merolae* 10D. Nature *428*, 653–657.

Matsuzaki, M., Kuroiwa, H., Kuroiwa, T., Kita, K., and Nozaki, H. (2008). A cryptic algal group unveiled: a plastid biosynthesis pathway in the oyster parasite *Perkinsus marinus*. Mol. Biol. Evol. *25*, 1167–1179.

Matsuzaki, M., Masuda, I., Kuroiwa, H., Kuroiwa, T., Nozaki, H., and Kita, K. (2009). A DNA-lacking plastid in the oyster pathogen *Perkinsus marinus*. Phycologia *48*, 82–83.

McFadden, G.I. (1993). Second-hand chloroplasts: Evolution of cryptomonad algae. In Advances in Botanical Research, J.A. Callow, ed. (London: Academic Press), pp. 189–230.

McFadden, G.I., and van Dooren, G.G. (2004). Evolution: Red algal genome affirms a common origin of all plastids. Curr. Biol. *14*, R514–6.

McFadden, G.I., Gilson, P.R., Hofmann, C.J., Adcock, G.J., and Maier, U.-G. (1994). Evidence that an amoeba acquired a chloroplast by retaining part of an engulfed eukaryotic alga. Proc. Natl. Acad. Sci. U.S.A. *91*, 3690–3694.

Mignot, J.P. (1965). Étude ultrastructurale de *Cyathomonas truncata* From (flagellé cryptomonadine. J. Microscopie *4*, 239–252.

Moestrup, O., and Sengco, M. (2001). Ultrastructural studies on *Bigelowiella natans*, gen. et sp. nov., a chlorarachniophyte flagellate. J. Phycol. *37*, 624–646.

Montane, M.H., and Kloppstech, K. (2000). The family of light-harvesting-related proteins (LHCs, ELIPs, HLIPs): Was the harvesting of light their primary function? Gene *258*, 1–8.

Moore, R.B., Obornik, M., Janouskovec, J., Chrudimsky, T., Vancova, M., Green, D.H., Wright, S.W., Davies, N.W., Bolch, C.J., Heimann, K., Slapeta, J., Hoegh-Guldberg, O., Logsdon, J.M., and Carter, D.A. (2008). A photosynthetic alveolate closely related to apicomplexan parasites. Nature *451*, 959–963.

Morrison, D.A. (2009). Evolution of the Apicomplexa: where are we now? Trends Parasitol. *25*, 375–382.

Moustafa, A., and Bhattacharya, D. (2008). PhyloSort: a user-friendly phylogenetic sorting tool and its application to estimating the cyanobacterial contribution to the nuclear genome of *Chlamydomonas*. BMC Evol. Biol. *8*, 6.

Moustafa, A., Beszteri, B., Maier, U.G., Bowler, C., Valentin, K., and Bhattacharya, D. (2009). Genomic footprints of a cryptic plastid endosymbiosis in diatoms. Science *324*, 1724–1726.

Nassoury, N., Cappadocia, M., and Morse, D. (2003). Plastid ultrastructure defines the protein import pathway in dinoflagellates. J. Cell Sci. *116*, 2867–2874.

Neilson, J., and Durnford, D.G. (2010). Evolutionary distribution of Light Harvesting Complex-like proteins in photosynthetic eukaryotes. Genome *53*, 68–78.

Nikolaev, S.I., Berney, C., Fahrni, J.F., Bolivar, I., Polet, S., Mylnikov, A.P., Aleshin, V.V., Petrov, N.B., and Pawlowski, J. (2004). The twilight of Heliozoa and rise of Rhizaria, an emerging supergroup of amoeboid eukaryotes. Proc. Natl. Acad. Sci. U.S.A. *101*, 8066–8071.

Nishiyama, T., Fujita, T., Shin-I, T., Seki, M., Nishide, H., Uchiyama, I., Kamiya, A., Carninci, P., Hayashizaki, Y., Shinozaki, K., Kohara, Y., and Hasebe, M. (2003). Comparative genomics of *Physcomitrella patens* gametophytic transcriptome and *Arabidopsis thaliana*: Implication for land plant evolution. Proc. Natl. Acad. Sci. U.S.A. *100*, 8007–8012.

Norris, B.J., and Miller, D.J. (1994). Nucleotide sequence of a cDNA clone encoding the precursor of the peridinin-chlorophyll *a*-binding protein from the dinoflagellate *Symbiodinium* sp. Plant Mol. Biol. *24*, 673–677.

Nosenko, T., and Bhattacharya, D. (2007). Horizontal gene transfer in chromalveolates. BMC Evol. Biol. *7*, 173.

Not, F., Valentin, K., Romari, K., Lovejoy, C., Massana, R., Tobe, K., Vaulot, D., and Medlin, L.K. (2007). Picobiliphytes: a marine picoplanktonic algal group with unknown affinities to other eukaryotes. Science *315*, 253–255.

Nozaki, H., Maruyama, S., Matsuzaki, M., Nakada, T., Kato, S., and Misawa, K. (2009). Phylogenetic positions of Glaucophyta, green plants (Archaeplastida) and Haptophyta (Chromalveolata) as deduced from slowly evolving nuclear genes. Mol. Biol. Evol. *53*, 872–880.

Nymark, M., Valle, K.C., Brembu, T., Hancke, K., Winge, P., Andresen, K., Johnsen, G., and Bones, A.M. (2009). An integrated analysis of molecular acclimation to high light in the marine diatom *Phaeodactylum tricornutum*. PloS One *4*, e7743.

Oeltjen, A., Krumbein, W.E., and Rhiel, E. (2002). Investigations on transcript sizes, steady-state mRNA concentrations and diurnal expression of genes encoding fucoxanthin chlorophyll *a*/*c* light harvesting polypeptides in the centric diatom *Cyclotella cryptica*. Plant Biol. *4*, 250–257.

Okamoto, N., Chantangsi, C., Horak, A., Leander, B.S., and Keeling, P.J. (2009). Molecular phylogeny and description of the novel katablepharid, *Roombia truncata* gen. et sp. nov., and establishment of the Hacrobia taxon nov. PLoS One *4*, e7080.

Osafune, T., Sumida, S., Schiff, J.A., and Hase, E. (1991). Immunolocalization of LHCP II apoprotein in the Golgi during light-induced chloroplast development in non-dividing *Euglena* cells. J. Electron Microsc. *40*, 41–47.

Palmer, J.D. (2003). The symbitotic birth and spread of plastids: how many times and whodunit? J. Phycol. 39, 4–11.

Parfrey, L.W., Lahr, D.J.G, and Katz, L.A. (2008). The dynamic nature of eukaryotic genomes. Mol. Biol. Evol. *25*, 787–794.

Patron N.J., and Waller, R.F. (2007). Transit peptide diversity and divergence: A global analysis of plastid targeting signals. BioEssays *29*, 1048–1058.

Patron N.J., Waller R.F., Archibald J.M., and Keeling P.J. (2005). Complex protein targeting to dinoflagellate plastids. J. Mol. Biol. *348*, 1015–1024.

Patron, N.J., Inagaki, Y., and Keeling, P.J. (2007). Multiple gene phylogenies support the monophyly of cryptomonad and haptophyte host lineages. Curr. Biol. *17*, 887–891.

Pearson, G.A., Hoarau, G., Lago-Leston, A., Coyer, J.A., Kube, M., Reinhardt, R., Henckel, K., Serrao, E.T., Corre, E., and Olsen J.L. (2009). An expressed sequence tag analysis of the intertidal brown seaweeds *Fucus serratus* (L.) and *F. vesiculosus* (L.) (Heterokontophyta, Phaeophyceae) in response to abiotic stressors. Mar. Biotechnol., in press.

Peers, G., Truong, T.B., Ostendorf, E., Busch, A., Elrad, D., Grossman, A.R., Hippler, M., and Niyogi, K.K. (2009). An ancient light-harvesting protein is critical for the regulation of algal photosynthesis. Nature *462*, 518–521.

Peeters, N., and Small, I. (2001). Dual targeting to mitochondria and chloroplasts. Biochim. Biophys. Acta *1541*, 54–63.

Perkins, F.O., Barta, J.R., Clopton, R.E., Peirce, M.A., and Upton, S.J. (2002). Phylum Apicomplexa Levine, 1970. In An Illustrated Guide to the Protozoa, J.J. Lee, ed. (Lawrence, KS: Allen Press), pp. 190–200.

Phillips, N., Burrowes, R., Rousseau, F., de Reviers, B., and Saunders, G.W. (2008). Resolving evolutionary relationships among the brown algae using chloroplast and nuclear genes. J. Phycol. *44*, 394–405.

Ralph, S.A., D'Ombrain, M.C., and McFadden, G.I. (2001). The apicoplast as an antimalarial drug target. Drug Resist. Updates *4*, 145–151.

Reith, M., and Munholland, J. (1995). Complete nucleotide sequence of the *Porphyra purpurea* chloroplast genome. Plant Mol. Biol. Rep. *13*, 333–335.

Reisinger, V., Ploscher, M., and Eichacker, L.A. (2008). LIL3 assembles as chlorophyll-binding protein complex during deetiolation. FEBS Lett. *582*, 1547–1551.

Rensing, S.A., Goddemeier, M., Hofmann, C.J., and Maier, U.-G. (1994). The presence of a nucleomorph hsp70 gene is a common feature of Cryptophyta and Chlorarachniophyta. Curr. Genet. *26*, 451–455.

Reumann, S., Inoue, K., and Keegstra, K. (2005). Evolution of the general protein import pathway of plastids. Mol. Membr. Biol. *22*, 73–86.

Reyes-Prieto, A., and Bhattacharya, D. (2007). Phylogeny of Calvin cycle enzymes supports Plantae monophyly. Mol. Phyl. Evol. *45*, 384–391.

Reyes-Prieto, A., Hackett, J.D., Soares, M.B., Bonaldo, M.F., and Bhattacharya, D. (2006). Cyanobacterial contribution to algal nuclear genomes a primarily limited to plastid functions. Curr. Biol. *16*, 2320–2325.

Reyes-Prieto, A., Weber, A.P.M., and Bhattacharya, D. (2007). The origin and establishment of the plastid in algae and plants. Ann. Rev. Genet. *41*, 147–168.

Reyes-Prieto, A., Moustafa, A., and Bhattacharya, D. (2008). Multiple genes of apparent algal origin suggest ciliates may once have been photosynthetic. Curr. Biol. *18*, 956–962.

Rice, D.W., and Palmer, J.D. (2006). An exceptional horizontal gene transfer in plastids: gene replacement by a distant bacterial paralog and evidence that haptophyte and cryptophyte plastids are sisters. BMC Biol. *4*, 31.

Richard, C., Ouellet, H., and Guertin, M. (2000). Characterization of the LI818 polypeptide from the green unicellular alga *Chlamydomonas reinhardtii*. Plant Mol. Biol. *42*, 303–316.

Rispe, C., and Moran, R. (2000). Accumulation of deleterious mutations in endosymbionts: Muller's ratchet with two levels of selection. Am. Natur. *156*, 425–441.

Robinson, T., and Katz, L.A. 2007. Non-Mendelian inheritance of paralogs of two cytoskeletal genes in the ciliate *Chilodonella uncinata*. Mol. Biol. Evol. *24*, 2495–2503.

Rodriguez-Ezpeleta, N., Brinkmann, H., Burey, S.C., Roure, B., Burger, G., Löffelhardt, W., Bohnert, H.J., Philippe, H., and Lang, B.F. (2005). Monophyly of primary photosynthetic eukaryotes: green plants, red algae, and glaucophytes. Curr. Biol. *15*, 1325–1330.

Rodriguez-Ezpeleta, N., Brinkmann, H., Roure, B., Lartillot, N., Lang, B.F., and Philippe, H. (2007). Detecting and overcoming systematic errors in genome-scale phylogenies. Syst. Biol. *563*, 389–399.

Rogers, M.B., Archibald, J.M., Field, M.A., Li, C., Striepen, B., and Keeling, P.J. (2004). Plastid-targeting peptides from the chlorarachniophyte *Bigelowiella natans*. J. Eukaryot. Microbiol. *51*, 529–535.

Rogers, M.B., Watkins, R.F., Harper, J.T., Durnford, D.G., Gray, M.W., and Keeling, P.J. (2007). A complex and punctate distribution of three eukaryotic genes derived by lateral gene transfer. BMC Evol. Biol. *7*, 89.

Rumpho, M.E., Dastoor, F.P., Manhart, J.R., and Lee, J. (2006). The kleptoplast. In The Structure and Function of Plastids, R.R. Wise, and J.K. Hoober, eds. (Dordrecht: Springer), pp. 451–473.

Sato, N., Ishikawa, M., Fujiwara, M., and Sonoike, K. (2005). Mass identification of chloroplast proteins of endosymbiotic origin by phylogenetic profiling based on organism-optimized homologous protein groups. Genome Informat. *16*, 56–68.

Savard, F., Richard, C., and Guertin, M. (1996). The *Chlamydomonas reinhardtii* LI818 gene represents a distant relative of the CabI/II genes that is regulated during the cell cycle and in response to illumination. Plant Mol. Biol. *32*, 461–473.

Schnepf, E., and Elbrachter, M. (1999). Dinophyte chloroplasts and phylogeny – A review. Grana *38*, 81–97.

Sepsenwol, S. (1973). Leucoplast of the cryptomonad *Chilomonas paramecium*. Evidence for presence of a true plastid in a colorless flagellate. Exp. Cell Res. *76*, 395–409.

Shalchian-Tabrizi, K., Minge, M.A., Cavalier-Smith, T., Nedreklepp, J.M., Klaveness, D., and Jakobsen, K.S. (2006). Combined heat shock protein 90 and ribosomal RNA sequence phylogeny supports multiple replacements of dinoflagellate plastids. J. Eukaryot. Microbiol. *53*, 217–224.

Shalchian-Tabrizi, K., Kauserud, H., Massana, R., Klaveness, D., and Jakobsen, K.S. (2007). Analysis of environmental 18S ribosomal RNA sequences reveals unknown diversity of the cosmopolitan phylum Telonemia. Protist *158*, 173–180.

Sidler, W., Nutt, H., Kumpf, B., Frank, G., Suter, F., Brenzel, A., Wehrmeyer, W., and Zuber, H. (1990). The complete amino-acid sequence and the phylogenetic origin of phycocyanin-645 from the cryptophytan alga *Chroomonas* sp. Biol. Chem. Hoppe-Seyler *371*, 537–547.

Six, C., Worden, A.Z., Rodriguez, F., Moreau, H., and Partensky, F. (2005). New insights into the nature and phylogeny of prasinophyte antenna proteins: *Ostreococcus tauri*, a case study. Mol. Biol. Evol. *22*, 2217–2230.

Slamovits, C.H., and Keeling, P.J. (2008). Plastid-derived genes in the nonphotosynthetic alveolate *Oxyrrhis marina*. Mol. Biol. Evol. *25*, 1297–1306.

Sláviková, S., Vacula, R., Fang, Z., Ehara, T., Osafune, T., and Schwartzbach, S.D. (2005). Homologous and heterologous reconstitution of Golgi to chloroplast transport and protein import into the complex chloroplasts of *Euglena*. J. Cell Sci. *118*, 1651–1661.

Standfuss, J., Terwisscha van Scheltinga, A.C., Lamborghini, M., and Kühlbrandt, W. (2005). Mechanisms of photoprotection and nonphotochemical quenching in pea light-harvesting complex at 2.5 Å resolution. EMBO J. *24*, 919–928.

Stauber, E.J., Busch, A., Naumann, B., Svatos, A., and Hippler, M. (2009). Proteotypic profiling of LHCI from *Chlamydomonas reinhardtii* provides new insights into structure and function of the complex. Proteomics *9*, 398–408.

Stegemann, S., Hartmann, S., Ruf, S., and Bock, R. (2003). High-frequency gene transfer from the chloroplast genome to the nucleus. Proc. Natl. Acad. Sci. U.S.A. *100*, 8828–8833.

Steiner, J.M., Yusa, F., Pompe, J.A., and Löffelhardt, W. (2005). Homologous protein import machineries in chloroplasts and cyanelles. Plant J. *44*, 646–652.

Stenzel, D.J., and Boreham, P.F. (1996). *Blastocystis hominis* revisited. Clin. Microbiol. Rev. *9*, 563–584.

Stiller, J.W., Reel, D.C., and Johnson, J.C. (2003). A single origin of plastids revisited: convergent evolution in organellar genome content. J. Phycol. *39*, 95–105.

Stiller, J.W., Huang, J., Ding, Q., Tian, J., and Goodwillie, C. (2009). Are algal genes in nonphotosynthetic protists evidence of historical plastid endosymbioses? BMC Genom. *10*, 484.

Sulli, C., and Schwartzbach, S.D. (1996). A soluble protein is imported into *Euglena* chloroplasts as a membrane-bound precursor. Plant Cell *8*, 43–53.

Tan, S., Ducret, A., Aebersold, R., and Gantt, E. (1997). Red algal LHC I genes have similarities with both chlorophyll *a/b*- and *a/c*-binding proteins: A 21 kDa polypeptide encoded by LhcaR2 is one of the six LHC I polypeptides. Photosynth. Res. *53*, 129–140.

Taylor, F.J.R. (1974). Implications and extensions of the serial endosymbiosis theory of the origin of eukaryotes. Taxon 23, 229–258.

Teles-Grilo, M.L., Duarte, S.M., Tato-Costa, J., Gaspar-Maia, A., Oliveira, C., Rocha, A.A., Marques, A., Cordeiro-da-Silva, A., and Azevedo, C. (2007a). Molecular karyotype analysis of *Perkinsus atlanticus* (phylum Perkinsozoa) by pulsed field gel electrophoresis. Eur. J. Protistol. *43*, 315–318.

Teles-Grilo, M.L., Tato-Costa, J., Duarte, S.M., Maia, A., Casal, G., and Azevedo, C. (2007b). Is there a plastid in *Perkinsus atlanticus* (phylum Perkinsozoa)? Eur. J. Protistol. *43*, 163–167.

Timmis, J.N., Ayliffe, M.A., Huang, C.Y., and Martin, W. (2004). Endosymbiotic gene transfer: Organelle genomes forge eukaryotic chromosomes. Nature Rev. Genet. *5*, 123–135.

Tonkin, C.J., Foth, B.J., Ralph, S.A., Struck, N., Cowman, A.F., and McFadden, G.I. (2008a). Evolution of malaria parasite plastid targeting sequences. Proc. Natl. Acad. Sci. U.S.A. *105*, 4781–4785.

Tonkin, C.J., Kalanon, M. McFadden, G. I. (2008b). Protein targeting to the malaria parasite plastid. Traffic *9*, 166–175.

Tyler, B.M., Tripathy, S., Zhang, X.M., Dehal, P., Jiang, R.H.Y., Aerts, A., Arredondo, F.D., Baxter, L., Bensasson, D., Beynon, J.L., Chapman, J., Damasceno, C.M.B., Dorrance, A.E., Dou, D.L., Dickerman, A.W., Dubchak, I.L., Garbelotto, M., Gijzen, M., Gordon, S.G., Govers, F., Grunwald, N.J., Huang, W., Ivors, K.L., Jones, R.W., Kamoun, S., Krampis, K., Lamour, K.H., Lee, M.K., McDonald, W.H., Medina, M., Meijer, H.J.G., Nordberg, K., Maclean, D.J., Ospina-Giraldo, M.D., Morris, P.F., Phuntumart, V., Putnam, N.H., Rash, S., Rose, J.K.C., Sakihama, Y., Salamov, A.A., Savidor, A., Scheuring, C.F., Smith, B.M., Sobral, B.W.S., Terry, A., Torto-Alalibo, T.A., Win, J., Xu, Z.Y., Zhang, H.B., Grigoriev, I.V., Rokhsar, D.S., and Boore, J.L. (2006). *Phytophthora* genome sequences uncover evolutionary origins and mechanisms of pathogenesis. Science *313*, 1261–1266.

van Dooren, G.G., Schwartzbach, S.D., Osafune, T., and McFadden, G.I. (2001). Translocation of proteins across the multiple membranes of complex patterns. Biochim. Biophys. Acta *1541*, 34–53.

Veith, T., Brauns, J., Weisheit, W., Mittag, M., and Büchel, C. (2009). Identification of a specific fucoxanthin-chlorophyll protein in the light harvesting complex of photosystem I in the diatom *Cyclotella meneghiniana*. Biochim. Biophys. Acta *1787*, 905–912.

Waller, R.F., and Jackson, C.J. (2009). Dinoflagellate mitochondrial genomes: stretching the rules of molecular biology. Bioessays *31*, 237–245.

Waller, R.F., Patron, N.J., and Keeling, P.J. (2006). Phylogenetic history of plastid-targeted proteins in the peridinin-containing dinoflagellate *Heterocapsa triquetra*. Int. J. Syst. Evol. Microbiol. *56*, 1439–1447.

Wilk, K.E., Harrop, S.J., Jankova, L., Edler, D., Keenan, G., Sharples, F., Hiller, R.G., and Curmi, P.M. (1999). Evolution of a light-harvesting protein by addition of new subunits and rearrangement of conserved elements: Crystal structure of a cryptophyte phycoerythrin at 1.63-Å resolution. Proc. Natl. Acad. Sci. U.S.A. *96*, 8901–8906.

Wilson, R.J.M.I., Denny, P.W., Preiser, D.J., Rangachari, K., Roberts, K., Roy, A., Whyte, A., Strath, M., Moore, D.J., Moore, P.W., and Williamson, D.H. (1996). Complete gene map of the plastid-like DNA of the malaria parasite *Plasmodium falciparum*. J. Mol. Biol. *261*, 155–172.

Wolfe, G.R., Cunningham, F.X., Durnford, D.G., Green, B.R., and Gantt, E. (1994). Evidence for a common origin of chloroplasts with light-harvesting complexes of different pigmentation. Nature *367*, 566–568.

Wong, J.T.Y., and Kwok, A.C.M. (2005). Proliferation of dinoflagellates: blooming or bleaching. Bioessays *27*, 730–740.

Yakushevska, A.E., Keegstra, W., Boekema, E.J., Dekker, J.P., Andersson, J., Jansson, S., Ruban, A.V., and Horton, P. (2003). The structure of Photosystem II in *Arabidopsis*: Localization of the CP26 and CP29 antenna complexes. Biochemistry *42*, 608–613.

Yoon, H.S., Hackett, J.D., and Bhattacharya, D. (2002). A single origin of the peridinin- and fucoxanthin-containing plastids in dinoflagellates through tertiary endosymbiosis. Proc. Natl. Acad. Sci. U.S.A. *99*, 11724–11729.

Yoon, H.S., Hackett, J.D., Ciniglia, C., Pinto, G., and Bhattacharya, D. (2004). A molecular timeline for the origin of photosynthetic eukaryotes. Mol. Biol. Evol. *21*, 809–818.

Yoon, H.S., Muller, K.M., Sheath, R.G., Ott, F.D., and Bhattacharya, D. (2006a). Defining the major lineages of red algae (Rhodophyta). J. Phycol. *42*, 482–492.

Yoon, H.S., Reyes-Prieto, A., Melkonian, M., and Bhattacharya, D. (2006b). Minimal plastid genome evolution in the *Paulinella* endosymbiont. Curr. Biol. *16*, R670-R672.

Zhu, S.H., and Green, B.R. (2008). Light-harvesting and photoprotection in diatoms: Identification and expression of L818-like proteins. In Photosynthesis. Energy from the Sun: 14th International Congress on Photosynthesis, J.F. Allen, E. Gantt, J.H. Golbeck, and B. Osmond, eds. (Dordrecht: Springer), pp. 261–264.

Zufall, R.A., McGrath, C.L., Muse, S.V., and Katz, L.A. (2006). Genome architecture drives protein evolution in ciliates. Mol. Biol. Evol. 23, 1681–1687.

Zybailov B., Rutschow H., Friso G., Rudella A., Emanuelsson O., Sun Q., van Wijk K.J. (2008). Sorting signals, N-terminal modifications and abundance of the chloroplast proteome. PLoS One 3, e1994.

# Index

## E



## F



## G



## H



## I



## J



## K



## L



## M

## W



## X